Molecular Neuroscience

LIVERPOOL
JOHN MOORES UNIVERSITY
AVRIL ROBARTS LRC
TITHEBARN STREET
LIVERPOOL L2 2ER
TEL. 0151 231 4022

Molecular Neuroscience

P. Revest
Department of Physiology, St Batholomew's and the Royal London School of Medicine and Dentistry, Queen Mary and Westfield College, London, UK

A. Longstaff
Department of Biosciences, School of Natural Sciences, University of Hertfordshire, Hatfield, UK

© BIOS Scientific Publishers Limited, 1998

First published 1998

All rights reserved. No part of this book may be reproduced or transmitted, in any form or by any means, without permission.

A CIP catalogue record for this book is available from the British Library.

ISBN 1 85996 250 5

BIOS Scientific Publishers Ltd
9 Newtec Place, Magdalen Road, Oxford OX4 1RE, UK
Tel. +44 (0)1865 726286. Fax +44 (0)1865 246823
World Wide Web home page: http://www.bios.co.uk/

DISTRIBUTORS

Australia and New Zealand
Blackwell Science Asia
54 University Street
Carlton, South Victoria 3053

India
Viva Books Private Limited
4325/3 Ansari Road, Daryaganj
New Delhi 110002

Published in the United States of America, its dependent territories and Canada by Springer-Verlag New York Inc., 175 Fifth Avenue, New York, NY 10010-7858, in association with BIOS Scientific Publishers Ltd.

Published in Hong Kong, Taiwan, Singapore, Thailand, Cambodia, Korea, The Philippines, Indonesia, The People's Republic of China, Brunei, Laos, Malaysia, Macau and Vietnam by Springer-Verlag Singapore Pte. Ltd, 1 Tannery Road, Singapore 347719, in association with BIOS Scientific Publishers Ltd.

Production Editor: Priscilla Goldby.
Typeset and illustrated by Florencetype Ltd, Stoodleigh, UK.
Printed by Redwood Books, Trowbridge, UK.

Front cover: Cultured left-upper quadrant neurons from *Aplysia californica* that have been microinjected with either lucifer yellow CH (*L-453*) or sulforhodamine 101 (*S-359*). These neurons display an extensive array of overlapping processes. Reprinted from Kleinfeld, D. and Salzberg, B. (1991) *J. Neurophysiol.* **66**: 316–333. With permission from the American Physiological Society.

Contents

Abbreviations

A	adenosine
AA	arachidonic acid
AC	adenylate cyclase
ACh	acetylcholine
ACPD	amino-cyclopentane dicarboxylate
AD	Alzheimer's disease
adRP	autosomal dominant retinitis pigmentosa
Aga	agatoxin
AGE	advanced glycosylation end-product
AHP	afterhyperpolarization
ALS	amyotrophic lateral sclerosis
AM	acetomethoxy
AMPA	α-amino-3-hydroxy-5-methyl-4-isoxazoleproprionic acid
ANP	atrial natriuretic peptide
AP	alkaline phosphatase
4-AP	4-amino-pyridine
AP-1	activator protein-1
AP2	assembly protein complex
AP4	2-amino-4-phosphonobutanoate
D-AP5 (D-APV)	D(–)-2-amino-5-phosphonovalerate
D-AP7	D(–)-2-amino-7-phosphonoheptanoate
apoE	apolipoprotein E
APP	amyloid precursor protein
AR	adrenergic receptor
βARK	β-adrenergic receptor kinase
arRP	autosomal recessive retinitis pigmentosa
ATP	adenosine triphosphate
βA	β-amyloid
L-BMMA	β-*N*-methylamino-L-alanine
bp	base pair
bRH	bacteriorhodopsin
α-BuTX	α-bungarotoxin
BZ	benzodiazepine
bZIP	basic leucine zipper
C	cytosine

CA	catecholamine
CACA	*cis*-4-aminocrotonic acid
CaM	calmodulin
CaMKI	calcium/calmodulin-dependent protein kinase I
CaMKII	calcium/calmodulin-dependent protein kinase II
cAMP	cyclic adenosine monophosphate
CAT	choline acetyltransferase
β-CCE	ethyl-β-carboline-3-carboxylate
β-CCM	methyl-β-carboline-3-carboxylate
β-CCP	*n*-propyl-β-carboline-3-carboxylate
cDNA	complementary DNA
cGMP	cyclic guanosine monophosphate
ChAT	choline acetyltransferase
CHO	Chinese hamster ovary
ChTX	cholera toxin
CL218872	3-methyl-6-[(3-trifluoropyridazine)phenyl]-1,2,4-triazolo [4,3β]pyridazine
cNOS	constitutive nitric oxide synthetase
CNQX	6-cyano-7-nitroquinoxaline-2,3-dione
CPZ	chlorpromazine
CRE	cAMP response element
CREB	cAMP response element-binding protein
CrTx	charybdotoxin
CTX	conotoxin
DA	dopamine
DAB	3,3′-diaminobenzene
DAG	diacylglycerol
DDF	*p*-(*N*,*N*-dimethylamino)-benzenediazonium fluoride
ddNTP	2,3′-dideoxynucleotide triphosphate
DHP	dihydropyridine
DMCM	methyl-6,7-dimethoxy-4-ethyl-β-carboline-3-carboxylate
DNA	deoxyribonucleic acid
DND	delayed neuronal death
dNTP	deoxynucleoside triphosphate
L-DOPA	L-3,4,-dihydroxphenylalanine
ds	double stranded
DTx	dendrotoxin
EAA	excitatory amino acid
EGF	epidermal growth factor
EGTA	ethyleneglycol-bis-(β-aminoethyl)-*N*,*N*′-tetraacetic acid
ELISA	enzyme-linked immunosorbent assay
EMSA	electromobility shift assay
epsp	excitatory postsynaptic potential
ES	embryonic stem
EST	expressed sequence tag
EtBr	ethidium bromide
FSH	follicle-stimulating hormone
FTX	funnel spider toxin
G	guanosine

GABA	γ-aminobutyric acid
GAD	glutamic acid decarboxylase
GC	guanylate cyclase
GDI	GDP dissociation inhibitor
GDP	guanosine diphosphate
GluR	glutamate receptor
GTP	guanosine triphosphate
HAP-1	huntingtin-associated protein-1
HD	Huntington's disease
HIV	human immunodeficiency virus
HPLC	high-performance liquid chromatography
HRP	horseradish peroxidase
5-HT	5-hydroxytryptamine, serotonin
HVA	high voltage-activated
IEG	immediate-early response gene
Ig	immunoglobulin
iGluR	ionotropic glutamate receptor
iNOS	inducible nitric oxide synthetase
IP_3	inositol 1,4,5-trisphosphate
ipsp	inhibitory postsynaptic potential
kbp	kilobase pair
lacZ	*E. coli* lactose Z gene
LDCV	large dense-core vesicle
LDL	low density lipoprotein
LEMS	Lambert–Eaton myasthenic syndrome
LGIC	ligand-gated ion channel
LTD	long-term depression
LTF	long-term facilitation
LTP	long-term potentiation
LVA	low voltage-activated
mAChR	muscarinic acetylcholine receptor
MAO	monoamine oxidase
MAP	microtubule-associated protein
MAPK	mitogen-activated protein kinase
MBTA	4-(*N*-maleimido)benzyltrimethylammonium
MCPG	α-methyl-4-carboxyphenylglycine
MEA	methanethiosulfonate ethylammonium
mepp	miniature end-plate potential
mepsp	miniature excitatory postsynaptic potential
mGluR	metabotropic glutamate receptor
MK801	dizocilpine
MPP^+	1-methyl-4-phenylpyridinium
MPTP	1-methyl-4-phenyl-1,2,3,4-tetrahydropyridine
mRNA	messenger RNA
NA	noradrenaline
nAChR	nicotinic acetylcholine receptor
L-NAME	L-*N*-nitroarginine
ncNOS	neuronal constitutive nitric oxide synthetase
NDGA	nordihydroguaiaretic acid

NF-H	high molecular weight neurofilament protein
NF-L	low molecular weight neurofilament protein
NFT	neurofibrillary tangle
NGF	nerve growth factor
NMDA	*N*-methyl-D-aspartate
NMR	nuclear magnetic resonance
NO	nitric oxide
NOS	nitric oxide synthetase
NSF	*N*-ethylmaleimide-sensitive factor
ori	origin of replication
PAF	platelet-activating factor
PBP	periplasmic amino acid-binding protein
PCP	phencyclidine
PCR	polymerase chain reaction
PD	Parkinson's disease
PDE	phosphodiesterase
PDGF	platelet-derived growth factor
PDS	paroxysmal depolarizing shift
PET	positron emission tomography
PGE_2	prostaglandin E_2
PHF	paired helical fragment
PIP_2	phosphatidylinositol-4,5-bisphosphate
PKA	protein kinase A, cAMP-dependent protein kinase
PKC	protein kinase C
PKG	protein kinase G, cGMP-dependent protein kinase
PLA_2	phospholipase A_2
PKU	phenylketonuria
PLC	phospholipase C
pNPP	*p*-nitrophenyl phosphate disodium
PRL	prolactin
PS-1	presenilin-1
PS-2	presenilin-2
PSD	postsynaptic density
PTK	protein tyrosine kinase
PTX	pertussis toxin
PZT	pentylenetetrazole
rDNA	recombinant DNA
RFLP	restriction fragment length polymorphism
RNA	ribonucleic acid
RNase	ribonuclease
rRNA	ribosomal RNA
SCLC	small-cell lung cancer
SDS	sodium dodecylsulfate
sFTX	synthetic funnel spider toxin
SH2	Src-homology 2
SNAP	soluble NSF attachment protein
SNL	spatial navigation learning
SOD	superoxide dismutase
SR	sarcoplasmic reticulum

ss	single stranded
SSV	small synaptic vesicle
STS	sequence-tagged site
SV40	simian virus 40
T	thymidine
TBPS	t-butylbicyclophorothionate
TEA	tetraethylammonium
TF	transcription factor
TM	transmembrane
3TM, 4TM	three transmembrane, four transmembrane, etc.
TPP	triphenyl-methylphosphonium
TRH	thyrotropin-releasing hormone
tRNA	transfer RNA
TTX	tetrodotoxin
UV	ultraviolet
VAChT	vesicular acetylcholine transporter
VDCC	voltage-dependent calcium channel
VDKC	voltage-dependent potassium channel
VDSC	voltage-dependent sodium channel
VMAT	vesicular monoamine transporter
xlRP	X-linked retinitis pigmentosa
ZK93426	ethyl-6-benzyloxy-4-methoxymethyl-β-carboline-3-carboxylate

Preface

This book, written largely over two summers (scholarly oases sandwiched between relentlessly busy academic sessions) has its roots in courses run at Queen Mary and Westfield College and the University of Hertfordshire. A third year B.Sc. neuroscience course, taken by both science and intercalated medical students at QMW, and final year neurophysiology and neuropharmacology modules coupled with material taught in molecular biology from UH, all shaped by the experience of delivery to several cohorts of students, have served as the precursors to this volume.

Our intention is that this text will prove useful to final year undergraduates and post-graduates interested in the impact that the new science of molecular biology is having on (the rather older) neuroscience. As such, we have assumed that most readers will have at least some familiarity with basic elements of neuron cell biology – action potentials, synapses and the like – and some knowledge of molecular biology, such as general features of the structures and functions of nucleic acids and proteins. However, we have included some 'reminders' at crucial places so we hope no one will be unable to follow an argument for lack of an essential fact. We assume no prior exposure to the techniques of genetic engineering; these are dealt with where appropriate as the text unfolds, although, inevitably, there is a higher dose of this type of material in the early chapters.

We have tried, as fits this level, to provide insights into general experimental strategies, and sometimes provide quite detailed accounts of experiments, so that the evidential basis for much of what we cite is clear. We make no apology for this approach, which invariably tends to highlight inconsistency and controversy, for science is a practical activity that proceeds by the not so steady accumulation of evidence that must be critically assessed. In the final year of a degree, it is not enough that students have the facility to read and comprehend a scientific paper, although it is one of our goals that the reader gains access to the molecular neuroscience literature. They should also begin to develop the mind set of the good scientist which requires the rather bizarre combination of highly analytical pedantry – allowing sensible critique of the literature – with the ingenuity and imagination needed to visualize the clarity of the concept despite the muddiness of the experimental waters. Real neuroscience is a messy business!

This book, like the courses which spawned it, cannot be comprehensive if it is to be of affordable length and sufficient depth. Of necessity, then, some topics have been left out which, in an ideal world, we would have liked included, but nonetheless we have attempted to capture something of the flavour of contemporary neuroscience in the choice of areas we have addressed.

Textbooks of physiology, pharmacology and medicine have traditionally focused, in examining neurotransmission, on the outposts of the peripheral nervous system, the neuromuscular junction and autonomic nervous systems, for good historic reasons. In exploring the roles of glutamate and GABA, we have chosen a rather different emphasis. This can be justified since so much exciting research currently being undertaken involves these transmitters, and also on quantitative grounds; simply put, most synapses in the brain use either glutamate or γ-amino butyrate.

The notion that three key revolutions will be preeminent in the science of the next century, namely in computing, understanding the genome and neurobiology, is well founded. In their very different ways, all are about information processing and are, in some senses, inextricably linked. It has been fun (mostly!) writing a book encompassing something of two of these extraordinary pursuits.

Several colleagues at UH, Virginia Bugeja, Heddwyn Jones and Robert Slater, provided material that was helpful in drafting part of Chapter 2, for which we are grateful. We thank Rachel Offord, Lisa Mansell and Priscilla Goldby at BIOS for the patience with which they dealt with our idiosyncrasies and David Hames for asking us to do the book in the first place. Needless to say, any errors or misrepresentations are entirely our fault, but, despite the many, sometimes spirited arguments about what to put in and what to leave out, as well as how to say what we did say, we are still friends.

Patricia Revest and Alan Longstaff

Chapter

Introduction

1

What should a book entitled Molecular Neuroscience be about? There are no hard and fast rules about this, but we have chosen to write a text which shows what molecular biology has done for neuroscience and the exciting problems which remain to be tackled. It is hoped that it will be useful to the student of neuroscience who may find tools to answer specific questions, and also of interest to the molecular biologist who might be tempted to apply their skills to the important questions neuroscience raises.

The desire to understand the workings of the human brain has stimulated diverse approaches from the mystical to the mechanistic. Neuroscience has always been multidisciplinary, embracing anatomy, physiology and biochemistry, as well as the clinical and behavioral sciences.

Mathematical modeling of individual neurons has been around for over half a century, and more recently an eclectic and heady mix of mathematics, computer science, artificial intelligence, connectionism and the like has spawned neural network models, some of which seem to offer a means to a conceptual (as opposed to a descriptive) approach to how neural systems work. What this is teaching us is that the brain is an information-processing device which works only by virtue of the connectedness of its components.

To this pot must now be added the molecular biology revolution. Molecular biology itself is ill-defined. In one sense it is concerned with how deoxyribonucleic acid (DNA) makes ribonucleic acid (RNA) makes protein, and as such is a subset of biochemistry. However, at another level, it is the science concerned with engineering these macromolecules. When this manipulation is done so as to discover something about how the nervous system works, we have molecular neuroscience.

The immense power of the molecular biology revolution to reveal hitherto unapproachable minutiae of cell functioning, with its inevitable emphasis on gene level explanations, is seductive but carries with it intellectual dangers. Molecular neuroscience is close to the lowest level on the reductionist hierarchy of analysis; only the sort of models which try to describe things like the electrostatic barriers which are thought to exist in the core of ion channels or the electron clouds which can surround an agonist-binding site are further from whole brain function. These reductionist approaches can provide descriptions of ion channels, receptors, enzymes, etc., and developments in modeling have enabled their properties to be simulated in higher level systems. These systems display appropriate higher level behaviors which could not necessarily be predicted from the characteristics of the individual elements, that is they have emergent properties. However, reductionism cannot explain why most humans like music and some become psychopaths.

One of the problems is that the practice of neuroscience, the day-to-day business of designing experiments, impaling cells, setting up a polymerase chain reaction (PCR) incubation, running gels and the like is highly reductionist. Indeed, many might argue that the more variables that can be controlled for in any given study, the better the science. However, the interpretation of the data we collect is another matter. It is untenable to argue, simply because concentrations of monoamines seem to be lower in the brains of depressed patients, that depression is caused by reduced brain monoamines. Firstly, because correlation is no proof of causation but, secondly, and more crucially, because the concept of depression is in an entirely different category of description from that of an altered concentration of neurotransmitter, and comparisons between them may not be meaningful. There are an unimaginable number of interactions that will intervene in the explanatory gap that separates neurotransmission from behavior which no experiment could take account of, even *in principle*, given the undoubted nonlinearity of the dynamic systems in the brain.

A related pitfall into which molecular biology can throw the incautious is that of *genetic determinism*, the notion that we are the slaves of our genotypes. Consider, for example, the inborn error of metabolism disorder, phenylketonuria (PKU). This arises from a single mutation in the gene coding for phenylalanine hydroxylase, the enzyme responsible for converting phenylalanine to tyrosine. Infants with PKU suffer severe disturbance to brain development in the first few months after birth. Untreated, they come to have serious cognitive deficits. Now, in this case, the link between genotype and phenotype would seem to be transparently straightforward. However, in many countries, PKU is now diagnosed at birth using a routine screening test. Affected infants are treated for life by being placed on a diet low in phenylalanine. Such individuals grow up with almost normal cognitive skills. Notice that this successful treatment has not altered the defective gene – the mutant genotype remains – but it has changed the phenotype. Moreover, the treatment works by manipulating the environment (diet). This example illustrates a very important general principle, that the phenotype arises from an interaction of genotype and environment. The same genotype can result in different phenotypes (owing to variations in the environment), or the same phenotype can be shared by organisms with a different genotype.

That such a complete disconnection of genotype and phenotype is possible in such a simple case surely highlights the difficulties inherent in postulating a schizophrenic genotype or a homosexual genotype. Molecular biology is exciting, but we must not imagine – like some modern Laplace – that once we know the locus and function of every gene we will have a complete understanding of the brain or be able to *predict* its behavior. With these provisos always in mind, we will not be too misguided by the remarkable discoveries that the molecular approach to neuroscience undoubtedly has in store.

This volume is a chimera containing both some neuroscience and some molecular biology; it is at the interface between these two disciplines. As such, it cannot be a comprehensive analysis of either. You will not find here detailed descriptions of neuroanatomical pathways, extensive examination of every putative neurotransmitter or an exhaustive review of neuropathology. Neither do we provide the detailed experimental protocols that would be needed to clone a gene, nor an exposition of the theory underlying, say, pulsed-field gel electrophoresis.

This text is designed to give the reader some insight into how neuroscience has been revolutionized at the molecular level in the past decade or so. We hope it will serve as a broad introduction to many of the ingenious approaches molecular biologists have used to fathom neural function, and some of the key findings in relation to aspects of neuroscience currently at the cutting edge and, importantly, which look like remaining high profile for a little while to come. As such, we trust that this book will provide sufficient familiarity with molecular biology techniques applied to neuroscience discovery that the exciting (and voluminous) literature in this area becomes accessible. A glance through the major journals, such as *Nature* and *Science*, shows that hardly an issue goes by that does not contain papers on molecular approaches to neuroscience. Moreover, most of these papers do not make easy reading. The density of technical vocabulary in most scientific journals has steadily increased over recent years, and the trend towards ever greater data compression in methods sections which are often now relegated to figure legends or notes at the end of an article makes the task of fathoming exactly how an experiment has been done hard, as the typical example below shows.

"The gene locus was targeted by homologous recombination using two isogenic flanking fragments of 5 kb (*Eco*RI–*Eco*RV) and 2.9 kb (*Apa*I) which were isolated from a mouse 129/SvEv genomic DNA library. *Neo*^{r} cassette and herpes simplex virus *tk* genes were used as positive and negative markers respectively. The linearized targeting construct was electroporated into AB2.1 ES cells and selection achieved using G418 and ganciclovir. The targeted alleles can be detected by the presence of a 12.6 kb fragment instead of 8 kb when DNA is digested with *XbaI* and probed with a 5′ diagnostic probe . . ."

To begin the demystification process, we start with two chapters which concentrate on the more important basic molecular biology methods as they have been applied to neuroscience. In subsequent chapters, the emphasis changes. In these we examine the major families of molecules which are at the interface between the extracellular and intracellular environments of the cell, the receptors and the ion channels which make excitable cells so responsive. We chose to illustrate how many of these molecules may act in an orchestrated way by looking in some detail at plasticity – the way in which the nervous system adapts to changes in its environment – which provides a wonderful excuse for exploring the molecular biology of signaling between neurons, signaling within neurons and the variety of ways in which membrane events can be coupled to changes in gene expression. Later in the volume, we touch upon a number of ways in which the system fails, in epilepsy and in some neuropathologies, and pose the question as to whether there are some common pathways which can link seemingly unrelated problems.

We introduce new methodologies in boxes so as not to disrupt the main text. Finally, we include a glossary of all words which appear in the text in **bold** as an *aide memoire,* and a number of appendices providing information which the reader might want to have to hand when trying to unravel a research paper in molecular neuroscience. There is also a reading list of selected key references and review papers.

There are a number of topics which have not been covered, and the most important of these is growth and development. It was a decision taken with some regret that in a text of this size we could either have covered this subject trivially or one of the other topics covered in the book would have had to be sacrificed. We leave it to either another larger edition or another volume to cover this interesting topic in the detail which it deserves.

One of the great legacies of molecular biology is that it has shown us superfamilies of molecules, members of which are related in terms of both mechanism of action and probably in evolutionary origin. The key question now is surely whether this extraordinarily detailed level of structural knowledge can translate into novel drug design or therapeutic strategies based on genetic engineering. Since drug–receptor interactions are a three-dimensional matter, only when it becomes possible to model how subtle mutations change the way proteins fold and alter the electronic properties of binding sites will it be possible to have a science of molecular biology-based drug design. As for genetic engineering, the ability to treat disorders by switching on or off the expression of particular receptor subunits at specific times and places within the brain is on the horizon in animals specially engineered and bred for the purpose. Achieving the same goal in a sick human is a different and far more difficult proposition.

Chapter 2

Receptor cloning

2.1 Introduction

This chapter describes how classical protein chemistry, combined with molecular biological techniques for manipulating nucleic acids, was used to clone and sequence the α-subunit of the nicotinic acetylcholine receptor (nAChR). This was the first time that this technology was harnessed to examine a neurotransmitter receptor. To appreciate how this was achieved, some of the basic techniques that molecular biologists use are introduced.

2.2 The amino acid sequence of a protein can be deduced by sequencing its DNA

As shown in Chapter 1, an aim of contemporary neuroscience is to learn about the structure of proteins important in nerve cell function, such as ion channels, receptors, neurotrophic peptides and the like. A starting point is to find the primary amino acid sequence of the protein, since from this it is possible to make informed guesses about higher order structure (disulfide bonds, α-helices, etc.) and how the protein relates to the rest of the cell; for example, how a receptor is orientated in the plasma membrane, which regions are extracellular and which intracellular. These guesses may then be tested experimentally in a variety of ways. The basic strategy is to deduce the amino acid sequence by cloning and sequencing the DNA that codes for the protein of interest. By **cloning** is meant making multiple copies of the DNA. This is necessary to provide sufficient material to allow the DNA to be sequenced.

The original cloning technique uses the natural replication machinery of bacteria. It works by introducing the DNA of interest into bacteria (usually a strain of the intestinal organism *Escherichia coli*) which reproduce to produce a colony containing millions of identical individuals or **clones**, each with identical copies of the DNA. However, foreign DNA will not normally replicate in bacterial cells. Neuroscientists are interested in eukaryotic DNA, and the DNA replication machinery of prokaryotes will not usually recognize eukaryotic DNA. The bacterial cells must be tricked into replicating the foreign DNA by first joining it to a **vector**, a nucleic acid molecule that possesses the ability to replicate in the host.

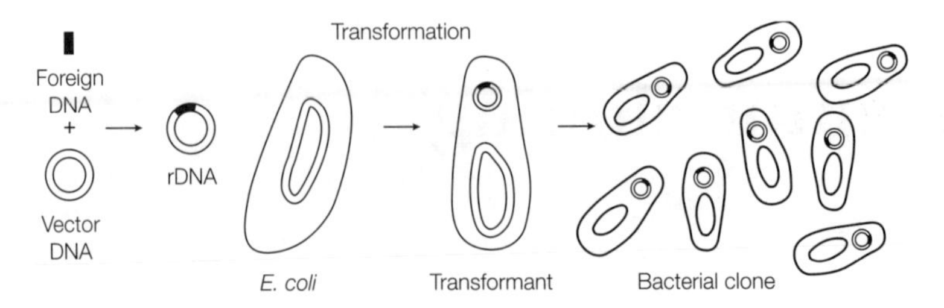

Figure 2.1. rDNA is engineered by joining the DNA of interest to a vector. Reproduction of transformed bacteria is accompanied by replication of rDNA.

The simplest vectors are derived from **plasmids**, double-stranded circular DNA molecules of several thousand base pairs that occur naturally in bacteria. The combination of foreign DNA and vector is a **recombinant DNA** molecule (rDNA) (*Figure 2.1*). To engineer rDNA, it is necessary to be able to cut and join, or **ligate** DNA molecules reproducibly in predictable ways. The rDNA is now introduced into bacteria, a process called **transformation**, which are then grown in culture using standard microbiological methods.

In a typical experiment, many different bacterial clones will be produced, each containing a distinct fragment of DNA, and it is necessary to be able to select the desired transformants, that is the clones containing the DNA of interest. Having identified the correct clones, the DNA can be extracted and sequenced.

In the sections which follow, each of the steps is examined in a little more detail.

2.3 Restriction enzymes can be used to cut DNA

Restriction enzymes or restriction endonucleases are invaluable tools for cutting DNA molecules at precise points. These enzymes recognize specific base sequences in DNA and cut the DNA at defined sites within the recognition sequence. Over 200 restriction enzymes with different sequence specificities have been characterized. In general, the recognition sequences are either four or six bases long, and sequences are rotationally symmetrical, that is the same series of bases is present on each strand of the DNA though arranged in opposite directions. Not all restriction enzymes cut at the center of the recognition sequence (*Figure 2.2*). Some enzymes, e.g. *Eco*RI, cut asymmetrically, leaving protruding 5′ or 3′ single-stranded tails often called '**cohesive**' or '**sticky**' ends. These are very useful when rejoining fragments. Other enzymes, e.g. *Hae*III, cut centrally and leave '**blunt**' or '**flush**' ends.

2.4 Separation of DNA fragments by size

The pieces of cut DNA called **restriction fragments** can be separated and analyzed by gel electrophoresis using agarose gels or, for small fragments, polyacrylamide gels. Gels separate molecules on the basis of size since they consist of a polymer mesh which impedes the migration of larger DNA fragments more than smaller ones. In general,

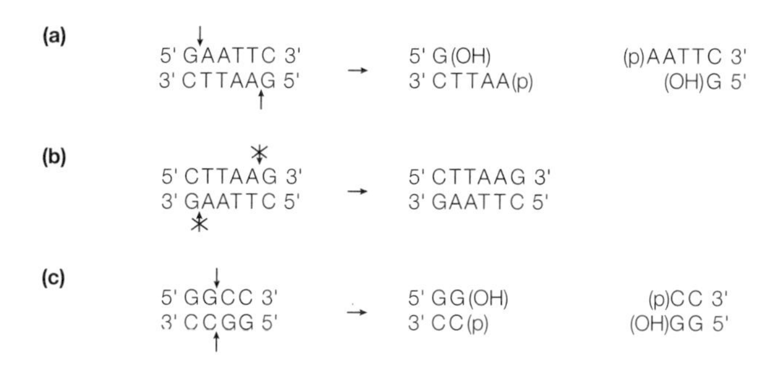

Figure 2.2. Recognition sequences of *Eco*RI and *Hae*III. (a) Vertical arrows indicate where *Eco*RI cuts to produce cohesive ends; (b) the same sequence in the opposite orientation is not recognized by *Eco*RI and so is not digested. (c) *Hae*III cuts to produce blunt ends.

the speed of migration is proportional to $\log_{10}$ of the size of the molecule. The driving force for the migration of fragments is provided by applying an electric field across the gel. DNA carries a net negative charge at neutral pH, due to the negatively charged phosphate groups in the DNA backbone, and so will migrate towards the positive electrode during electrophoresis. The nucleic acids on the gel are often detected by staining with ethidium bromide (EtBr), a molecule which binds to DNA or RNA and fluoresces when excited by ultraviolet (UV) light. Gels are examined under a UV light source to reveal discrete bands. The intensity of the fluorescence is related to the amount of EtBr bound, and this is determined by the number of base pairs. Gel staining can thus be used to quantify the amount of nucleic acid in any band. As little as 1 ng of DNA per band can be detected reliably in this way.

To see how restriction fragments may be identified, consider the digestion of a small circular double-stranded DNA molecule such as a plasmid vector by a restriction enzyme, e.g. *Eco*RI (*Figure 2.3*). In this example, the enzyme attacks at four sites, creating fragments A–D which, being different sizes, can be separated by gel electrophoresis.

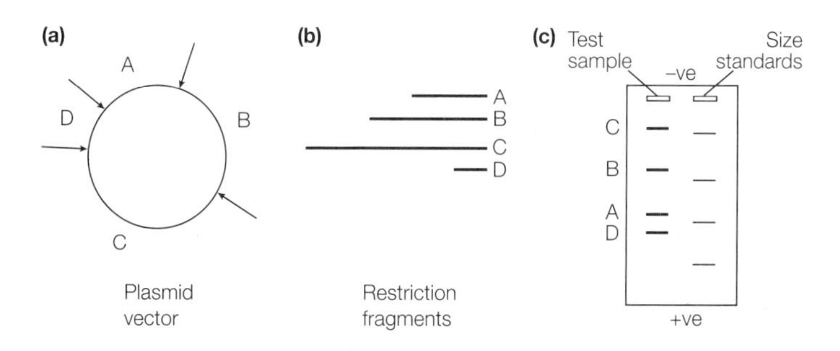

Figure 2.3. Digestion of a circular DNA molecule by a single restriction enzyme. (a) Plasmid vector (the arrows indicate recognition sites for *Eco*RI; (b) restriction fragments; (c) restriction fragments are separated by size using gel electrophoresis. The use of size standards allows the sizes of the test sample restriction fragments to be estimated.

Table 2.1. Fragment length depends on the size of enzyme recognition sequences

Restriction sequence length (bp)	Exemplar enzyme	Fragment length (bp)
4	*Hae*III	$4^4 = 256$
6	*Eco*RI	$4^6 = 4096$
8	*Not*I	$4^8 = 65\,536$

It is not obvious how to work back from the bands on the gel and place them in the correct order on the map. In fact it is necessary to use more than one type of restriction enzyme. Each enzyme is used separately and in pair-wise combinations. A series of logical steps solves the puzzle, and a restriction map, showing the positions of attack of each enzyme, is prepared.

The average size of restriction fragments depends on the length of the enzyme recognition sequence, assuming that the four bases making up DNA are distributed randomly throughout the genome (*Table 2.1*). This is because the number of restriction sites depends on the probability of a particular sequence occurring. Shorter recognition sequences occur more often and so result in shorter fragments. Thus, with four bases to chose from, a 6-bp sequence will occur, on average 4096 bp apart. Enzymes with 6-bp recognition sequences are used most commonly because they produce fragments that are small enough to handle easily but are sufficiently large to be useful.

In fact, the assumption about random distribution of bases is flawed. Mammals have a paucity of C and G bases in their DNA, and even these are often methylated, a modification which blocks digestion by many restriction enzymes. Consequently, the frequency of CG-containing recognition sequences is low in the mammalian genome.

2.5 DNA molecules can be joined using ligases

DNA fragments are joined or **ligated** into vectors such as plasmids by DNA ligases which catalyze the formation of a covalent bond between juxtaposed 5′-phosphate and 3′-hydroxy groups in the DNA backbone. Ligation may be used to join complementary sticky ends to each other, where Watson–Crick base pairing occurs between the complementary sequences of the sticky ends, to join blunt ends or to seal nicks (breaks in a single strand of a double-stranded molecule).

A DNA fragment can be inserted into an open vector if both molecules have complementary single-stranded DNA tails. Cohesive ends left by restriction enzymes provide suitable complementary regions which **anneal** (stick together) (*Figure 2.4*). The hydrogen bonds involved are not enough to form a stable hybrid, but the DNA ligase completes the covalent bond between the molecules.

There are several possible outcomes from a ligation reaction, but generally the most likely outcome is simple resealing of the vector molecule without any inserted foreign DNA. To try to optimize hybrid formation there are several strategies that can be used.

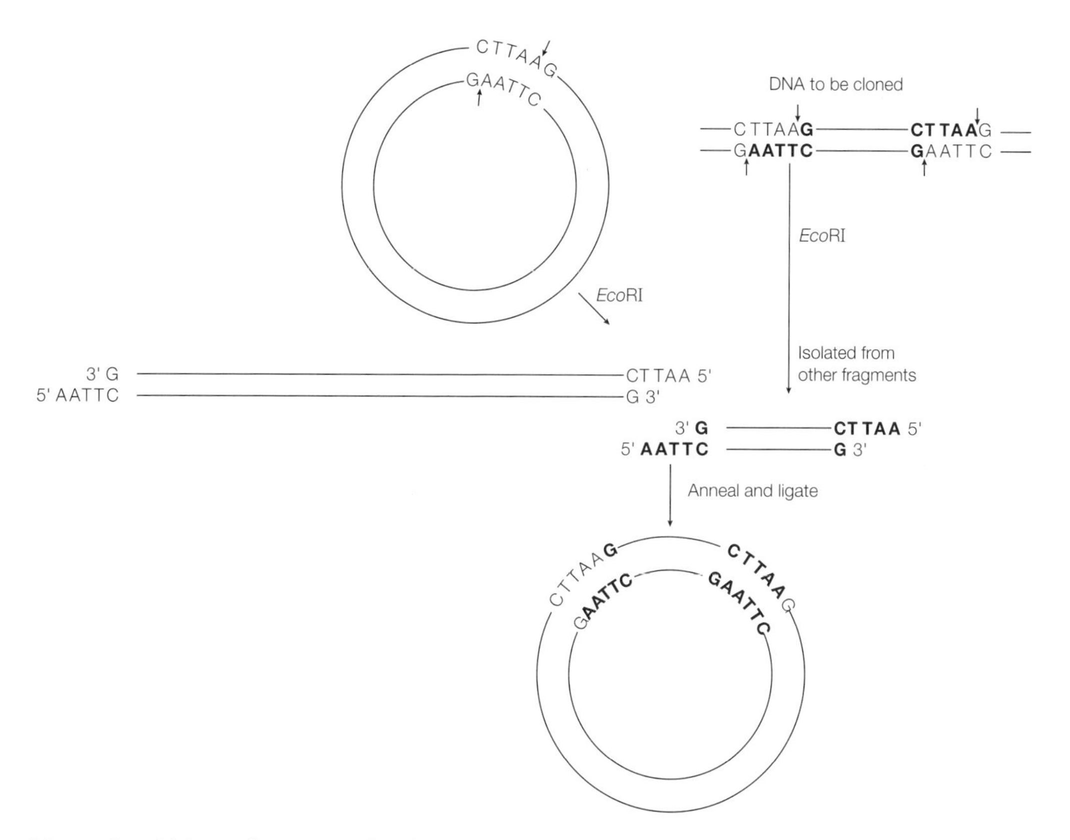

Figure 2.4. Using a ligase to splice DNA fragments into a plasmid vector.

Alkaline phosphatase treatment frequently is used to remove the 3′-terminal phosphate groups from the ends of the vector molecules so that the resulting structure cannot ligate and recircularize. The foreign DNA to be inserted is not treated and hence retains the ability to be ligated into the vector. As a consequence, the vector must incorporate an insert in order to be recircularized successfully. The net result is that the only circular plasmids generated are those containing inserts.

There are situations where blunt-ended molecules (i.e. no single-stranded tails) are to be inserted into a vector. The enzyme T4 DNA ligase is capable of joining blunt ends, but only at a low frequency. A useful method for joining blunt-ended molecules is called homopolymer tailing. The method uses an enzyme called terminal transferase which adds nucleotides to the 3′ ends of double-stranded DNA. Thymidine (T) bases, for example, may be added to the ends of the opened plasmid, and adenosine (A) added to the ends of the insert, thus providing complementary sticky ends.

Other methods of putting sticky ends onto blunt-ended molecules involve the use of **linkers**, synthetic double-stranded oligonucleotides that incorporate one or more restriction sites, and the use of **adaptors**. Adaptors are short synthetic oligonucleotides that have one blunt end and one sticky end. The blunt end of the adaptor is ligated to the blunt end of the DNA fragment producing a new molecule with sticky ends.

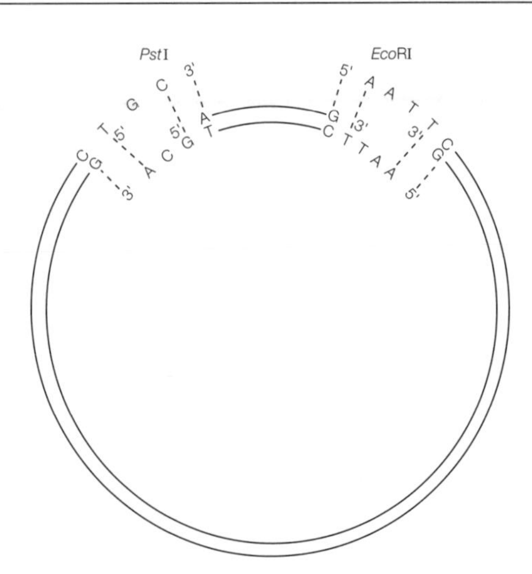

Figure 2.5. Sticky end ligation with different cohesive ends permits directional cloning.

The linker is present in high concentration because, in the absence of complementary tails between the foreign DNA and the linker (both of which have blunt ends), there is no Watson–Crick base pairing for initial annealing. Ligation thus occurs with low efficiency. In order to protect any *Eco*RI recognition sites in the insert DNA from digestion by *Eco*RI, the insert DNA can be pre-treated with *Eco*RI methylase which methylates the restriction site, preventing the action of *Eco*RI.

The orientation that the DNA fragment takes within a vector is often crucial. To ensure that the foreign DNA inserts only in the correct orientation, both vector and foreign DNA are digested with two restriction enzymes producing different sticky ends (*Figure 2.5*).

2.6 rDNA is introduced into bacterial cells for cloning

Introducing genetically engineered DNA into bacteria is called **transformation**. Bacteria will take up DNA naturally, but only with a very low efficiency, so various strategies have been developed to increase DNA uptake. One widely used method incubates actively growing cells plus the DNA at 4°C with a hypotonic solution of $CaCl_2$ or $Ca_3(PO_4)_2$. The salt causes the DNA to be precipitated out of solution as a large macromolecular aggregate. A brief heat shock (42°C for 2 min) encourages the bacterial cells to endocytose the DNA. However, only about one in a thousand cells transform under these conditions and it is necessary to select out the transformants. This is made possible by suitable design of the vector used to create the rDNA.

2.7 Vectors are constructed using genetic engineering techniques

Vectors are the vehicles used to ensure that eukaryotic DNA is cloned in a prokaryotic host cell. Most are based on plasmids or on viruses which infect bacteria known as **bacteriophages**, usually referred to simply as **phages**.

2.7.1 Plasmid vectors

Apart from its main chromosomal DNA which contains about 4×10^6 bp, an *E. coli* cell contains large numbers of circular mini-chromosomes or plasmids, each with several thousand base pairs. The number of copies of a plasmid – the **copy number** – depends on the plasmid and the host cell. **Relaxed control** plasmids replicate to produce 10–200 copies per cell, whereas **stringent control** plasmids have very low copy numbers. Clearly it is relaxed control plasmids which are most useful for DNA cloning. Plasmids carry genes which confer resistance to specific antibiotics. It is this property which makes them useful in selecting transformants.

In general, small DNA molecules are easier to manipulate than large ones, and naturally occurring plasmids are rather large. Hence, plasmid-derived vectors are artificial constructs made by cutting out and joining together only those regions of a natural plasmid that are useful for gene cloning. The crucial features of a constructed plasmid are listed below.

(i) It should be as small as possible and hence easy to handle. Small closed circular DNA molecules are readily isolated by differential centrifugation.
(ii) It must be capable of autonomous replication. This requires a short sequence of DNA called the **origin of replication** (ori). Moreover, this should be a relaxed origin to allow high copy numbers to be achieved. This may be considerably increased (100-fold) by inhibiting replication of the main chromosomal DNA with chloramphenicol. This prevents cell division but the plasmids continue to replicate.
(iii) Marker genes, such as those for antibiotic resistance, should be present, enabling identification of transformed cells.
(iv) It should have single recognition sites for several restriction enzymes to allow foreign DNA to be inserted.

A commonly used family of plasmid vectors are the pUC vectors (*Figure 2.6*). These are derived from an earlier version (pBR322), originally manufactured from bits of three naturally occurring plasmids. pUC plasmids are small (~2.7 kb), have an origin of replication which can give a high copy number (500–700) and contain an ampicillin resistance gene (*amp*^{r}). In addition, they contain a **polylinker**, a short stretch of DNA containing the recognition sequences for several different restriction enzymes.

LIVERPOOL
JOHN MOORES UNIVERSITY
AVRIL ROBARTS LRC
TITHEBARN STREET
LIVERPOOL L2 2ER
TEL. 0151 231 4022

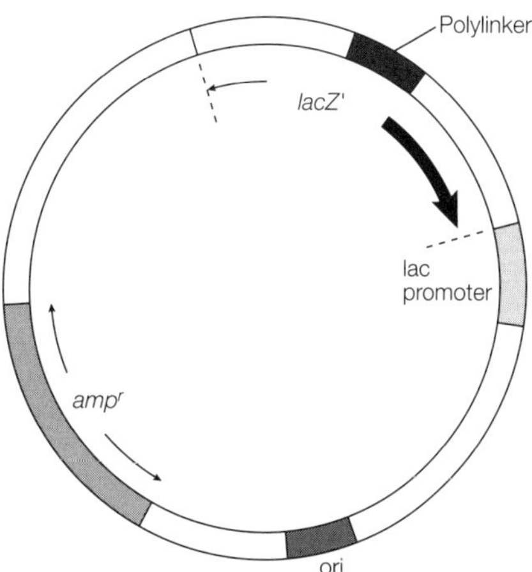

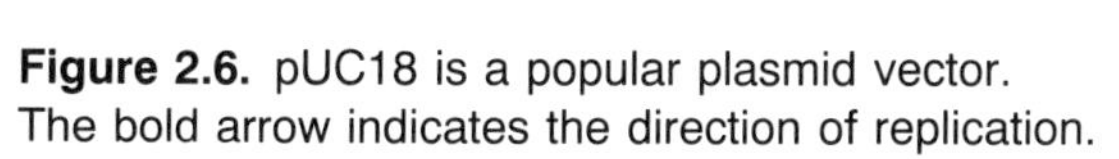
Figure 2.6. pUC18 is a popular plasmid vector. The bold arrow indicates the direction of replication.

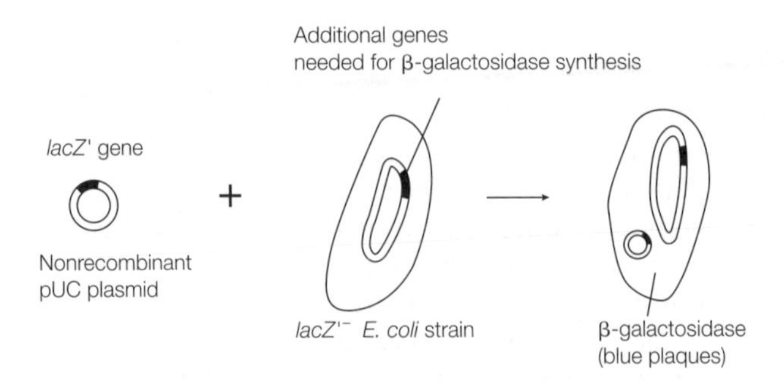

Figure 2.7. Only *lacZ′⁻ E. coli* transformed by the nonrecombinant *lacZ′*-containing plasmid will secrete β-galactosidase.

The pUC plasmid allow selection of transformants in two ways. Firstly, because the vector carries the ampicillin resistance (*amp*^{r}) gene, those *E. coli* cells that have acquired the plasmid will gain resistance to the antibiotic. Growing the cells in a medium containing ampicillin will kill the bacteria which fail to transform, sparing the successful transformants. The second type of selection is for cells transformed by a plasmid which contains the foreign DNA insert. There is no point, after all, in cloning cells that contain only the vector. This selection works because the plasmid has a fragment of the *E. coli* lactose Z (*lacZ′*) gene built into it which, together with the lactose promoter gene immediately upstream, instructs the manufacture of the α-subunit of β-galactosidase (an enzyme which hydrolyzes lactose) in any *E. coli* transformed by nonrecombinant plasmids which do not contain the insert. The *E. coli* used are *lacZ′*⁻, that is lacking the *lacZ′* gene; they cannot synthesize the β-galactosidase α-subunit. However, they can make the rest of the enzyme. Hence, it is the combination of plasmid and *E. coli* that is needed to produce fully functional enzyme (*Figure 2.7*).

β-Galactosidase can be assayed using a colorless synthetic substrate, X-gal. The substrate is cleaved by β-galactosidase to give galactose and an indoxyl derivative. This subsequently oxidizes to give a blue-colored product. *E. coli* colonies synthesizing β-galactosidase show up as blue plaques in a Petri dish.

Although the pUC vector has the polylinker spliced into the *lacZ′* gene, this does not, of itself, prevent the production of functional (though altered) enzyme. However, if a large fragment of foreign DNA has been ligated into the polylinker, the *lacZ′* gene is disrupted and a functional α-subunit of β-galactosidase is not produced. *E. coli*

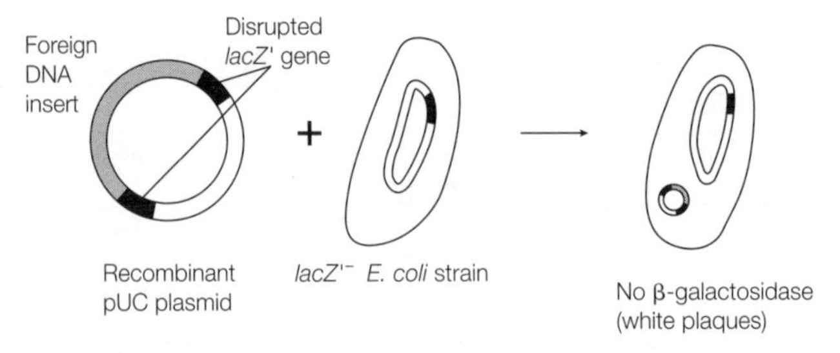

Figure 2.8. Recombinant plasmids have disrupted *lacZ′* genes and fail to make the α-subunit of β-galactosidase. The resulting transformants fail to secrete functional enzyme and remain white in media containing X-gal.

transformed by recombinant plasmids will fail to produce β-galactosidase and will thus appear as white colonies (*Figure 2.8*).

2.7.2 *Phage vectors*

Bacteriophages are viruses which infect bacteria. Many phage vectors are based on λ-phage which naturally infects *E. coli*. λ-Phage has a 49-kbp double-stranded DNA chromosome located within a protein head. A protein tail, attached to the head, allows the phage to dock onto the surface of the bacterium and inject its DNA. One of two things may now happen. Either the phage DNA replicates in the host and hijacks its protein synthetic machinery to manufacture phage proteins, so producing multiple copies of itself and rupturing the cell (the lytic pathway), or the phage DNA is incorporated into the *E. coli* chromosome (the lysogenic pathway). The lytic and lysogenic pathways are controlled by distinct phage genes, but it is only the lytic pathway which is used to produce cloning vectors.

When it infects a bacterium, a phage injects a linear double-stranded DNA molecule bearing cohesive ends, called **cos** sites, that can anneal with each other to form a circular molecule. In this state, the DNA undergoes repeated cycles of replication. The replicated DNA copies do not recircularize but instead anneal to each other via cos sites to form very long stretches of DNA containing many repeats of the viral genome. These extensive molecules are called **concatamers**, and they direct transcription and translation of phage proteins. Packaging of the DNA into new phage requires the linked cos sites in the concatamer. These bind specific phage proteins, permitting the assembly of the head around the DNA between successive cos sites, which are then cut to liberate the DNA primed head. The tail is then added subsequently. Newly produced viruses are liberated by cell lysis and can then infect other bacteria.

For use as a cloning vehicle, it is necessary to retain about 75% of the λ-phage genome to ensure manufacture of new phage particles. However, genes controlling lysogeny can be deleted with impunity to make room for foreign DNA inserts. Typically, a phage vector is engineered to have one or two restriction enzyme sites.

To produce an rDNA molecule using the phage λgt10, it is cut with *Eco*RI (*Figure 2.9*). The foreign DNA is also engineered to have *Eco*RI sticky ends. Now, in the presence of DNA ligase, the foreign DNA is spliced into the λ-phage DNA which assembles to form concatamers via the cos sites. Lysates prepared from λ-phage-infected *E. coli* are added to the DNA which thus becomes packaged into viable new phage particles. The recombinant phages can then be used to infect fresh *E. coli* cells which are spread on agar plates. The result will be phage plaques, each derived from a single recombinant phage.

Phage vectors have some advantages over plasmid vectors. Larger (up to 18 kbp) fragments of DNA can be inserted, and the transformation efficiency of phages is vastly increased over that of plasmids.

2.8 Several methods have been devised to sequence nucleic acids

Once the clone containing the DNA of interest has been identified (precisely how this is achieved is described later), the DNA can be purified and sequenced. Methods for

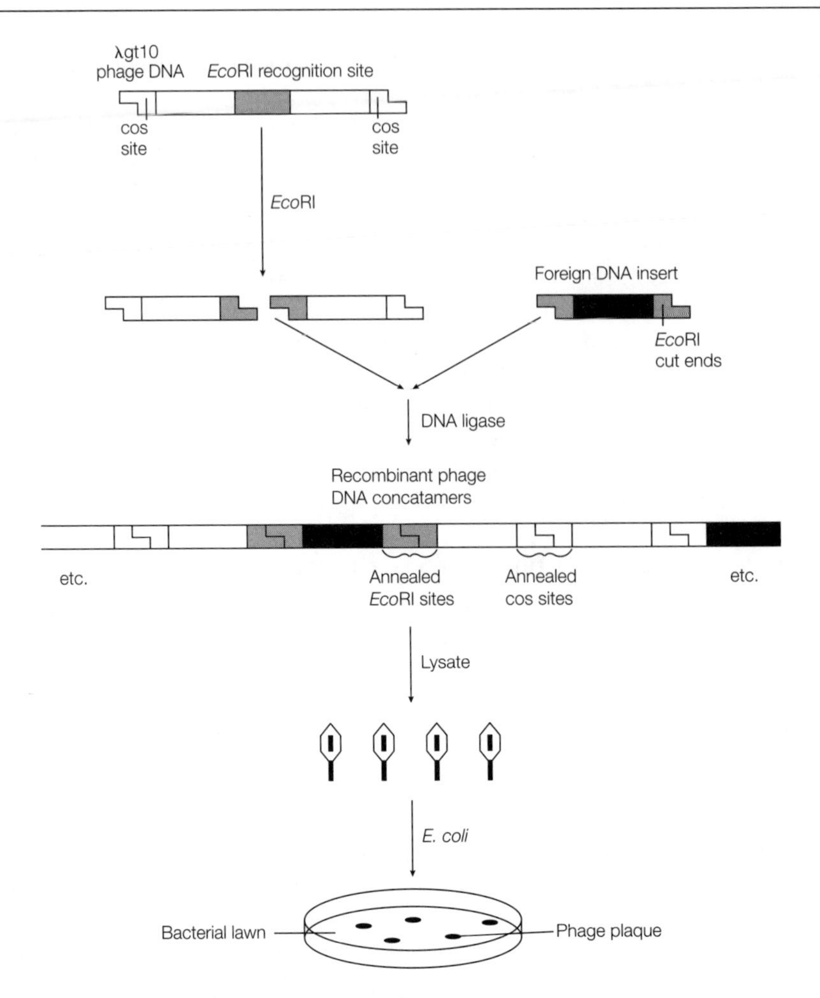

Figure 2.9. Production of rDNA using λgt10 phage.

sequencing both RNA and DNA have been invented. A method of sequencing DNA devised by Fred Sanger called appropriately **Sanger sequencing** currently is the most popular method. The DNA to be sequenced is used as a template for the synthesis of complementary DNA (cDNA) by DNA polymerase I. This requires a primer which anneals to the 3′ end of the template. You may recall that *in vivo* DNA replication always requires a primer because DNA polymerases only recognize double-stranded nucleic acids. *In vivo* replication uses RNA primers, whereas DNA repair *in vivo* uses DNA primers. Furthermore, DNA synthesis proceeds by the stepwise addition of nucleotides to the 3′ end of the growing molecule. In other words, DNA elongation is in the 5′→3′ direction. The sequencing reaction mix contains the four deoxynucleoside triphosphates (dNTPs) that are the precursors for DNA synthesis. These are radiolabeled. The key to Sanger sequencing is the addition of low concentrations of 2′,3′-dideoxynucleoside triphosphates (ddNTPs) to the reaction mix. These molecules are incorporated readily into the 3′ end of a growing DNA via their 5′-triphosphate groups. However, lacking a 3′-hydroxyl residue, they cannot make phosphodiester bonds with any incoming nucleoside triphosphate, with the result that DNA elongation is halted. In practice, four separate reactions are run in parallel, each with a different ddNTP present, together with all four labeled dNTPs, primer and the template DNA for sequencing (*Figure 2.10a*). The concentration of ddNTP is carefully adjusted so that there is a reasonable probability that chain termination occurs at all possible sites. So, for example, the template DNA will contain many A residues. One of the four reactions contains dideoxythymidine triphosphate (ddTTP) at a concentration that ensures ddTTP is incorporated in some growing strands for every position of A in the template. The result is a series of labeled strands, the lengths of which depend on the locations of a particular base (A, say) from the 3′ end of the template DNA (*Figure 2.10b*). These strands are separated from the DNA template by denaturing, and then separated on the basis of size by polyacrylamide gel

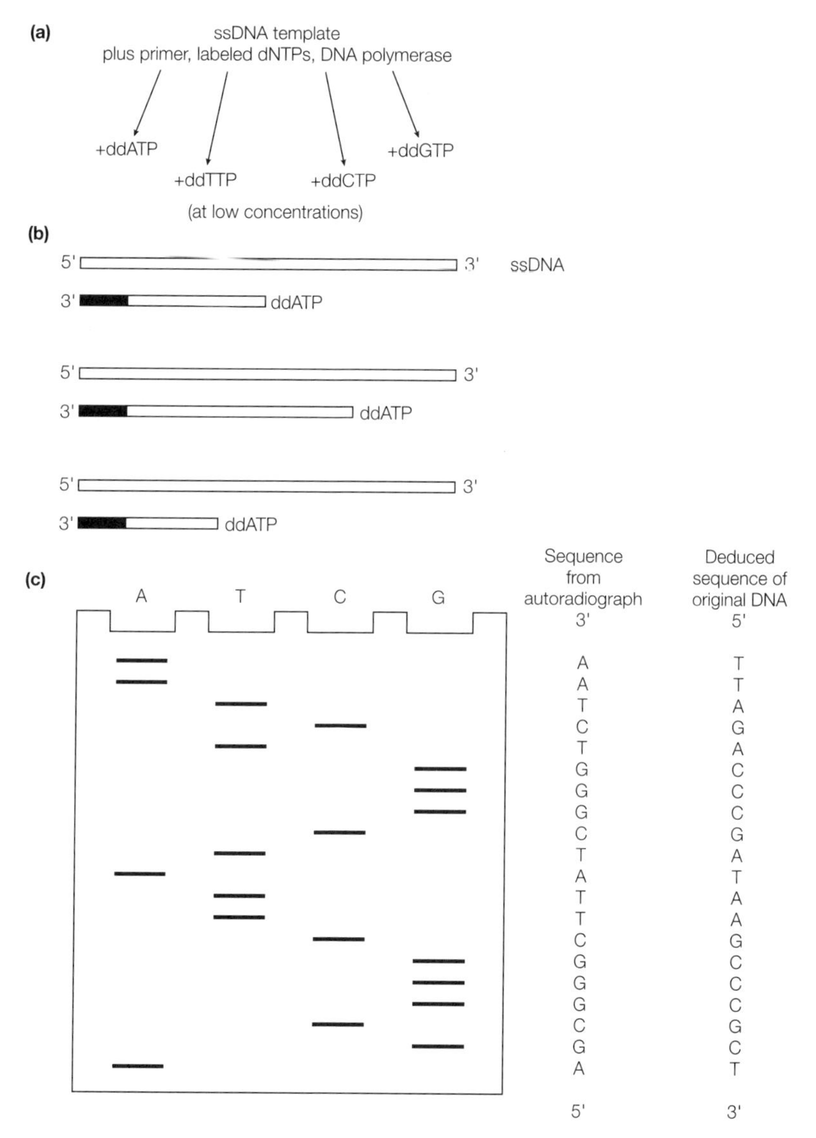

Figure 2.10. Sanger sequencing. (a) The four sequencing reaction mixes. (b) In each reaction mix, chains are produced of differing lengths. In this example, each chain is terminated whenever ddATP replaces dATP. The primer is shown as a dark bar. (c) Autoradiograph of a gel with labeled chains showing the deduced DNA sequence.

LIVERPOOL JOHN MOORES UNIVERSITY
LEARNING SERVICES

electrophoresis. Autoradiography of the gel reveals a pattern of bands for each of the four reactions, from which the sequence can be deduced (*Figure 2.10c*).

A minor variation of the procedure labels the primers rather than the added dNTPs. The importance of DNA sequencing lies in its ease and speed. Whilst RNA sequencing has been developed (also by Fred Sanger), RNA is less stable than DNA, exquisitely susceptible to digestion by contaminating ribonucleases (RNases) and so is appreciably more difficult to work with. Protein sequencing (for example by **Edman degradation**), whilst having an important role in cloning, as we shall see, is laborious, may take months or even years, and can give ambivalent results for some amino acids. In consequence, it is now far easier to clone and sequence the DNA coding for a protein than to sequence the protein itself.

2.9 A strategy for cloning a desired sequence

There are now several approaches to cloning either the DNA coding for a given protein, or a whole gene, including its control elements, such as the promoter, which are of course not transcribed. The first to be developed involved choosing a suitable source of DNA, cutting this into manageable fragments and ligating it into vectors which are then used to transform *E. coli* so as to amplify the amount of DNA. Finally, it is necessary to find the clone(s) containing the desired sequence.

To clone the DNA that encodes a particular protein, there are two potential sources of nucleic acids. One is the total cell DNA, **genomic DNA**. However, the major problem in cloning a desired sequence is identifying it from all the rest of the DNA; it is the ultimate 'needle in a haystack' hunt. The human genome is some 3×10^9 bp long. Clearly proteins of interest to neuroscientists cover a huge size range – from the tripeptide thyrotropin-releasing hormone (TRH) to dystrophin, which weighs in at a massive 3685 amino acids; but, more typically, the α-subunit of the nAChR is 437 amino acids which would be coded by 1311 bases along the DNA coding strand. This represents just 0.0000004 of the total genomic DNA. So, cutting the total genome into restriction fragments, ligating these into vectors, transforming *E. coli* and then hunting for the clone that contains the DNA coding for the α-subunit of nAChR is a truly gargantuan task. It is made worse by knowing that only about 3% of the total genome codes for proteins – most of the rest has no obvious function – so the vast majority of clones produced will contain DNA of no interest. Nonetheless **genomic libraries** – the collection of vectors containing the entire DNA – have been prepared for a number of organisms and are used as source material to hunt for desired sequences. Indeed it is the only useable source if a complete gene is required.

However, the scale of the hunt can be minimized by using the fact that cells transcribe only coding DNA. Moreover, while each nucleated human cell may have about 100 000 genes, a given cell expresses only a fraction of these. Thus, the liver probably expresses only 5000 or so, the brain is estimated to express some 30 000–50 000, the large number presumably reflecting the diversity of neuronal and glial cell types. The pattern of gene expression in a cell will, of course, be reflected in the messenger RNA (mRNA) that it synthesizes. Hence, mRNA purified from specific tissues is a suitable starting material to clone a desired sequence. The mRNA is used as a template to synthesize cDNA, which is essentially a copy of the DNA coding sequence. This cDNA is inserted into vectors to generate a **cDNA library**. Note that it is not possible to clone an entire gene from a cDNA library since it contains only DNA that codes for proteins; no control elements are represented.

Both genomic and cDNA libraries can be screened to discover the clones that contain the desired DNA. One powerful and extensively used strategy is based on the principle that complementary single strands of DNA will hybridize by Watson–Crick base pairing. For this technique to work, it is necessary to know, or to be able to best guess, a short part of the sequence. Radiolabeled oligonucleotides complementary to this sequence are synthesized chemically and used to probe all the clones. If the desired sequence is present, the probe will hybridize to it and can be detected by the label. Alternatively, sometimes bacterial clones will transcribe and translate the protein and so can be screened using antibodies raised against the protein. However, it is worth pointing out that this alternative is rather 'hit and miss' as there are a number of reasons why eukaryotic proteins may not be expressed in prokaryotes.

2.9.1 cDNA is made from total mRNA

The first step in preparing cDNA is to isolate total RNA from the sample of interest. The tissue is homogenized, shaken with acidic phenol and centrifuged so that it rapidly separates into aqueous and organic layers. Phenol denatures proteins, which enter the organic phase. At pH 7, the diester phosphates in nucleic acids are negatively charged. However, the phosphates in DNA are neutralized more easily than those in RNA (DNA phosphates have a greater pK_a) and so at acid pH the DNA goes into the organic phase but the RNA remains in the aqueous layer.

In order to purify mRNA from total RNA, which also includes transfer RNA (tRNA) and ribosomal RNA (rRNA), use is made of the fact that all eukaryotic mRNA is polyadenylated at its 3′ end, the so-called poly(A) tail. Passing the total cell RNA extract down a column packed with deoxythymidine linked to cellulose [oligo(dT)-cellulose] results in mRNA being retained on the column by hydrogen bonding via the poly(A) tails whilst the rest of the RNA is washed through. The mRNA can now be eluted from the column by increasing the ionic strength of the buffer.

The central dogma of molecular biology asserts that information flow in cells is from DNA to RNA to protein, and this is certainly the case generally. However, there are RNA viruses which on infecting cells make DNA copies of their genome. One such virus is the human immunodeficiency virus (HIV). These viruses produce an enzyme, **reverse transcriptase**, which catalyzes the synthesis of DNA from an RNA template. This enzyme is harnessed to make cDNA. It recognizes only double-stranded nucleic acids so a primer is required to reverse transcribe the single-stranded message. An oligo(dT) primer is the obvious choice as it hybridizes to the poly(A) tail of the mRNA (*Figure 2.11*, steps 1 and 2). However, for very long mRNA molecules, the enzyme may not make it all the way to the 5′ end. A trick to get round this is random priming cDNA synthesis in which a large number of short oligonucleotides are synthesized, in the hope that some will by chance be complementary to mRNA sequences scattered along the message, and so act as primers for reverse transcriptase. The end result is

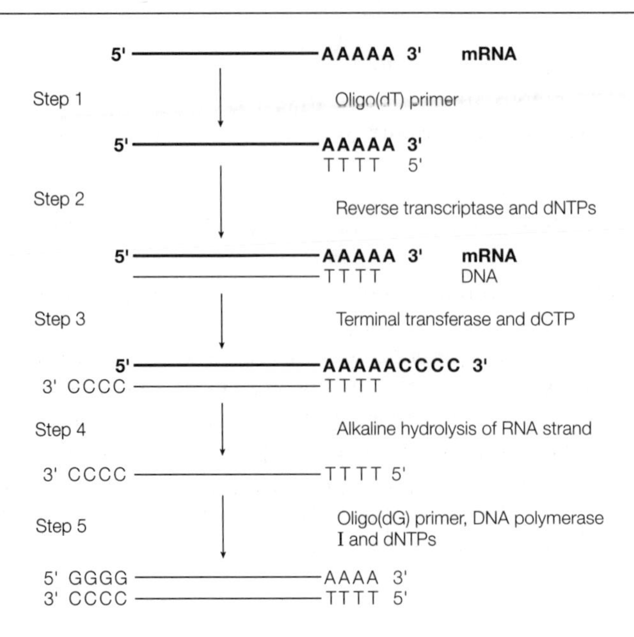

Figure 2.11. cDNA is synthesized from an mRNA template.

an RNA–DNA hybrid, and several procedures exist to transform this into double-stranded DNA that can be used to produce rDNA vectors for cloning.

One method first uses terminal transferase to link cytidylate residues to the 3′ end of the DNA and RNA strands (*Figure 2.11*, step 3). Next, the RNA is digested either by alkaline hydrolysis (step 4) or RNase, and the second DNA strand synthesized by DNA polymerase I using an oligo(dG) primer which is complementary to the 3′ poly(C) tail (step 5).

A second, more recent technique (*Figure 2.12*) uses an *E. coli* enzyme, RNase H, which creates nicks in the RNA molecule cutting it into fragments. These fragments serve as primers for DNA polymerase I to synthesize the second strand. Actually, synthesis is incomplete since nicks remain in the second strand, but these are readily closed by having DNA ligase present in the reaction.

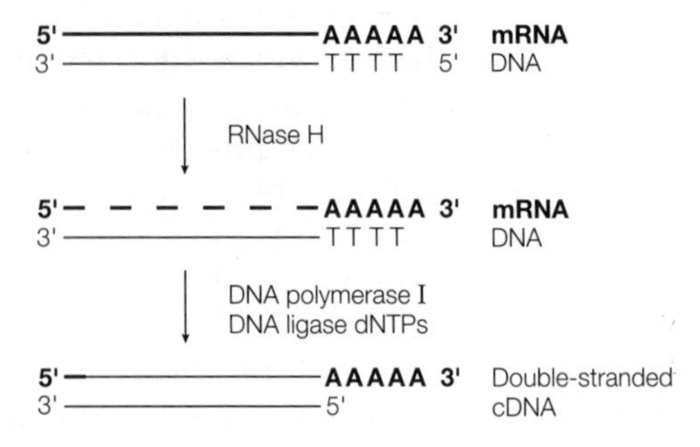

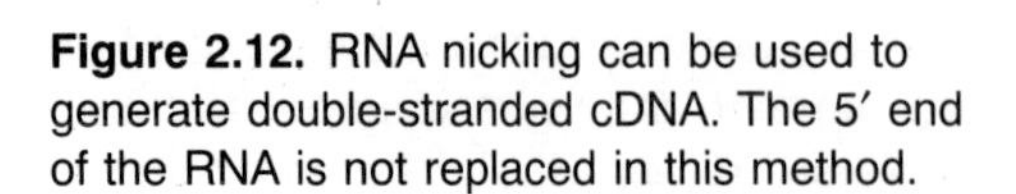
Figure 2.12. RNA nicking can be used to generate double-stranded cDNA. The 5′ end of the RNA is not replaced in this method.

The cDNA is now ligated into vectors, usually by modifying the ends using appropriate linkers as described previously (Section 2.5), so as to create a cDNA library. Transforming bacteria with the library will result in up to a million clones, in the form of either phage plaques or bacterial colonies. These must now be screened to find few – or may be just one – clones that contain the desired sequence.

2.9.2 A cDNA library can be screened using synthetic oligonucleotide probes

If part of the sequence of a protein is known, an oligonucleotide probe can be synthesized chemically which is complementary to the cDNA coding for several consecutive amino acids. The design of the probe needs careful consideration. The minimum length oligonucleotide to recognize a unique sequence in a eukaryotic cDNA library is 15–16. Generally, oligomers with 17–20 nucleotides are used. This requires a knowledge of at least six contiguous amino acids. If it is too short – say a hexamer (6-mer) – it will bind ubiquitously, because the sequence it recognizes is likely to occur frequently. A very long probe will in all probability recognize the crucial DNA uniquely, but would require an extensive knowledge of the protein sequence, which is exactly what the DNA cloning is attempting to discover! The main problem is with codon degeneracy. Recall that the genetic code has several key characteristics. It is **universal**, in that it is common to all organisms (although this is not strictly true since there are minor differences between the prokaryotic, mitochondrial and eukaryotic codes), **specific**, in that a codon uniquely specifies only one amino acid, and – important in the context of probe construction – it is **degenerate**. This means that a given amino acid may be coded for by more than one codon. A glance at the genetic code (see Appendix 1) shows that while methionine and tryptophan have only a single codon, some amino acids have two, three or even four codons. Hence a knowledge of amino acid sequence does not uniquely specify the corresponding DNA coding sequence. In other words, there are several possible DNA sequences which code a run of amino acids, and the more amino acids there are in the sequence the greater the number of possible DNA coding sequences. *Figure 2.13* illustrates this point. Any of 32 sequences could code for this peptide of five amino acids.

Ways to circumvent this are to choose sequences with the minimum of degeneracy (those high in methionine or tryptophan) and to produce degenerate probes; a mixture containing all of the possible complementary sequences. An additional trick is to make the probe length one base less than that needed for an exact number of codons, e.g. a 17-mer rather than an 18-mer. This obviates the need to take account of the degeneracy in the last codon since it is only the 3′ base of any codon which is degenerate (a property known as wobble!). Furthermore, codon usage varies between species, making it possible to design simpler probes, since in a given species some of the degenerate alternatives are very improbable. These probes are often called guessemers! Probes must then be labeled so that the DNA sequences they recognize can be identified (***Box 2.1*** p. 28).

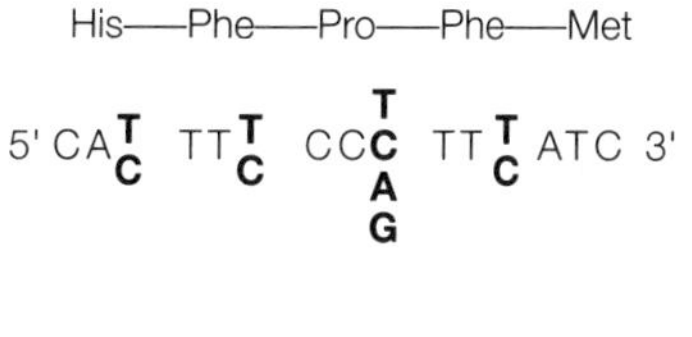

His—Phe—Pro—Phe—Met

5′ CA(T/C) TT(T/C) CC(T/C/A/G) TT(T/C) ATC 3′

Figure 2.13. An example of coding degeneracy.

2.9.3 *Hybridization screening can be used to localize the required clone*

To probe the cloned cDNA library, nitrocellulose filters are placed briefly over the agar surface on which the clones (either bacterial colonies or phage plaques) are growing (*Figure 2.14*) Bacteria or phage particles from each clone are thus transferred to the filter which then becomes a replica of the original plate, which is stored at 4°C. The filters are treated to lyse the bacterial cells or phage particles and the DNA denatured to single strands which bind avidly to the nitrocellulose via the sugar–phosphate backbone. The filters are now incubated with the labeled probe, washed to remove unbound probe, and subjected to autoradiography to localize clones which have bound the probe. Comparing the autoradiogram with the original plate will allow the required clone to be selected. Its DNA may now be sequenced.

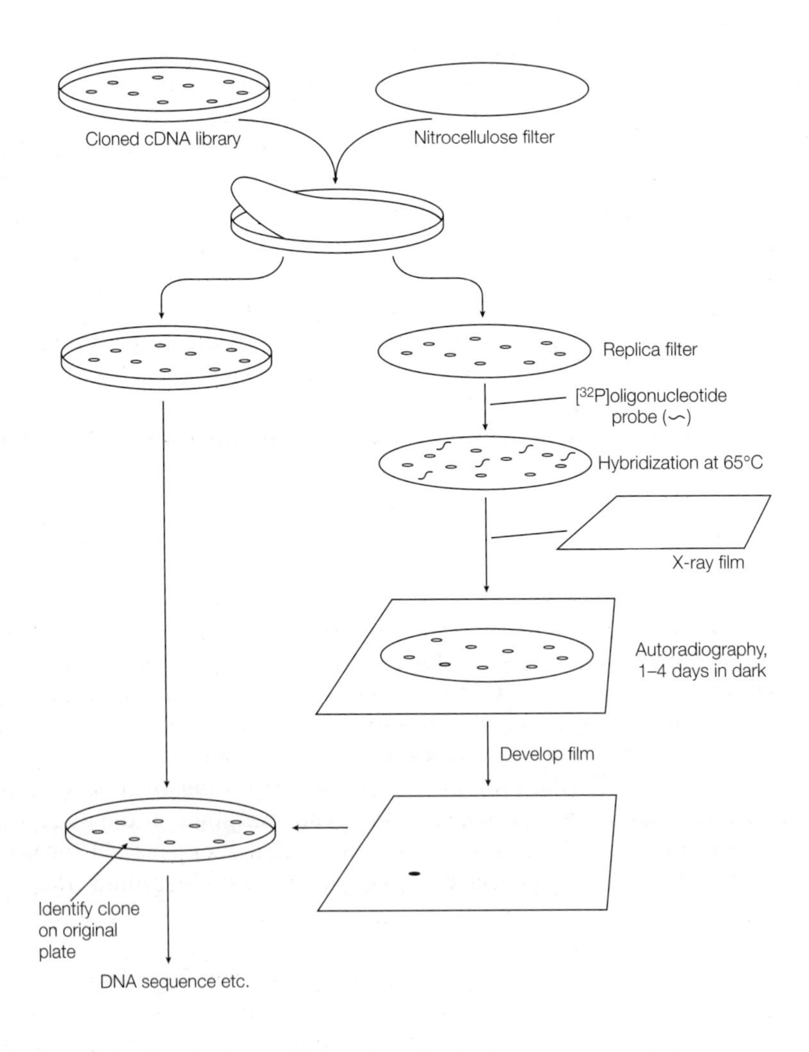

Figure 2.14. Hybridization screening of a cDNA library.

In the circumstances described above, in which the amino acid sequence of a part of the protein was known exactly, the hybridization screening is done under high-stringency conditions, that is both the temperature and the salt concentration of the incubation medium are high (65°C and millimolar concentrations are typical). This ensures that the probe anneals only to DNA sequences that are highly complementary. High-stringency conditions reduce the risk of false positives. However, there are situations in which low-stringency hybridization – a lower temperature (42°C) and lower salt concentration – is desirable. Most ion channels, receptors, growth factors, cell adhesion molecules, etc. that interest neuroscientists fall into families. Since individuals within a family share considerable homology (by definition), a probe designed on the basis of one family member is likely to have some complementarity with the other members of the same family. By relaxing the stringency of the hybridization screen, annealing will occur between the probe and DNA with which it is only partially complementary. This **heterologous gene probe** technique has become very important in identifying novel receptor subtypes in particular.

Clones identified by oligonucleotide probing do not always contain the required DNA. With degenerate probes there is always a risk of false positives, since many of the oligomers in the mixture may hybridize to unwanted sequences. Also, it commonly occurs that the clone contains only part of the desired sequence, the rest having ended up in other clones simply because of the vagaries of the way in which the DNA was cut and spliced into the vectors. However, the DNA in the identified clone may be cut out of the vector, labeled and used as a probe to screen other clones for the rest of the molecule. Where possible, it is useful to use two (or more) probes directed to widely different regions of the target DNA. A clone which hybridizes to both (or all) probes is more likely to have all (or at least a bigger chunk) of the desired DNA.

2.10 The nicotinic cholinergic receptor was sequenced by probing a cDNA library

The strategy and techniques introduced above are essentially those used to obtain the first ever DNA sequence for a neurotransmitter receptor, the nAChR. These receptors mediate transmission at vertebrate neuromuscular junctions, generating end-plate potentials that are the primary triggers for muscle contraction. Characterization, purification and eventually sequencing of the nAChR was greatly facilitated by finding extremely rich sources of the receptor in membranes of the electric organs of the marine ray, *Torpedo* sp. (a marine elasmobranch) and the eel, *Electrophorus electricus* (a freshwater teleost). This was fortuitous, since many neurotransmitter receptors are in low abundance, which means that not only is receptor protein in short supply, but receptor mRNA – the source material for cDNA – is scarce. Furthermore, a snake peptide neurotoxin, α-bungarotoxin (α-BuTX), which has a very high affinity (K_D ~10^{-10} M) for the receptor was available for visualizing the receptor by autoradiography (using radiolabeled toxin) and for purification of the receptor from solubilized *Torpedo* electric organ membranes by **affinity column chromatography** (*Figure 2.15*). Binding of the receptor to the column is so tight that a competitive antagonist is used to elute the receptor from the column. Because of this, another toxin from the cobra, *Naja naja siamensis*, called najatoxin, which has a lower affinity for the receptor, has been used as it is easier to elute the purified receptor protein.

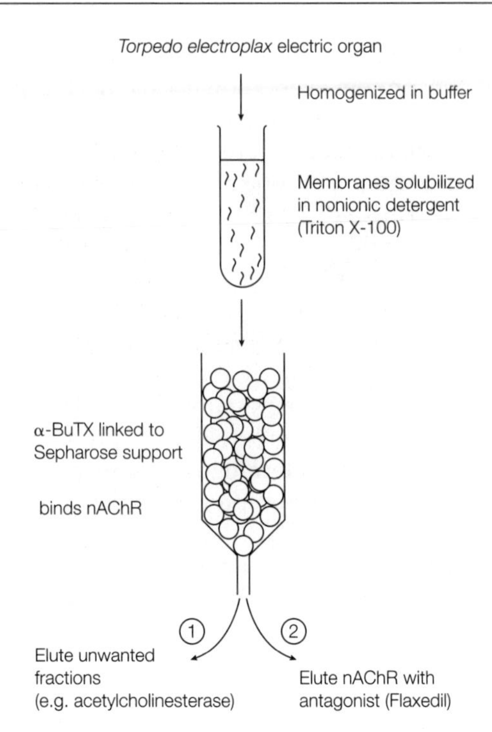

Figure 2.15. Nicotinic receptor can be purified by α-BuTX affinity chromatography.

Running the isolated receptor on denaturing polyacrylamide gels established that it is comprised of four distinct subunits, designated α, β, λ and δ, and that the α-subunit is present in a stoichiometry of 2:1 with each of the other subunits. All the subunits were shown to be extensively glycosylated and capable of phosphorylation. The implication is that the native receptor is a pentamer with $\alpha_2\beta\lambda\delta$ quaternary composition. Physical biochemistry, including X-ray diffraction, had established the size and shape of the receptor, its transmembrane location and that large portions of the protein project from both outside and inside faces of the membrane. Moreover, when viewed from above by electron microscopy, it resembled a rosette surrounding a central pore. Hence, even before the new science of molecular biology had been brought to bear, a fairly comprehensive view of the nicotinic receptor had been built up (*Figure 2.16*).

Purification of the receptor subunits was followed by heroic efforts to sequence the proteins by Edman degradation and, in 1980, one laboratory published the primary sequence of the N-terminal 50 or so amino acids for each of the subunits, showing that they shared considerable homology and hinting at a common evolutionary origin. With this sequence information, Shashoka Numa and colleagues at Kyoto University, Japan were able to construct probes for two stretches of the N-terminal end of the α-subunit. These were used to probe a cDNA library generated from *Torpedo* electric organ mRNA.

The original library consisted of 2×10^5 clones. This was probed with a degenerate probe corresponding to residues 25–29 (*Figure 2.17a*) to yield 57 positive clones. Using a second degenerate probe directed at DNA coding for residues 13–18 (*Figure 2.17b*) resulted in 20 clones. Digesting the DNA in these clones with several restriction enzymes gave identical restriction fragments, showing that they all represented a single

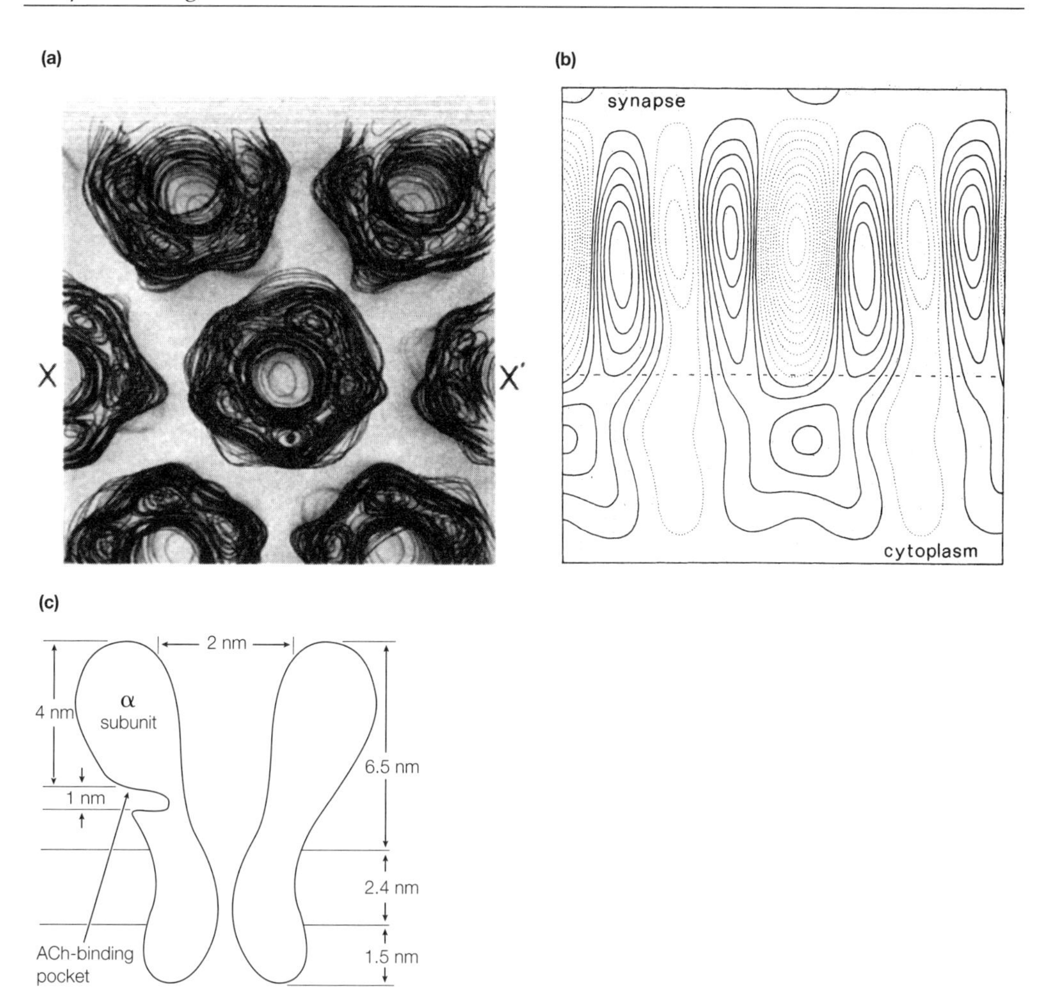

Figure 2.16. (a) Electron density map of *Torpedo* nAChR crystals showing how native receptor would appear in the synapse; (b) section along the line XX′. (c) Cartoon showing how the nAChR lies in the membrane. (a) and (b) reprinted with permission from *Nature,* Brisson, A. and Unwin, P.N.T., Quaternary structure of the acetylcholine receptor, **315**, pp. 474–7. Copyright 1985 Macmillan Magazines Limited.

mRNA species. The two clones with the largest cDNA inserts were selected for DNA sequencing. That the sequence of nucleotides 1–162 corresponded exactly to the α-subunit N-terminal 54 amino acids known directly from the protein sequencing was good evidence that the cDNA identified really did correspond to the α-subunit. The strategy used to clone the nAChR α-subunit is summarized in *Figure 2.18*.

2.10.1 Subunit structure was inferred from the amino acid sequence

Armed with a complete primary sequence, it was possible to make intelligent (and now testable) guesses about the higher order structure of the subunit. The authors deduced that the N-terminal end was extracellular for the following reasons:

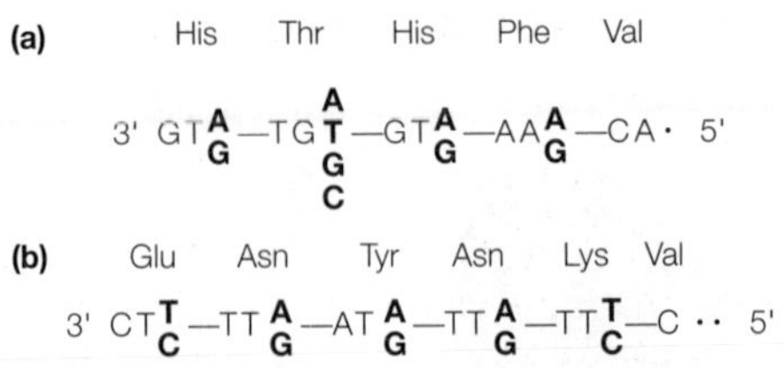

(a) His Thr His Phe Val

3' GT(A/G)—TG(A/T/G/C)—GT(A/G)—AA(A/G)—CA· 5'

(b) Glu Asn Tyr Asn Lys Val

3' CT(T/C)—TT(A/G)—AT(A/G)—TT(A/G)—TT(T/C)—C ·· 5'

Figure 2.17. Probes used to identify clones containing nAChR cDNA. The degenerate probe used for the sequence 25–29 (a) and for the sequence 13–18 (b). Each of the probes is complementary to the strand in the cDNA which is the coding strand in the original gene. Note that the 5′ ends of the probes are truncated.

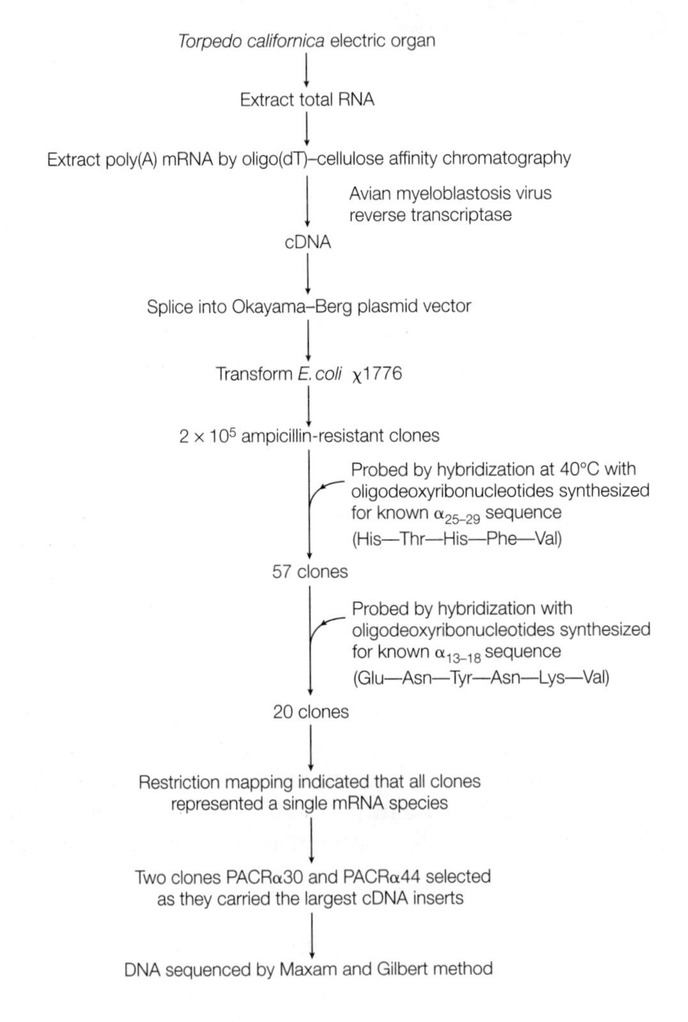

Figure 2.18. Cloning and sequencing the nAChR α-subunit; a similar strategy was used subsequently for the other subunits.

(i) The 5′ end of the cDNA coded for 24 hydrophobic amino acid residues which resembled the signal sequence which ensures that secretory peptides are translocated through the membrane. In the mature α-subunit, this pre-sequence is cleaved off.
(ii) An asparagine residue at 141 appeared to be an *N*-glycosylation site.
(iii) Pharmacology experiments suggested that nAChRs are activated by binding two molecules of acetylcholine (ACh) and, since the biochemistry indicated an $\alpha_2\beta\lambda\delta$ structure, this suggested that the α-subunit carried the ligand-binding site. The N terminus contained a region which bore all the hallmarks of being the binding site for ACh. This was thought to consist of an anionic site for binding the positively charged nitrogen of the choline and an esteratic site which recognized the carbonyl group.

A photoactivatable competitive antagonist of ACh, *p*-(*N*,*N*-demethylamino)-benzenediazonium fluoroborate (DDF), that covalently reacts with native receptor upon UV irradiation, had been shown to label two adjacent cysteine residues (Cys192 and Cys193) in the N terminus. Moreover, a second competitive antagonist, *4*-(*N*-maleimido) benzyltrimethylammonium (MBTA), bonds covalently with sulfhydryl groups exposed on Cys192 and Cys193 after reduction of the receptor with dithiothreitol. The implication of these experiments is that the anionic binding site for ACh must lie within about 1 nm of these cysteine residues. These adjacent cysteines are now known to form a disulfide linkage. Within the neighborhood, stabilized by a β-pleated sheet secondary structure, are acidic amino acid residues and a histidine residue, candidates for anionic and carbonyl binding sites.

Clearly, this candidate ligand-binding site would be extracellular. We point out that implicit in the above model is the assumption that *only* the α-subunit has amino acids capable of influencing the binding of ACh. The advent of more modern genetic engineering technology shows that this is probably not the case and is a matter taken up again in Chapter 5.

Amino acids differ in their hydropathicity (see Section 4.2.2). Some such as phenylalanine are hydrophobic whereas those with a charged side chain (e.g. aspartate or arginine) are hydrophilic. Hence, regions with amino acids that were relatively hydrophobic were thus hypothesized to be membrane-spanning segments (M1–M4, see *Figure 2.19*)

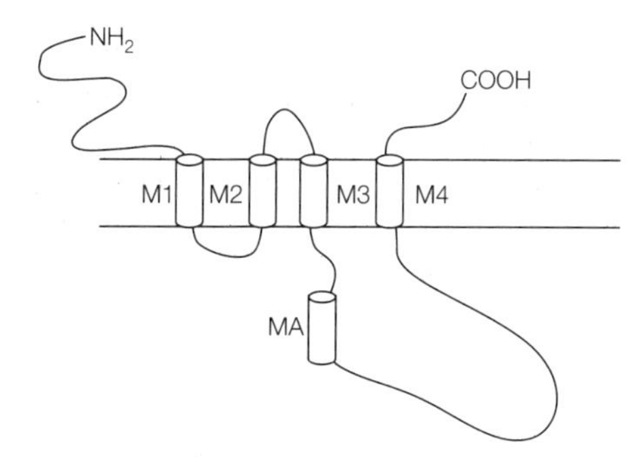

Figure 2.19. Model of nAChR α-subunit structure.

2.10.2 *High homology was demonstrated between subtypes*

Within a year of the first sequence being published, all of the subunits of *Torpedo* nAChR had been sequenced and, soon after, subunit sequences of nAChRs of other species including chick, calf, mouse and human were being reported. Comparisons showed that there is extraordinarily high homology between the same subunit of different species (for the α-subunit there is 97% homology between human and calf and even 80% homology between human and *Torpedo)* and fairly high homology between different subunits.

2.10.3 *Functional receptor can be expressed in* Xenopus *oocytes*

Cloning and sequencing the cDNA is only the first step in the molecular biological analysis of a protein. At some stage, it is crucial to be able to transcribe and translate the cDNA, that is to express it, in some suitable *in vitro* system in which it may be characterized by appropriate biochemical, physiological and pharmacological assays. At a trivial level, this provides the definite proof that the cloned cDNA actually does code for the desired protein. More importantly, it allows different combinations of subunits to be expressed together so that the roles of individual subunits can be discerned. Furthermore, techniques exist to mutate the cDNA in precise ways. When expressed, mutant receptors can be functionally assayed to provide invaluable information about which bits of the molecule are involved in which functions. The *sine qua non* for this **site-directed mutagenesis** (see ***Box 4.3***, p. 74), one of the most important tools of the molecular neurobiologist as seen in subsequent chapters, is that the protein can be expressed.

One of the most useful, and earliest expression systems developed was the use of the *Xenopus* oocyte. *Xenopus laevis* is a South African frog which has large (~1 mm diameter) egg cells (oocytes) that can be harvested by a simple surgical procedure. When microinjected with eukaryotic poly(A)$^+$ RNA, it will often (though not always) translate it, expressing sufficient protein after a few days to enable successful characterization. *Xenopus* oocyte expression is well suited to the analysis of ligand-gated ion channels such as the nAChR, since the oocytes do not naturally synthesize these proteins, and they can be subjected to all the usual battery of electrophysiological procedures developed to study nerve and muscle cells; intracellular recording, voltage clamping and patch clamping.

There are now a number of techniques for transcribing the cDNA to obtain the mRNA for injection into oocytes. In the case of the nicotinic receptor, the cDNAs coding for the four subunits were cut out of their recombinant plasmids and ligated into new constructs containing the origin of replication and promotor for simian virus 40 (SV40) early genes. This was used to **transfect** COS monkey cells. The rationale for this approach is that any plasmid containing the SV40 gene control elements would be expected to replicate to high copy numbers inside monkey cells. The method for expressing nAChR subunits is summarized in *Figure 2.20.*

Expression of all four subunits of *Torpedo* nAChR in voltage-clamped *Xenopus* oocytes showed inward currents that could be blocked by the nicotinic receptor antagonist,

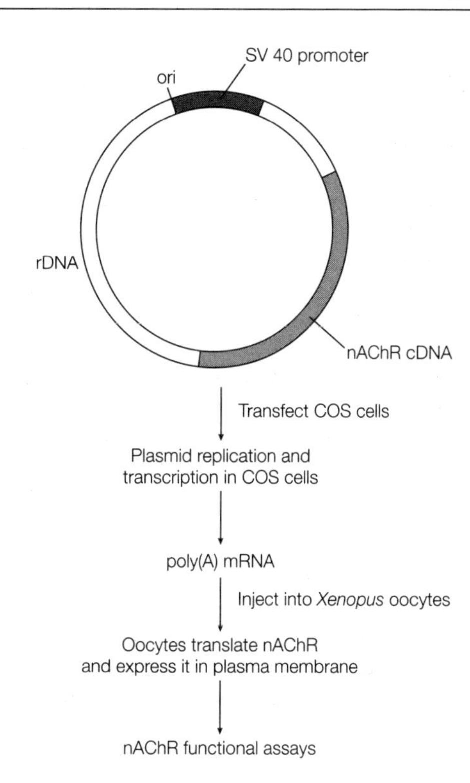

Figure 2.20. Expression of nAChR subunits using COS cells and *Xenopus* oocytes.

D-tubocurarine, but not atropine, a muscarinic antagonist. The oocytes bound [^{125}I]α-BuTX with much the same affinity as purified native receptor, and log dose–response curves for the ACh agonist, carbamylcholine, were the same in oocytes injected with nAChR mRNA or purified native receptor protein. These experiments demonstrated that the original cDNA really did code for nAChR. Subsequent experiments in which only three of the four possible subunits were expressed together revealed that all four subunits are needed to obtain functioning receptor.

Many receptors, ion channels and other proteins of importance in the nervous system have been cloned and sequenced by variations on the theme reviewed above. More recently, the advent of other techniques, such as the polymerase chain reaction (PCR), has revolutionized DNA cloning, and in subsequent chapters we shall see that this has been of immense utility for molecular neuroscience.

Box 2.1. Labeling of oligonucleotide probes

Labeling can be either radioactive or non-radioactive, and relies on the inclusion of a reporter molecule in the probe in order to visualize the position of the probe. Three methods for radiolabeling (listed below) are in common use, and probes labeled with radioisotopes such as ^{3}H, ^{35}S and ^{32}P can be visualized by autoradiography. However, these methods require a relatively long exposure time and, in techniques which require the simultaneous visualization of two or more probes (see Chapter 3), autoradiography is not able to distinguish between different probes in the same samples.

(i) Nick translation works on the principle that however carefully DNA has been prepared, some nicks will be present. DNA polymerase I will catalyze the repair and the DNA will be labeled if [^{32}P]deoxynucleoside triphosphate (e.g. [^{32}P]dATP) precursors are provided (*Figure 1a*).

(ii) End labeling can be used to label the ends of double-stranded DNA providing they are sticky. The required catalyst is the Klenow fragment of DNA polymerase I, which has the polymerase activity but lacks the nuclease activity of native enzyme. This fills in the sticky ends which once again will be labeled if radiolabeled precursors are provided (*Figure 1b*).

(iii) Random priming is predicated on the principle that in a sufficiently varied assortment of random hexamers, some will recognize part of the probe. The Klenow fragment is then used to fill in the gaps with labeled dNTPs (*Figure 1c*).

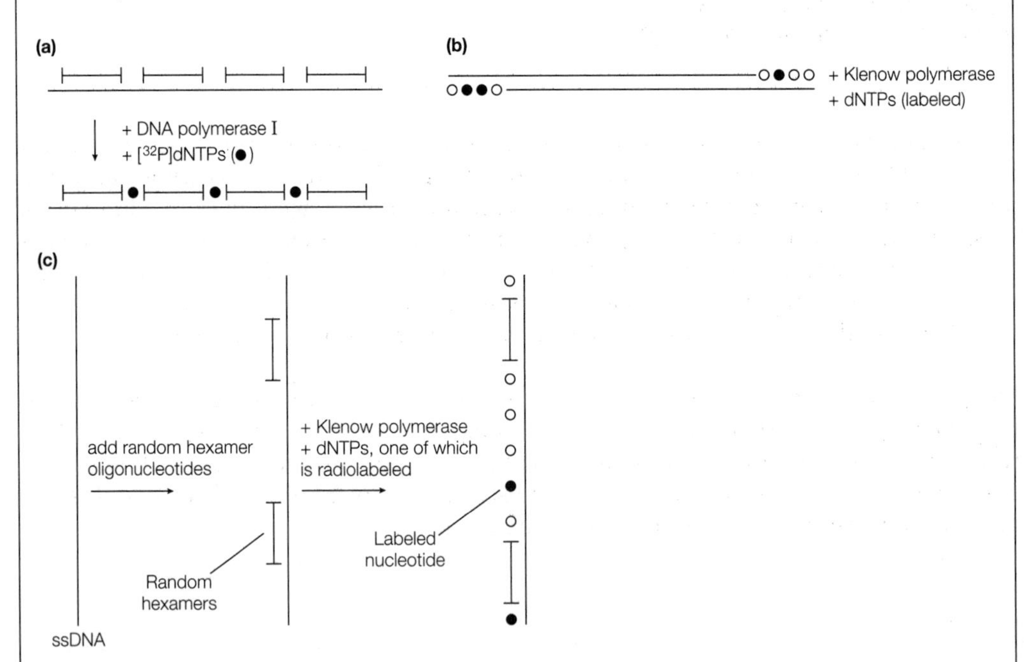

Figure 1. Methods for radiolabeling of oligonucleotide probes.

Continued

Nonradioactive methods of probe labeling have the advantages of fast visualization and high resolution. There are a number of different nonradioactive methods which involve the incorporation of a reporter molecule such as digoxigenin-UTP into the probes. This can then be detected with an antibody specific for the reporter molecule. This antibody can be conjugated to the enzyme alkaline phosphatase and, in the presence of substrate, a blue precipitate is formed (*Figure 2a*).

A method of directly labeling the probe with alkaline phosphatase involves the incorporation, during the synthesis of the probe, of a modified nucleic acid which carries a linker arm with a terminal primary amine group. The alkaline phosphatase is cross-linked to the amine group attaching the enzyme directly to the probe (*Figure 2b*). This method has the advantage of very low background, which gives a high contrast between positive and negative samples.

Another nonradioactive method is chemiluminescent labeling. In this method, the enzyme horseradish peroxidase is linked to the probe by glutaraldehyde. On addition of luminol and hydrogen peroxide, light will be produced at the site of the hybridized probe (HP) (*Figure 2c*).

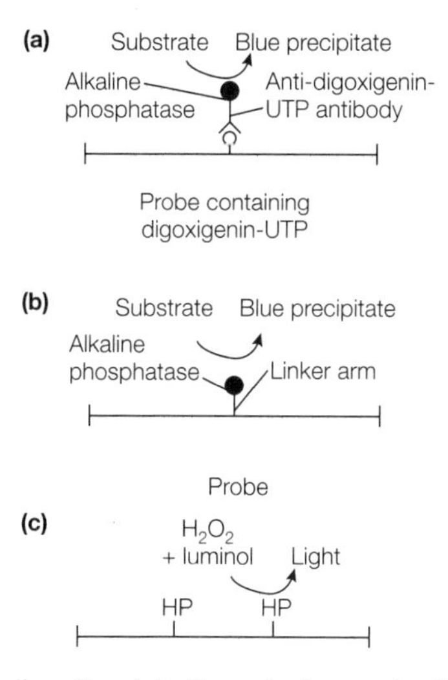

Figure 2. Methods for nonradioactive labeling of oligonucleotide probes.

Chapter

3

Molecular anatomy of the nervous system

3.1 Introduction

The human brain may contain up to 10^{12} neurons, of many different types, from the large pyramidal cells of the cerebral cortex to the small granule cells of the cerebellar cortex. In addition, there are possibly up to 10 times that number of glial cells; astrocytes, oligodendrocytes and microglia. How this diversity is derived and controlled is a major problem in developmental neurobiology and in plasticity, but, before any investigations can be carried out at the molecular level into how a particular function is performed in the brain, it is important to identify both the specific molecules involved and their locations.

An upper estimate of the number of genes in the human genome is about 100 000, but not all of these are expressed in all cells, although, of course, all nucleated cells contain the entire genome. There are many genes which are thought to be expressed only in the nervous system, and estimates of the number of different genes expressed in brain range from 30 000 to 50 000. The brain-expressed genes also seem to be larger, with an average mRNA size of 5 kb, which is twice the size of abundant mRNAs from nonbrain tissues.

3.2 Many neurotransmitters can be located by histochemistry

Since the ideas of chemical transmission at synapses were first hypothesized in the early 1900s by Elliot, Dixon and Dale, and elegantly shown in 1921 by Otto Loewi with the demonstration of cholinergic transmission in frog hearts, many methods have been devised to locate specific neurotransmitters to particular synapses.

In the substantia nigra, it is possible to identify visually the presence of neuromelanin, which is associated with the neurotransmitter, dopamine, by the natural dark color of the pigment which gives the brain region its name. However, most molecules in the brain have to be labeled in some way in order to locate them. One way in which this can be done is by the histological staining of thin tissue sections.

Some neurotransmitters, such as the catecholamines, can be identified chemically by exposure of the tissue sections to formaldehyde, which gives rise to compounds which are fluorescent when exposed to UV light. For other compounds, it is possible to use **immunocytochemistry**, where antibodies have been raised either against the neurotransmitter itself or the enzymes involved in its synthesis or degradation. In this way, cholinergic neurons have been identified using antibodies raised against choline acetyltransferase (CAT) and GABAergic neurons identified using glutamic acid decarboxylase (GAD), both of which are present in the synapses.

A variety of immunological techniques can be used in these types of assay, but a method which is both sensitive and specific involves the use of a secondary antibody (*Figure 3.1a*). The tissue sections are first incubated with the primary antibody, usually a monoclonal immunoglobulin G (IgG), which is not modified in any way, raised against the specific target in one species (e.g. mouse anti-CAT). A second polyclonal antibody, raised against IgGs of the first species (e.g. rabbit anti-mouse IgG) which is labeled in some way, is then applied and the label visualized. The label may be a radioactive label which can be visualized using **autoradiography** or a fluorescent molecule which is attached to the secondary antibody. Another alternative method uses the vitamin, **biotin**, which is attached to the antibody. This biotinylated protein is then incubated with labeled avidin or streptavidin which have a very high affinity for the biotin. In fact, the binding is so tight ($K_a > 10^{15}\ M^{-1}$) that the binding is essentially irreversible. Avidin and streptavidin also have multiple binding sites for biotin, allowing more than one molecule to bind to each biotin molecule, thus amplifying the signal.

In the case where the label is an enzyme such as horseradish peroxidase (HRP) or alkaline phosphatase (AP), the assay is called an enzyme-linked assay (*Figure 3.1b*). In this case, the sections will be incubated with a chromogenic enzyme substrate in order to locate the antibody. A variety of substrates are available for each of these enzymes which can produce colored precipitates, fluors or **chemiluminescence**. An advantage of this enzyme labeling is that the signal is amplified, but care must be taken to avoid nonspecific binding which could mask the specific signal.

The peptide neurotransmitters are particularly suitable for detection by the immunological methods as their large size makes the production of suitable antibodies relatively easy. These methods are also widely used to detect a wide range of proteins, including

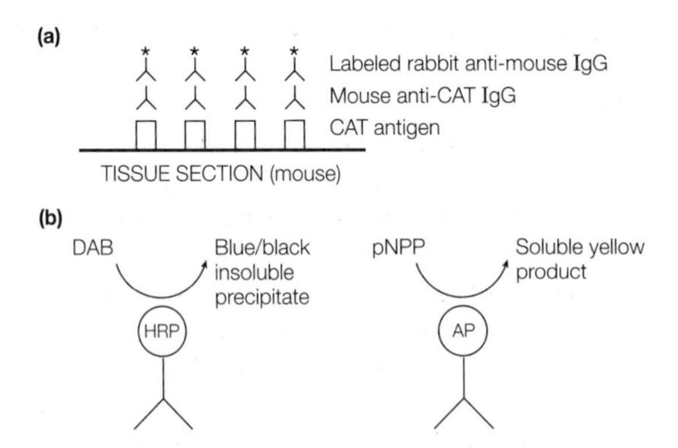

Figure 3.1. Immunocytochemical localization of CAT. (a) A labeled secondary antibody binds to the anti-CAT (primary) antibody. (b) Enzyme-linked antibodies are incubated with the appropriate substrate, e.g. 3,3′-diaminobenzene (DAB) and *p*-nitrophenyl phosphate disodium (pNPP).

neurotransmitter receptors and neurotransmitter-associated enzymes. The availability via gene cloning and *in vitro* gene expression of large amounts of receptor protein has enabled antibodies to be raised against many different receptors and receptor subtypes, allowing localization of the pathways using these neurotransmitters.

3.3 ELISA tests can detect a wide range of molecules

The most widely used immunoassays are enzyme-linked immunosorbent assays or **ELISAs**, which use similar techniques to the enzyme-linked immunocytochemistry. The main difference is that the antigen or antibody is absorbed directly onto plastic microtiter plates, usually 96-well plates or 12-well strips (*Figure 3.2*). Because the assay is carried out in a fixed medium, this method lends itself well to automation and is now widely used in analytical laboratories. There are now a large number of commercial ELISA assay kits available for a wide range of molecules. This method can detect either antigen directly absorbed onto the plate or in a refinement, called 'antigen capture' assays, which can detect extremely low levels of antigen. The plate is first coated, by absorption, with an antibody, after which the antigen is added and 'captured' by the antibody. The antigen is then detected using a second antibody which is raised against a different **epitope** of the antigen to the first antibody.

3.4 Some neurotransmitters can be located by specific uptake mechanisms

Due to the involvement of glutamate in the metabolism of all neurons and glia, the enzymes responsible for its synthesis are not located solely in glutamatergic neurons so they cannot be identified specifically using immunocytochemical methods. One method which has been used to try and identify pathways using glutamate exploits the fact that the action of glutamate is terminated by high-affinity sodium-dependent

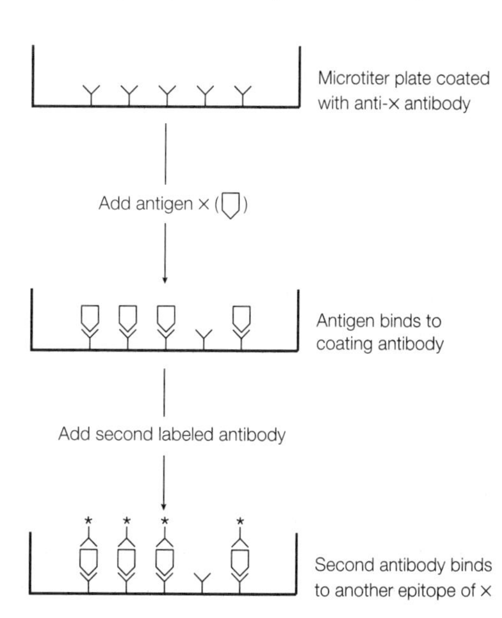

Figure 3.2. Antigen capture ELISA assays can detect very low levels of antigens.

reuptake systems present in the nerve terminals and surrounding glial cells. Using radiolabeled D-aspartate, which whilst being a good substrate for the transporter is not metabolized, it is possible to identify glutamatergic synapses. This technique has been successful in identifying numerous pathways by the retrograde axonal transport of labeled D-aspartate.

In all of these histological studies, a method of confirming the presence of a specific neurotransmitter pathway is by lesioning the putative pathway and, after allowing time for the synapses to disintegrate and be removed, measuring a reduction in the marker used. However, this may not be as effective for uptake studies, as gliosis at the site of the disintegrating synapses may increase glial uptake, even though neuronal uptake is reduced. Other methods used to identify glutamatergic neurons use the fact that the glutamate released as a neurotransmitter is in a separate pool from that involved in cell metabolism and that this pool can be labeled using [^{3}H]acetate. It has since been discovered that at glutamatergic synapses, released glutamate is taken into both presynaptic terminals and glia. The glial glutamate is converted to glutamine by glutamine synthetase, this glutamine is released from the glial cells and taken up by the glutamatergic neurons, where glutaminase converts the glutamine back into glutamate (*Figure 3.3*).

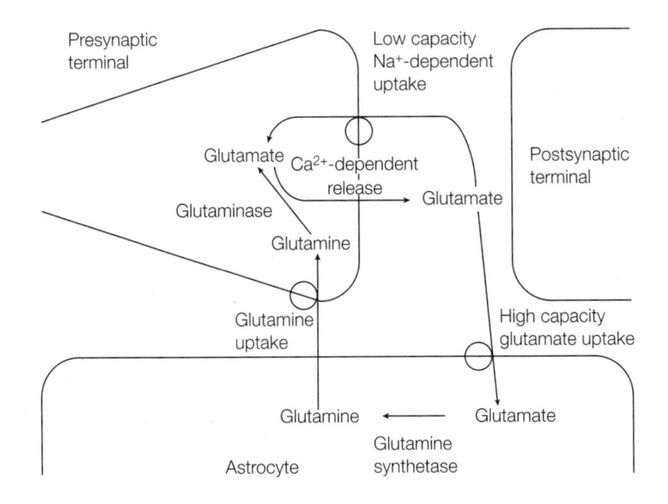

Figure 3.3. Pathways for the uptake and conversion of neurotransmitter glutamate.

3.5 Receptors can be counted using receptor ligand-binding assays

Receptor binding assays make it possible to locate receptors in a quantitative as well as a qualitative fashion. The method consists of incubating the preparation, which could be tissue slices, homogenized tissues or isolated membrane fractions, with a radiolabeled ligand which has a high affinity for the receptor, until equilibrium is reached. After this, the remaining free ligand is separated from the receptor–ligand complexes and the amount of bound radioactivity counted. There are many variations of this method and there are a number of technical difficulties which can make the

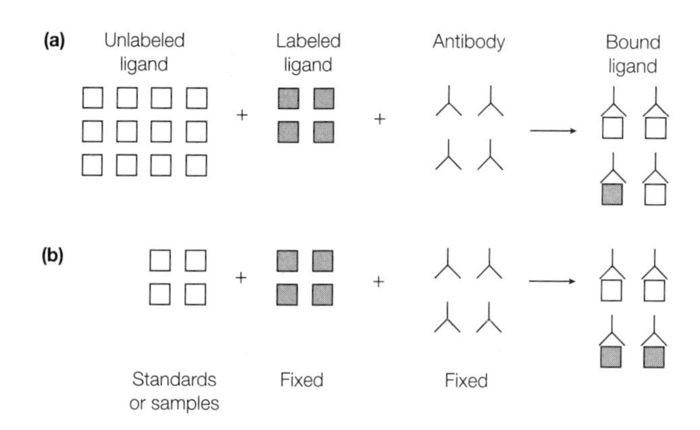

Figure 3.4. In radioimmunoassays, the radiolabeled ligand bound to the antibody varies inversely with the concentration of the unlabeled ligand. (a) High concentrations of unlabeled ligand lead to low concentrations of bound labeled ligand, and (b) vice versa.

interpretation of the results difficult; however, the major problem with these methods is that they require quite large amounts of material and a sufficiently high affinity ligand.

Radioimmunoassays are methods for quantifying receptors where a limited amount of antibody binds to saturating amounts of the ligand (i.e. receptor) (*Figure 3.4*). The amount of free and bound ligand is detected by the addition of a very small (tracer) amount of radiolabeled ligand. In this way, the labeled and unlabeled ligand compete for binding to the antibody. The bound and unbound (or 'free') ligand are then separated, often by adsorption of the free ligand onto coated charcoal, and the radioactivity of the bound fractions counted. Standard curves can be calculated by using known quantities of ligand.

3.6 In situ hybridization can determine patterns of gene expression

The techniques of *in situ* hybridization use the fact that single-stranded nucleic acids will hybridize by base pairing with their complementary sequences (see also Section 2.9.3). So, in order to locate particular mRNA in a tissue section, it can be incubated with a labeled nucleic acid which is complementary to the mRNA of interest. Only those cells, or regions within cells, which contain the mRNA will be labeled and, under the correct conditions, the labeling will be proportional to the amount of the specific mRNA present in the section. The labeled probes can be made from various types of nucleic acids: single-stranded cRNA, single-stranded cDNA or synthetic complementary oligonucleotides. The probes can be labeled either radioactively, using ^{3}H, ^{32}P, ^{35}S or ^{125}I, or nonradioactively (see ***Box 2.1***, p. 28). Using small oligonucleotide probes, for example a 48-mer, it is possible to design experiments not only to distinguish the distribution of the mRNA coding for a particular protein, but to distinguish, at the single-cell level, which particular subtypes of that protein are being expressed. In this way, the distribution of protein kinase C (PKC) subtypes has been investigated in the hippocampus, where PKC-α is most prominent in CA3 pyramidal cells, PKC-β in CA1 cells and PKC-ε in the dentate gyrus. Due to the different activation properties of the different isoforms (see Chapter 8), this differential distribution may have important consequences for the role of PKC in plasticity in the hippocampus.

Using *in situ* hybridization, it is possible not only to identify which neurons are producing a particular protein but also to examine changes in gene expression over time or in response to specific stimuli, although these methods are not sensitive to changes in low-abundance mRNAs.

3.7 Different kinds of blotting can identify DNA, RNA and proteins

In the different blotting techniques, the nucleic acids or proteins are first removed from the tissue and separated by gel electrophoresis (see Section 2.4). They are then transferred out of the gel onto a nitrocellulose filter by a wicking action. The gel is placed on a sponge, with the filter on top, and absorbent paper is placed on top of the filter. Salt buffer solution is drawn up through the gel and the filter, transferring the DNA from the gel to the filter. These techniques make the subsequent manipulation of the molecules easier than when they are incorporated in a fragile gel. After blotting, the molecules can be identified with appropriate probes.

The original blotting technique, now known as **Southern blotting** after its inventor, Edwin Southern, involves the transfer of DNA, whilst **Northern blotting** is used to detect RNA (*Figure 3.5*). After transfer, the nucleic acids are identified by hybridization histochemistry (see above). **Western blotting** involves the transfer of protein from a sodium dodecylsulfate (SDS)–polyacrylamide gel and its detection using antibodies.

Further elaborations include Southwestern blotting where DNA-binding proteins are identified by probing with nucleic acids which mimic the binding site on the nuclear DNA. Similarly, Northwestern blotting is used to detect RNA-binding proteins. These techniques can be used to investigate the proteins which can act to regulate the transcription and translation of DNA and RNA.

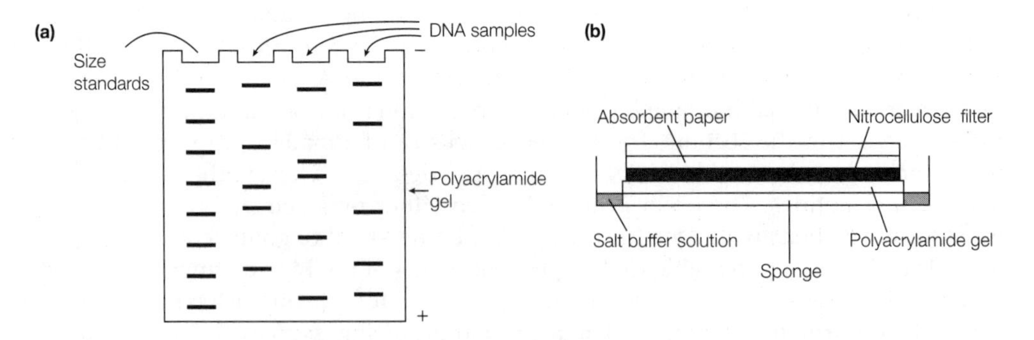

Figure 3.5. Southern blotting. (a) DNA fragments are separated by gel electrophoresis and (b) the separated DNA is transferred to the nitrocellulose filter. It can then be detected by probe hybridization.

If there is the possibility that a particular molecule belongs to a family of related molecules, then it is possible to use low-stringency probes to discover other structurally related members of the same family. This method can also be used to detect homologous molecules in different species. If two similar proteins exist in different species then a low-stringency probe designed to detect one of the proteins may detect the other as well.

3.8 Differential hybridization and subtracted cDNA libraries can measure differences in gene expression

Differential hybridization is a method which can be used to identify genes which are switched either on or off by particular conditions, for example cells grown with and without certain growth factors. In this technique (*Figure 3.6*), a cDNA library is constructed in a λ-phage vector from poly(A)$^+$ mRNA from the cell population with the putative induced gene(s). After infection in *E. coli*, these are then plated out, and identical replicates transferred to two sets of nitrocellulose filters. One set of filters is then probed with labeled cDNA probes constructed by reverse transcription of the mRNA from the treated cells and the other set with probes prepared from the mRNA of untreated cells. After visualization of the probes, genes which are induced will appear as clones on the 'treated' filters but not on the 'untreated' ones. These clones can then be identified and the cDNA isolated.

However, this method is not sensitive enough to identify low-abundance mRNAs. In order to do this, the method of subtraction cloning is used (*Figure 3.7*). One example

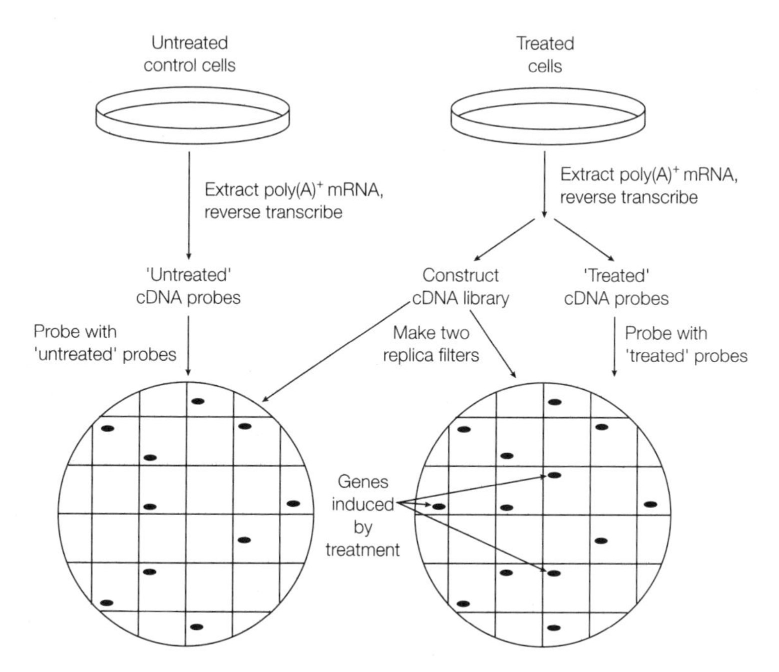

Figure 3.6. In differential hybridization, clones corresponding to induced genes will appear on filters hybridized with probes derived from treated cells but not on those from control cells.

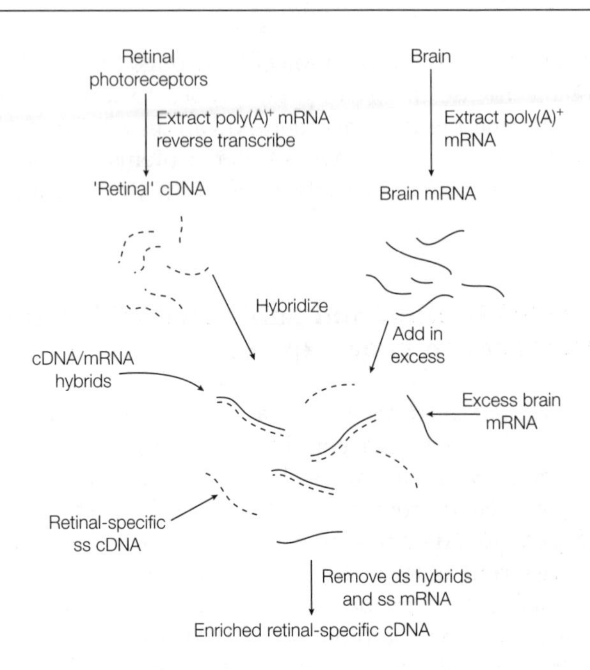

Figure 3.7. Subtraction cloning allows very low-abundance mRNAs to be enriched.

of this is the identification of genes which are expressed in photoreceptors but not in brain. Poly(A)$^+$ mRNA extracted from retinal photoreceptors is used to make cDNA. This is then hybridized with a large excess of poly(A)$^+$ mRNA from brain. Retinal cDNA from genes expressed in both brain and retina forms DNA–RNA hybrids with the brain mRNA, whilst retinal-specific cDNA remains single-stranded as there is no corresponding brain mRNA. In order to separate the double- and single-stranded nucleic acids, the mixture is passed down a hydroxyapatite column. Under the appropriate conditions, the ds DNA–RNA hybrid remains bound to the column and free cDNA can be washed through. This cDNA can then be used to construct a library which will be highly enriched in retinal-specific clones.

3.9 The polymerase chain reaction can produce large amounts of a specific DNA

The polymerase chain reaction (PCR), devised by Kerry Mullis in 1983, is a method for synthesizing virtually unlimited amounts of DNA. During the replication of DNA, DNA polymerase uses single-stranded DNA as a template for the synthesis of the second strand. However, in order to initiate replication, the enzyme requires a small section of double-stranded DNA. Taking the double-stranded DNA to be amplified, in order to provide single-stranded DNA all that is required is to heat it to about 95°C to separate the strands (*Figure 3.8a*). The initiation site can be provided by **primers**, which are short sections of single-stranded DNA which are complementary to the 3′ and 5′ ends of the sequence to be amplified. After cooling the sample, the primers will bind to the single DNA strands and, in the presence of DNA polymerase and dNTPs, new DNA strands will be synthesized on each of the single strands. The cycle can be repeated by reheating the sample in order to separate these newly synthesized double strands of DNA. The only difference is that in the second cycle there are now twice the number of single DNA strands. This means that starting with a single double-stranded copy of the DNA to be amplified, after one cycle there will be two copies, after two cycles there will be four, and so on.

As the primers determine the 5′ and 3′ ends of the desired sequence, after the first cycle of PCR there will be four strands, two of which have ends which correspond to the two primers (*Figure 3.8b*). When the cycle is repeated, the two original strands will be copied in the same way. However, the two new strands will terminate at the ends of the desired sequence. In this way, after two cycles, new strands will be made which consist *only* of the desired sequence, and after each cycle the proportion of strands of the desired sequence will increase. Theoretically, after only 20 cycles, there will be more than 1 million copies of the original DNA.

In the original experiments, the polymerase enzyme used was from *E. coli* and, because it was denatured in the heating step, it had to be added freshly at each cycle. Now, the enzyme used is a heat-stable polymerase, called *Taq* polymerase, which was derived

(a)

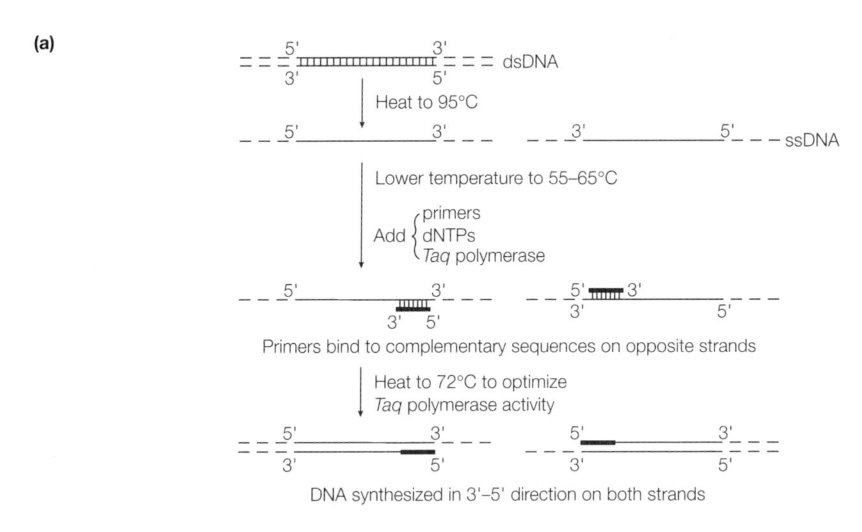

(b)

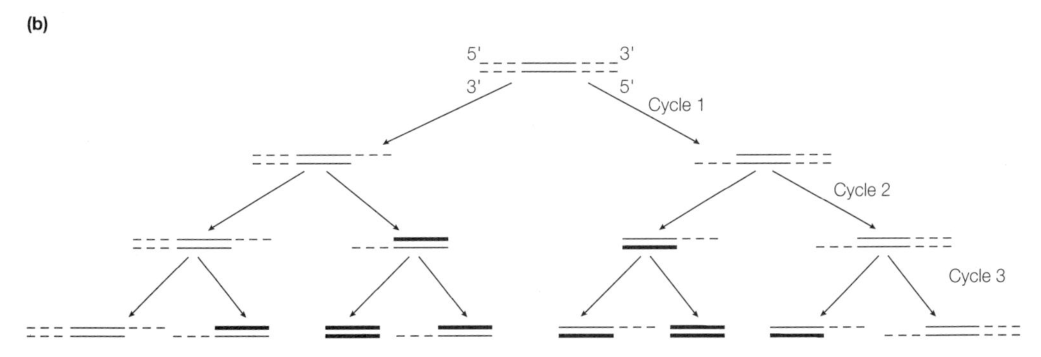

Figure 3.8. The PCR cycle. (a) A single PCR cycle; the desired sequence is shown as a solid line and primers are shown as black bars. (b) As the number of cycles increases, the desired sequence (bold type) is copied more and more times.

originally from the bacterium *Thermus aquaticus*, which lives in hot springs, but is now produced by genetic engineering. This polymerase works optimally at about 72°C.

A key issue in amplification of the correct DNA is the choice and construction of the primers which are required to initiate the DNA synthesis. If the sequences of the 3′ and 5′ ends of the DNA are known, then the synthetic oligonucleotides can be made exactly. If the only information available is an amino acid sequence, then primers have to be designed with the degeneracy of the genetic code taken into account (see Section 2.9.2). If short (5+) amino acid sequences have been determined accurately, then the construction of a number of degenerate probes which will allow all possible codons to occur will ensure that in the mixture there is at least one oligonucleotide which matches the DNA sequence exactly. If the amino acid is longer but less accurately known, then the use of best guess oligonucleotides of 30 or more bases is the most appropriate. In this method, amino acid sequences are selected, if possible, that contain amino acids with a low degeneracy, such as methionine or tryptophan. An analysis exists of the preferred codons in all the DNA sequences available in the GenBank database, which enables the preferred codon to be selected according to the species. Using this information, it is possible to construct an oligonucleotide which will represent the most likely codons for each amino acid. Although there will be regions of this probe which will not anneal with the target sequence, the overall homology will be large enough to locate the appropriate sequence. Another possibility which has been tried is to use inosine in the third position of the ambiguous codon as it can form equally stable base pairs with either adenosine, cytosine, guanine or thymidine.

In order to anneal successfully with the desired section of DNA, primers need to be of sufficient length and with a sufficiently high GC composition. An optimal length is about 20 bp and with a minimum 50% GC content. Because the G–C pairing occurs via three hydrogen bonds, it is stronger than the A–T bond which only has two hydrogen bonds.

In order to use PCR for producing DNA for cloning, it is advantageous to start with as much DNA as possible in order to reduce the number of PCR cycles required. This is because *in vitro* there are not the mechanisms which are present *in vivo* which correct errors which occur during DNA synthesis. *In vivo*, this correction mechanism reduces the error to about one mismatched nucleotide in 10^9. However, without this, the *Taq* polymerase produces about one error in 2×10^4 nucleotides, so fewer cycles will produce less incorrect DNA.

At high annealing temperatures, the specificity of binding of the primers is greatly increased. In this way, variations of the PCR technique rely on the activity of the polymerase being inhibited until the reaction mixture is hot, hence the name 'hot start PCR'. This may be achieved by simply adding the polymerase to the heated mixture or by blocking the activity of the enzyme by a heat-sensitive mechanism, such as anti-polymerase antibodies which bind to the enzyme and inhibit it but which are denatured at higher temperatures, thus releasing the enzyme. In this way, the initial cycle will only be initiated by highly specific binding of the primers.

However, there are cases when nonspecificity can itself produce interesting results. As will be seen in the following chapters, it has become evident that receptors occur in families and superfamilies of receptors which are related by the similarities of their structures. By using primers derived from the sequence of one member of a family, it is possible using low-stringency conditions to amplify related sequences in the same sample. These can then be sequenced and expressed in order to identify them. This technique has been so successful at identifying receptors that there are now a number of sequences which are obviously receptors of a given family for which there is no known endogenous ligand, and these receptors appropriately have been called **orphan receptors**.

3.10 Single-cell PCR

With the combination of the techniques of patch clamping (for details of this see ***Box 4.2***, p. 72), in order to define the biophysical properties of a single neuron, and using mRNA to generate cDNA that can be amplified by PCR and sequenced, it is now possible to compare precisely electrophysiological properties with molecular biology in single neurons. This can be useful in allocating particular properties to a particular channel or receptor subtype. In single-cell PCR, after recording the electrical activity of the neuron, the cytoplasm of the cell can be aspirated into the recording pipette and this can then be processed for PCR. Because there is only a very small amount of RNA present, great care must be taken to avoid both degradation of the RNA present and contamination from external sources.

Before carrying out the PCR, it is necessary to synthesize double-stranded cDNA using reverse transcriptase. After this the cDNA can be amplified in the usual way. It is normally necessary to carry out two separate rounds of PCR, each with about 30–40 cycles, in order to obtain sufficient material. After the first round of PCR, the DNA can be separated by gel electrophoresis and the required band selected for further amplification. In the second round, it is possible to use primers which have restriction sites appropriate to the vector at their 5′ termini. This enables the PCR fragments to be cloned into the appropriate vector.

These methods have been used to investigate the properties of different α-amino-3-hydroxy-5-methyl-4-isoxazoleproprionic acid (AMPA)-type glutamate receptors in different types of neurons. These type of glutamate receptors when expressed *in vitro* can be either homo- or hetero-oligomers (see Section 5.9.1) and, depending on which subtypes of the receptor are present, have large differences in their Ca^{2+} permeability. However, although the results of these experiments showed that those cells with the higher Ca^{2+} permeability had more of the mRNA coding for the Ca^{2+}-permeable receptor subtype, this type of mRNA was not found exclusively in the Ca^{2+}-permeable cells. It was suggested that differences in translation of the different mRNAs could also account for differences in the levels of native receptors.

3.11 mRNA differential display uses PCR to study changes in gene expression

As mentioned above, at high temperatures, primers will only anneal to highly specific sequences, but, at lower temperatures, less specific sequences can be primed. This is the basis of the techniques of **RNA fingerprinting**. These techniques use PCR to amplify the RNA expressed in a cell without knowing in advance anything about the genes being expressed. There are a number of different related protocols for RNA fingerprinting, one of these is called **mRNA differential display** (*Figure 3.9*).

In this technique, firstly the total poly(A)$^+$ mRNA is extracted and reverse transcribed into cDNA in three batches. Each batch is reverse transcribed using an oligo(dT) primer which ends in either C, G or A. This will result in each batch transcribing approximately one-third of the total mRNA, for example the batch primed with oligo(dT) plus A will only transcribe those mRNAs where the base immediately following the poly(A) tail is T (*Figure 3.9*). After this, the amplification of each batch of cDNAs by PCR is carried out using the oligo(dT) primer used for the reverse transcription and a primer with a random sequence, called an arbitrary primer, often a 13-mer. At low annealing temperatures (40°C), the arbitrary primer will anneal to an approximately homologous sequence in some of the cDNAs. It has been found that under these conditions each batch generates about 100 distinct labeled bands when the PCR products are separated on a gel. This is then repeated with different arbitrary primers to amplify other cDNAs. By using 32 arbitrary primers, each producing 100 bands, from the three batches then about 10 000 ($32 \times 100 \times 3$) mRNAs will be amplified. Given that there is evidence to suggest that each of these bands may be a mixture of at least two different cDNAs and assuming that a eukaryotic cell contains between 12 000 and 15 000 different mRNAs, it is probable that the majority of mRNAs from a given cell are amplified by this technique.

An example of this technique in action is the identification of an mRNA which is induced in rat brain by acute cocaine administration. Comparing the bands on gels prepared from poly(A)$^+$ mRNA derived from both control and cocaine-treated rats, there was a band present on the 'treated' gel which corresponded to an induced gene. The induced PCR product was then used to probe cDNA libraries constructed from different regions of rat brain. A complete sequence of the cDNA clone was isolated and sequenced. From the sequence of the predicted protein product, it was suggested that it is a secreted protein, and comparisons of expression in different brain regions showed that this mRNA was only induced in the striatum where cocaine could increase its levels fourfold.

An interesting variation of RNA fingerprinting uses primers which are not totally random but which are based on sequences or motifs which are known to be relatively conserved between genes with related functions. This method has been successful in isolating new members of a group of molecules which contain a so-called zinc finger region (see Chapter 8).

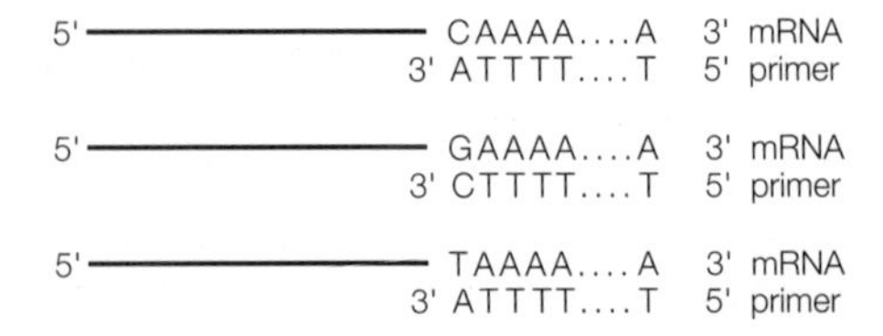

Figure 3.9. In mRNA differential display, the total poly(A)$^+$ mRNA is reverse transcribed into three batches, depending on which base (A, G or C) follows the poly(A) tail.

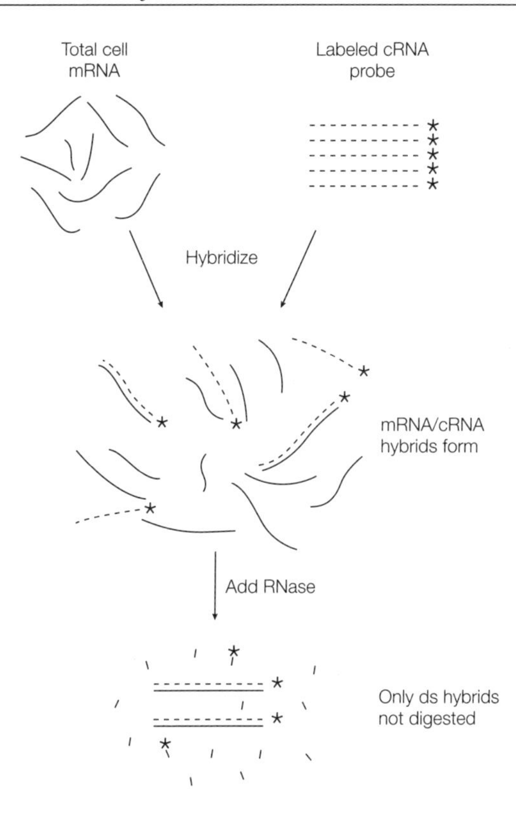

Figure 3.10. Nuclease protection assay allows the quantification of specific mRNAs.

A method which can be used to quantify a specific mRNA in a mixture of mRNAs is the nuclease protection assay (*Figure 3.10*). A large amount of a radiolabeled complementary RNA probe is incubated with mRNA extracted from the cells. This mRNA will contain many sequences, including the mRNA of interest. The labeled probe will hybridize with matching sequences forming double-stranded RNA. The sample is then digested with RNase A and RNase T1 which will digest only single-stranded RNA. Thus any noncomplementary mRNA and any excess probe will be digested, leaving only the mRNA–probe hybrids which can be run on a gel and analyzed. This method enables very small amounts of mRNA to be counted (<2 μg of total mRNA) and can be very specific as the hybridization conditions can ensure that the probe binding is only to completely complementary sites.

3.12 A large number of genes expressed in human brain have been identified

The human genome project has at its heart the aim of sequencing the entire human genome of about 3×10^9 bp. However, it has been estimated that only about 3% of the entire genome codes for expressed genes. The remaining 97%, which has been called 'junk' DNA, consists of noncoding regions found both within and between the genes. The intragenic DNA is found as introns, between the coding sequences, the exons, within the gene. The extragenic DNA is found between genes, and consists of both unique sequences of unknown function (~70%) and repeat sequences of varying degrees of repetition (30%). The function of the extragenic DNA is unknown, although it has been suggested that this is in some way the 'leftovers' of evolution and that the presence of a certain amount of unused DNA enables changes to the organism to occur with less deleterious effects.

A method which was developed in order to obtain unique landmarks in the genome to aid mapping was sequence-tagged sites (STSs). An STS is a short stretch of DNA of 100–200 bp that is a unique sequence, i.e. it is found only once in the entire genome. Parts of the STS may be found elsewhere, but the ends of the sequence are unique. If PCR is carried out on the entire genomic DNA using primers complementary to the unique ends of the STS, then there is only one product, which can be tested by separating the DNA on a gel. If the primers have amplified a true STS then there should be a single intense band corresponding to the amplified STS.

It has been argued that because most of the human genome has no known function then it would be more beneficial to first sequence those portions of the genome which are expressed, that is by extracting all of the mRNA, converting it to cDNA and sequencing the cDNA. A drawback of this approach is that there would be no information as to how the genes are regulated. Even so, the entire transcript of most full-length mRNAs will also contain untranslated 5′ and 3′ sequences. Expressed sequence tags (ESTs) are related to STSs in that they are short sequences, not of genomic DNA but of cDNA. Using automated partial DNA sequencing, a large number of these ESTs have now been sequenced. The sequences obtained have been compared with the sequences of previously known human genes, and some of them have been identified. However, the major interest in the method is the discovery of a large number of genes which are expressed in the brain and, although these may not all be brain specific, the use of subtraction techniques using libraries derived from nonbrain tissues will significantly increase the number of brain-specific genes identified. From the first paper in 1991 describing the method, which identified 337 new genes, to the end of 1993 over 10 000 EST sequences had been deposited in public databases. This number represents a significant proportion of human genes, and gives a real possibility that within a short time at least the part of the human genome which is expressed will be fully identified, even if we will have to wait a bit longer for the entire genome.

Chapter 4

Voltage-gated ion channels

4.1 All excitable cells contain voltage-sensitive ion channels

In all excitable cells there are protein channels in the cell membrane which, when open, allow the passage of ions. The opening of some of these channels is determined by the voltage across the membrane, hence the name **voltage-gated ion channels**.

As most of these channels have a significantly higher permeability for one ion over all others, they have been identified and classified according to their ionic selectivity. There are channels which are selective for sodium, calcium, potassium or chloride, as well as other less selective cation channels.

The first of these channels to be characterized were the voltage-dependent sodium (VDSC) and potassium (VDKC) channels of nerve cells which are responsible for the electrical events underlying action potentials. These were first described by Alan Hodgkin and Andrew Huxley in squid axons where their role was studied using a combination of electrophysiology, particularly **voltage clamping** (***Box 4.1***, p. 71), and pharmacology.

Hodgkin and Huxley showed that the action potential was made up of an early current flowing into the cell through sodium selective channels which could be blocked by the toxin of the pufferfish, tetrodotoxin (TTX), and a later potassium current which could be blocked by the large cation, tetraethylammonium (TEA) (*Figure 4.1*).

Their analysis was the first demonstration of the existence of separate permeability pathways for sodium and potassium which could be manipulated independently. Their work, for which they were awarded a Nobel prize, led to the idea of single protein species which were responsible for the permeation of single ions, and so the hunt was on to identify and eventually isolate these channels.

Since then, ion channels have been found in all eukaryotic cells and in some prokaryotes. It has been suggested that all cells may possess ion channels. These channels have a wide diversity of ion permeation and modes of regulation, but there are some common principles.

One of the reasons that Hodgkin and Huxley used squid axons for their experiments was the technical constraints which required the use of large cells. Since then,

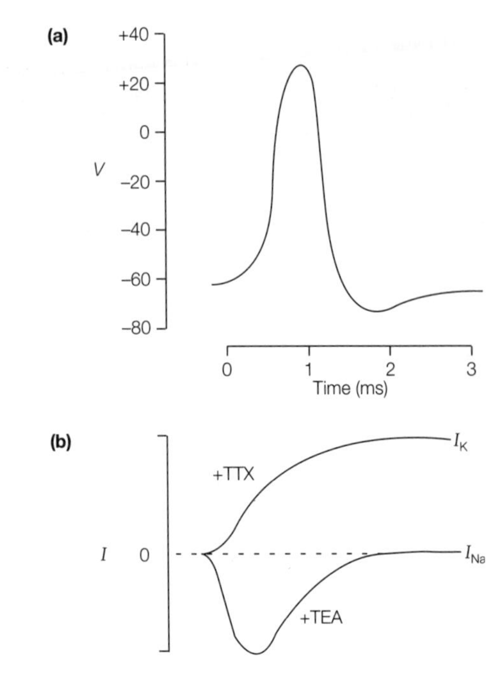

Figure 4.1. (a) Action potential. (b) Voltage clamping experiment showing two components identified using TTX and TEA.

techniques have been developed to record electrical signals from much smaller cells, and in 1976 the first paper was published by Bert Sackmann and Erwin Neher describing a method to record the signals produced by the opening of a single ion channel, a technique called **patch clamping** (*Box 4.2*, p. 71). This has invigorated the electrophysiological study of ion channels and, combined with molecular biology, has allowed the study of channels at the molecular level.

4.1.1 Different channels can have very different properties

Voltage-gated ion channels have a number of important properties which determine their activity. In order to respond to the voltage, these channels must have some kind of voltage sensor which, when triggered, can open the channel, and this rate of activation may vary depending on the type of channel. The early sodium current and late potassium current of the action potential occur in the order they do because of the different channel activation properties. The sodium channels respond immediately, that is their activation is fast, but the activation of the potassium current is delayed by about 0.5 msec after the membrane is depolarized, so the potassium current occurs after the sodium current. The timing of the closure of the channel is also important. After about 1 msec, the open sodium channels close. These closed channels cannot be reopened immediately and in this state are called inactivated. Inactivated sodium channels remain closed until the membrane voltage has returned to the resting potential, when the channel reverts to the closed but potentially activateable state. Hodgkin and Huxley developed a mathematical model of the voltage dependence of activation and inactivation which provides an empirical framework for the testing of hypotheses of channel function.

Voltage-gated channels are also selective, to varying degrees, as to which ions are allowed to flow through them. Sodium channels are over 10 times more permeable to sodium than to potassium, and some potassium channels prefer potassium over sodium by 100-fold.

Both voltage dependence and ionic selectivity must be related to the specific structure of the channels involved, and it has been possible to show that specific regions (and even specific amino acid residues) of channels are crucial determinants of these properties.

4.2 Sodium channels were the first to be identified

The first ion channel to be characterized at the molecular level was the sodium channel, and it was isolated from the electric organ, the electroplax, of the eel *Electrophorus electricus* by Shosaku Numa and colleagues in 1984. This was the tissue used to isolate the acetylcholine receptor as described in Chapter 2. When the 'battery' of the electroplax is discharged, the depolarization produced by the activation of large numbers of acetylcholine receptors triggers the opening of the voltage-dependent sodium channels, greatly increasing the current flow across the membrane.

The pufferfish toxin, TTX, used by Hodgkin and Huxley to block the sodium current in squid axons, was used to purify the sodium channel protein by column chromatography. The column contained beads coated with TTX and, when a crude membrane extract was added to the top of the column, the toxin bound the sodium channel protein so that it was retained in the column. The unwanted proteins and membrane lipids could then be washed out. To remove the sodium channel protein, a large excess of toxin was added to the column and the purified protein bound to the excess TTX eluted. The sodium channel from the eel electroplax was found to be a single glycosylated protein of molecular weight 260 kDa containing about 29% carbohydrate.

4.2.1 The electroplax sodium channel was cloned using cDNA libraries

The primary cDNA sequence of the sodium channel was determined in a similar way to that of the acetylcholine receptor. Purified channel protein was partially digested with trypsin and the amino acid sequences of six short peptides were determined by **Edman degradation**.

A cDNA library was constructed using pUC8 plasmids in *E. coli* using poly(A)$^+$ RNA from the eel electroplax. The clones were then probed for the presence of the sodium channel protein by testing them with rabbit antiserum raised against the sodium channel protein. It is very unusual and interesting that the prokaryotic cells of *E. coli* translated the mRNA and inserted immunologically recognizable proteins into the cell membrane. Normally, special expression vectors are needed for this to occur; this was a lucky accident. The positive clones were then tested for the presence of the cDNA by Southern blotting hybridization with a synthetic ^{32}P-labeled oligonucleotide probe.

The sequence of the oligonucleotide probe was calculated from the amino acid sequence of part of one of the peptide fragments. The probe used had a mixture of 32 nucleotide sequences because of the degeneracy of the genetic code (see the example in *Figure 2.13*).

One of the clones which had tested positive to the antiserum was labeled by the ^{32}P-labeled probe. However, calculations based on the known molecular weight of the channel showed that this clone could not contain the cDNA coding for the whole protein. A second cDNA library was constructed and, using part of the cDNA sequence of the previously identified clone, they probed the second library to find a further clone containing more of the sodium channel sequence. This procedure recovered more of the sequence, but there was still not enough to code for the entire protein. The partial sequence had obviously reached the 3′ end of the coding sequence, because of the presence of the stop codon (TAA) and a poly(A) tail. Using oligonucleotide probes targeted at the 5′ end of the clones, a search was made for clones containing the 5′ end of the channel DNA. Eventually, after using different probes and different libraries, the whole cDNA was found contained within seven overlapping clones (*Figure 4.2*).

An interesting technical variation was used to construct the final library of clones. In order to enrich this library with clones containing the cDNA of the sodium channel, instead of using oligo(dT) primers which would have synthesized any poly(A)$^+$ mRNA, they used a primer which was complementary to a region at the 5′ end of the partial sequence already determined. This meant that the cDNA cloning would synthesize the missing 5′ end of the coding strand. They could identify the start of the coding sequence by the presence of the start codon (ATG) with flanking sequences known to occur at the start of eukaryotic mRNAs.

After sequencing the different clones and lining up the overlaps, the entire coding sequence and its corresponding amino acid sequence could be examined. The entire nucleotide sequence was 7230 bases long, and a good check that no bases were missing was that all the peptide fragments were coded in the same reading frame. The strategy outlined above highlights the fact that cloning is not always a straightforward matter and calls for ingenuity and perseverance. The protein sequence thus turned out to be 1820 amino acids long and the calculated molecular weight of 208 kDa agreed well with that of the purified channel, allowing for the additional carbohydrate which most membrane-spanning proteins have on their extracellular face.

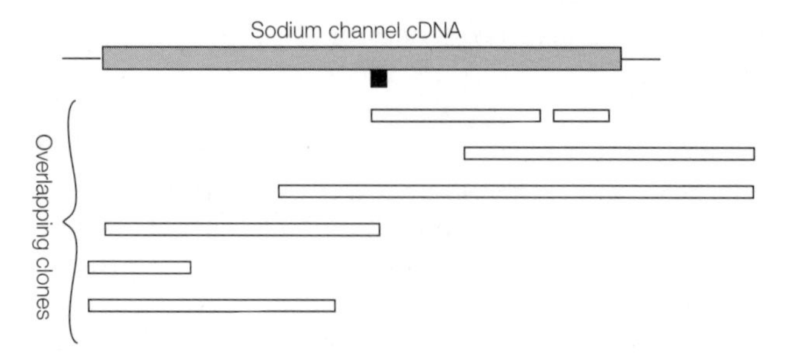

Figure 4.2. Diagram of sodium channelDNA (shaded bar), showing overlapping clones and oligo primer site (dark bar).

The complete amino acid sequence showed the presence of four internal repeats of approximately 300 amino acids. These regions had identical or similar amino acids at about 50% of the equivalent positions in each repeat, and this homology was highly statistically significant, indicating that the four repeats may have arisen from an original single domain which by gene duplication is now repeated four times.

4.2.2 Aspects of the secondary structure can be inferred from the amino acid sequence

The secondary structure of a protein and its distribution in the membrane can be partly inferred from a technique called **hydropathy analysis** of the amino acid sequence. Hydropathy analysis is based on the fact that amino acids can be hydrophilic or hydrophobic to varying degrees according to the identity of their side chains. For example, charged amino acids such as arginine and aspartate are hydrophilic, whereas phenylalanine and methionine with nonpolar side chains are hydrophobic. Thus, in solution, amino acids will dissolve (partition) preferentially in water or lipid according to their hydrophobicity. Values can be assigned to each amino acid according to their partition in oil:water, and this is known as the hydropathicity index.

When amino acids are linked together in a polypeptide, each single amino acid makes a contribution to the hydropathicity profile of the whole region but influenced by the hydropathy of its neighbors. To take account of this, the hydropathicity index of any given amino acid is averaged with those of its neighbors to give a running average. In practice, the averaged hydropathicity index is calculated for each residue of the entire protein region including nine residues flanking each amino acid.

When this hydropathy analysis was carried out on the sodium channel, it showed that each of the internal repeats or domains contained a possible six hydrophobic sequences of 18 or more residues (*Figure 4.3*). The importance of this lies in the fact that the common secondary protein structure, the **α-helix**, can form a membrane-spanning region if there are approximately 20 amino acids in the hydrophobic section.

This led Numa to suggest that the structure of the sodium channel consisted of four identical domains each containing six transmembrane (TM) segments (*Figure 4.4*). This model of the sodium channel, first proposed in 1984, has since been refined but is essentially that initially proposed.

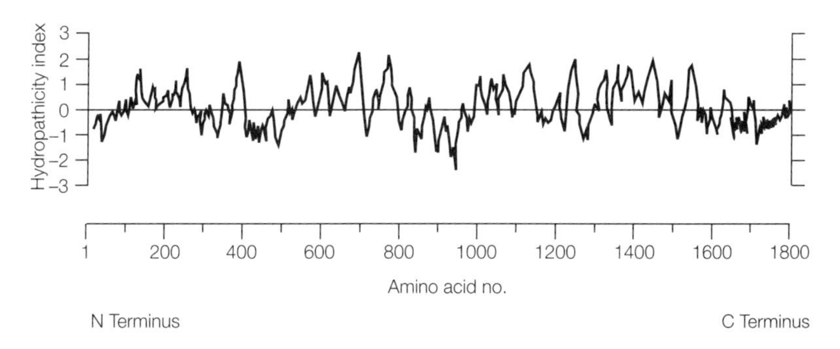

Figure 4.3. Hydropathy profile of the Torpedo sodium channel. Adapted from Noda *et al.* (1984) Primary structure of *Electrophorus electricus*, *Nature* **312**: 121–127.

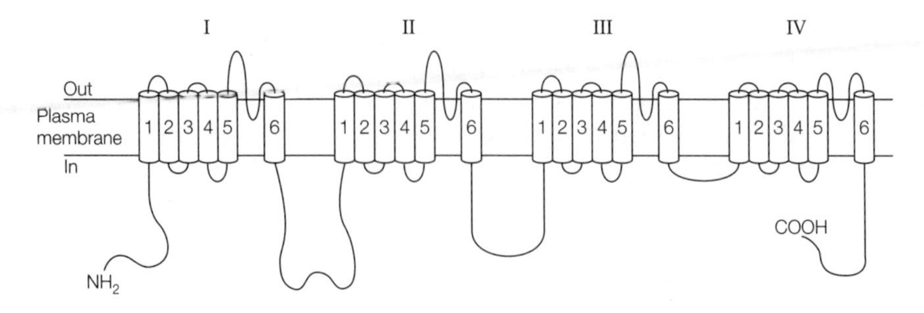

Figure 4.4. Model of sodium channel.

When the electroplax mRNA was injected into *Xenopus* oocytes and the sodium channel activity measured using patch clamping, the expressed protein was able to behave as a TTX-sensitive, voltage-gated sodium channel, showing that in this case the single mRNA was sufficient to produce a functional channel.

4.2.3 Mammalian sodium channels contain additional subunits

When sodium channels were purified from mammalian tissues (mainly rat and human), it was found that the extracted channel consisted of three proteins, the original 260 kDa protein plus two smaller proteins of 36 and 33 kDa. These new subunits were named β1 and β2 and the original subunit designated α.

The cDNA sequence of the electroplax sodium channel was then used to identify the rat brain sodium channels. A cDNA library from rat brain was screened by hybridization with a probe derived from electroplax sodium channel cDNA. This method isolated the cDNA of three different sodium channel α-subunits in rat brain called I, II and III. Remarkably, there was over 60% sequence homology between the rat and electroplax sodium channels. It was later found, by using the **alternative splicing** of two exons from the gene (*Figure 4.5*), that the type II existed in two different forms. By this method, slightly different proteins, called splice variants, can be produced from the same gene.

Type II sodium channels are expressed early in development in embryonic and neonatal brain, whilst the variant IIA is most prominent in adult brain. Other, similar, sodium channel α-subunits were also found in adult skeletal muscle (μ1) and heart (h1).

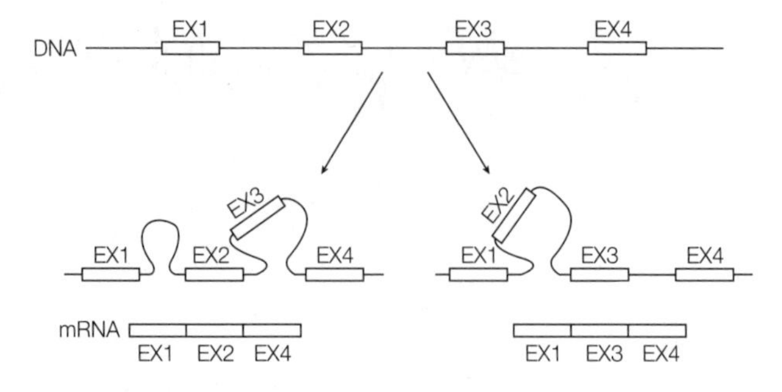

Figure 4.5. Diagram of alternative splicing.

The β1- and β2-subunits have also been cloned, and both have a single putative TM segment. Both β-subunits are glycosylated, showing that they have part of the protein exposed on the extracellular face of the membrane. The β1-subunit is noncovalently bound to the α-subunit whilst the β2-subunit is linked via disulfide bonds.

When the mRNA of the α-subunits from mammalian tissues is injected into oocytes, while functional ion channels are expressed some of their properties are different from those of native channels, especially the rate and voltage dependence of activation. When α- and β1-subunits are coexpressed, that is both α and β1 mRNA are injected into the same oocytes, the expressed sodium channels have properties which are similar to those of the normal channels.

Armed with the knowledge of the entire sequence of the sodium channel and a model of the secondary structure, it became possible to study which parts of the protein were important in defining channel properties such as activation, inactivation and ion selectivity. A technique which has been revolutionary in studies of channels at the molecular level is called **site-directed mutagenesis** (***Box 4.3***, p. 74).

4.3 Structure and function can be related at the molecular level

4.3.1 Channel activation involves the movement of gating charges

From early studies on the voltage dependence of channel activation and its association with the movement of electrical charges within the channel protein, called the gating current, it was calculated that activation is associated with a current equivalent to the movement of six charges across the entire width of the membrane (or a greater number of charges moving a smaller distance).

The fact that channel activation must involve the movement of charge focused attention on regions of the channel with charged amino acid residues within the hydrophobic portion of the membrane. Of the six TM segments of each domain of the sodium channel, one of these, the S4 segment, which is about 20 amino acids long, has a primary structure with a repeated sequence of a positively charged residue (Arg or Lys) followed by two hydrophobic amino acids (*Figure 4.6a*). As the α-helix has a pitch of about 3.5 amino acids per complete turn, the S4 segment is thought to form an α-helix with a spiral of positive charges on the outside (*Figure 4.6b*).

The involvement of these charged residues in activation has been shown clearly using a variety of mutants generated by site-directed mutagenesis. Numa, Stuhmer and colleagues constructed cDNA mutants which would produce sodium channels in which the positively charged residues at the various sites in the S4 region of domains I and II were replaced with neutral or negatively charged residues. The cDNA was transcribed to produce mRNA which was injected into *Xenopus* oocytes. Patch clamping was used to analyze the characteristics of the sodium channels expressed. Fitting the recorded currents to the equations derived by Hodgkin and Huxley to model activation and inactivation, it was shown that a reduction in the net positive charges of S4 in domain I causes a decrease in the apparent gating charge and a reduction in the steepness of the activation curve. This strongly suggested

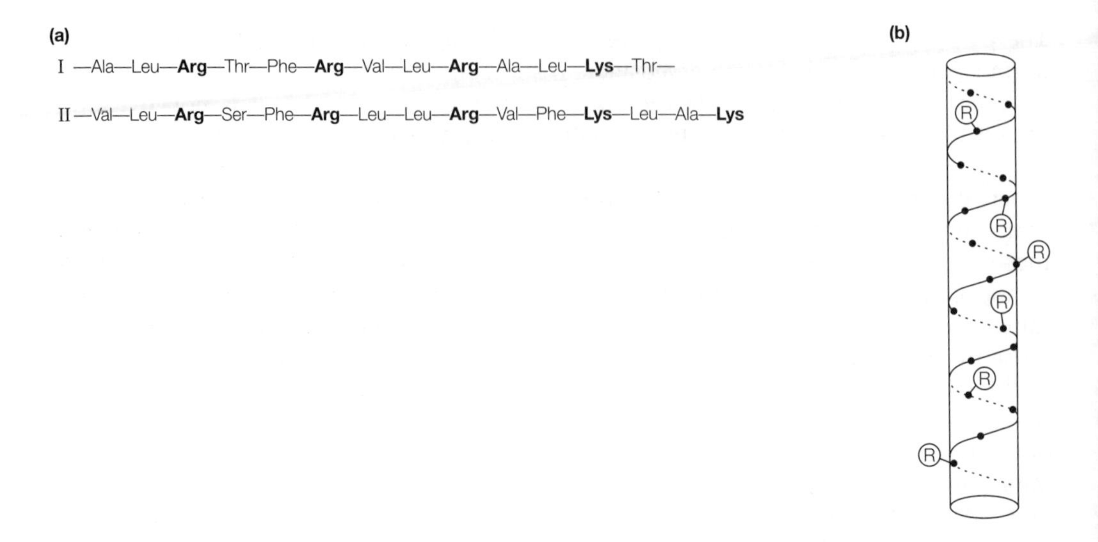

Figure 4.6. (a) The sequence of the S4 segments of domain I (215–227) and domain II (848–862). (b) The α-helical arrangement of amino acids; only arginine (R) side chains are shown.

that these positively charged residues of the S4 region may form part of the gating charge.

A model has been proposed to account for the action of S4 as a voltage sensor, where the S4 segment acts as a rotating helix. In response to depolarization, ionic bonds between the positively charged residues of the α-helix and negative charges in the other surrounding TM segments are broken. The S4 helix then rotates upwards towards the extracellular face, thus moving charge across the membrane. An attraction of this model is that if each of the four domains contributes the equivalent of only 1.5 charges, then the gating charge of 6 can be accounted for. However, both this model and a later one, which requires a large conformational change in the S4 segment from an α-helix to a β-pleated sheet, have to account for the large energy barrier which would have to be overcome to generate the movement.

4.3.2 Channel inactivation is due to intracellular residues

The region of the sodium channel thought to be responsible for inactivation was shown to be in the cytoplasmic portion of the channel, since treatment of the inside of the squid axon with proteolytic enzymes, which would remove parts of the protein on the cytoplasmic face, abolished inactivation. This was confirmed by experiments where oocytes injected with wild-type mRNA and then treated with trypsin showed an almost complete removal of inactivation. Mutants constructed lacking cytoplasmic portions of the sodium channel in between domains III and IV showed a strong reduction in the rate of inactivation. Single-channel recordings from these mutant channels showed that the channel openings were much longer, with the mean channel open time being increased by at least a factor of 10.

4.3.3 *Pore-lining residues and ion conductance can be identified using TTX*

The ion flow through voltage-gated ion channels is proposed to occur through a transmembrane water-filled pore, although other models based on the binding of ions directly to the channel protein, which then acts as an ion conductor, are under discussion. The identity of the region which lines the pore of the channel and controls both the size of the current (single-channel conductance) and the identity of the permeant ion (selectivity) has been studied using a variety of compounds which can block the pore. As all externally applied compounds which are not lipid soluble can only interact with exposed residues, the use of site-directed mutagenesis to change the effectiveness of the blockade provides information on the site of binding in the pore.

The toxin originally used to isolate the sodium channel protein, TTX, is known to bind with very high affinity to the extracellular mouth of the pore. The amino acid sequence and predicted structure of the sodium channel show that between the TM segments S1/S2 and S3/S4 the extracellular portions are quite short; however, the S5/S6 extracellular portion is longer and contains a number of negatively charged amino acids. This region is thought to form a hairpin loop structure in the membrane, variously called the H5 or P loop, and consists of two short sections called SS1 and SS2 (*Figure 4.4*). Evidence for the existence of this loop first came from studies of potassium channels. Indeed, the similarities between many different types of ion channel that have emerged has proved useful since testable inferences can be made about one channel on the basis of experiments with others.

A mutation, in which one of the negatively charged amino acids, a glutamate residue found at position 387 in SS2 (Glu387) of domain I, was converted to glutamine reduced the affinity for TTX by more than 10 000 times. Identification of similar residues in each domain showed that acidic amino acids in all four domains were required for high-affinity TTX binding, as well as other residues in SS2, showing clearly that these regions of the molecule form part of the external mouth (*Figure 4.7*).

Other molecules, such as some local anesthetics, block sodium channels from the intracellular face and act preferentially when the channel is open. Identification of aromatic residues at the end of the S6 region which are necessary for binding of local anesthetics, such as etidocaine, suggests that residues in the S6 segments line the pore on the intracellular face.

Sodium channel conductance is also dependent on the acidic residues which are involved in TTX binding. Mutations with neutral amino acids substituted at these positions (e.g. 387) have a reduced sodium conductance, as measured by whole-cell patch

Domain I	L(380) M T Q Ⓓ F W Ⓔ N L Y(390) Q L T
Domain II	V L C(940) G Ⓔ W I Ⓔ T M W Ⓓ C(950) M
Domain III	V A T(1420) F K G W M Ⓓ I M Y A(1430) A
Domain IV	I(1710) T T S A G W Ⓓ G L L(1720) A P I

Figure 4.7. Negatively charged amino acids (circled) are found in the SS2 segment.

clamping, which may be reduced by as much as 90%. The involvement of amino acid residues in ion conductance and selectivity will be discussed further after the section on calcium channels.

4.4 Voltage-gated calcium channels are responsible for diverse functions

Intracellular calcium is the trigger for **exocytosis** in many cell types, and in all excitable cells voltage-dependent calcium channels are responsible for the influx of calcium which couples excitation to neurotransmitter release. Voltage-gated calcium channels also allow calcium influx in all muscle types, and are especially numerous in the transverse tubules of skeletal muscle where they have an additional role in excitation–contraction coupling. The identification of a number of different types of calcium channel was made originally on the basis of their different electrophysiology and sensitivity to a diversity of toxins and drugs.

4.4.1 *Calcium channels are distributed differently and can be blocked by toxins*

Two types of calcium channels are found in muscle. The voltage-dependent calcium channel (VDCC) found in skeletal muscle is slow to activate, requires a large depolarization to activate it and shows no inactivation. These channels are called L channels (long-lasting) and have a single-channel conductance of about 24 pS. In cardiac muscle, a second calcium current is also found which can be activated by small depolarizations and is rapidly inactivated in a voltage-dependent manner. These channels are called T (transient) and have a much smaller single channel conductance (8 pS).

These calcium channels are classified further as either high or low voltage activated on the basis of the depolarization step required to activate them. Thus L channels are high voltage-activated (HVA) type requiring large (high) depolarizations, and T channels are low voltage-activated (LVA) type activated by small (low) depolarizations.

The L channels are blocked by **phenylalkylamines** like verapamil, but different **dihydropyridines** (DHPs) have opposite effects which can be observed in single-channel recordings. Nifedipine blocks calcium channels, whereas the closely related compound BAY K 8644 acts as an agonist, greatly increasing the mean channel open times.

Other calcium channels are found in neurons. Initial studies on neurons demonstrated the presence of three types of calcium channels. Two of these were identified as the L and T types already seen in cardiac muscle, but a third channel called N (neuronal) was an HVA with an inactivation that was intermediate between L and T and was insensitive to DHPs.

A further three types of calcium channel have since been identified in neurons by the use of toxins which can block particular components of the calcium current. When HVA currents are activated in the presence of DHPs, the proportion of the current blocked by DHPs can be attributed to L channels. N channels can be blocked by a toxin, ω-conotoxin GVIA (CTX), from the cone shell mollusc *Conus geographus*. In the presence of CTX and DHP, both L and N channels are blocked, and the current flowing through N channels can be calculated by subtraction. However, in neurons, much of the calcium current is insensitive to both DHPs and CTX. Hence, this method has been continued with a number of other toxins (*Table 4.1*), each defining a new part of the calcium current. They were named:

P, because they are the predominant calcium current in Purkinje cells;
Q, a current found in cerebellar granule neurons (and Q comes after P);
R, what remains after blockade of all other calcium currents (part of this current is thought to be HVA and part LVA).

The discovery that many sodium and calcium channels can be blocked by toxins such as TTX and CTX has stimulated a wide search to find other toxins which can be used as channel blockers. The toxin which blocks P channels, ω-CTX MVIIC, was discovered by probing a cDNA library constructed from the venom ducts of the marine snail *Conus magus* using probes based on known *C. magus* toxins. However, problems have been highlighted regarding the specificity of the block and degree of blockade of these toxins, so conclusions regarding quantitative aspects of toxin use must be approached with caution.

4.5 Calcium channels have been cloned

In a manner similar to that used to clone the voltage-gated sodium channel, the first cloned calcium channel was purified from a tissue rich in calcium channels, namely skeletal muscle, using a specific high-affinity ligand, a 1,4-dihydropyridine. The skeletal muscle L channel exists as a complex of five polypeptides, α_1 (175 kDa) α_2 (150 kDa) β (52 kDa), γ (32 kDa) and δ (25 kDa).

Table 4.1. Properties of voltage-gated calcium channels

Channel	Type	Channel blockers
L	HVA	Dihydropyridines (both antagonists and agonists) Phenylalkylamines Benzothiazepines
T	LVA	No *specific* blockers
N	HVA	ω-CTX GVIA ω-CTX MVIIC
P	HVA	ω-Aga IVA ω-CTX MVIIC FTX
Q	HVA	ω-CTX MVIIC ω-Aga IVA (>1000 μM)
R	Mixture of HVA/LVA	No *specific* blockers. Possibly not a single channel type

ω-CTX, ω-conotoxins from the cone snails, *Conus geographus* and *Conus magus*; ω-Aga, ω-agatoxin; and FTX, funnel spider toxin (polyamine-like) from the funnel web spider, *Agelenopsis aperta*.

4.5.1 The α_1-subunit from skeletal muscle was the first to be cloned

Using primers derived from the amino acid sequences of polypeptides cleaved from the 175-kDa purified protein, the cDNA coding for the α_1-subunit of the skeletal muscle L channel was cloned in a series of steps similar to those used to clone the voltage-gated sodium channel. It was found to code for a protein of 1873 amino acids with a molecular weight of 212 kDa. The difference in molecular weight from that of the purified protein is due to a truncation of the translated protein possibly due to post-translational processing.

The predicted structure of the calcium channel shows a close similarity to the voltage-gated sodium channel, with four domains each containing six TM segments and an H5 loop (see *Figure 4.4*). Overall, the sequence homology between the sodium and calcium channels showed that 35% of amino acids were identical, and this rose to 55% when the presence of similar residues was allowed for. This is called **conservative substitution**, when, for example, either glutamate or aspartate are considered to be equivalent, both having negatively charged side chains of roughly similar size.

The sequence shows a similar pattern to the sodium channel of amino acids in the S4 segments, giving it a similar role as a voltage sensor, and further similarities were found in the position of the pore lining and internal and external domains.

The L channel antagonists are important classes of drugs, used in the treatment of a number of cardiovascular diseases, because of their actions on cardiac and smooth muscle. The phenylalkylamines, such as verapamil, bind to the intracellular face of the α_1-subunit and bind preferentially to the open channel. **Photoaffinity labeling**, where a photoreactive analog is equilibrated with the channel and then covalently bound to it by photoactivation using UV light, has been used to identify which areas of the protein are involved in phenylalkylamine binding. After photolabeling, the α_1-subunit was cleaved, at lysine and arginine residues by trypsin, to make a number of peptide fragments. The fragments were then separated on SDS gels and the labeled fragment was identified, by immunoprecipitation, using anti-peptide antibodies. These antibodies had been raised against the different fragments of the α_1-subunit. This method identified a 9.5-kDa fragment containing residues 1339–1400. Reducing the size of this fragment using another protease defined a smaller 5-kDa fragment consisting of residues 1349–1391. In the complete subunit, this corresponds to a region which is centered around the S6 TM segment of domain IV, suggesting that the binding site for phenyalkylamines lies at the intracellular end of S6.

The binding of DHPs, such as nifedipine and amlodipine, is from the extracellular side of the membrane, and seems to be located in a hydrophobic region close to the outer surface of the membrane. Although binding is not enhanced by channel opening, it is increased by prolonged depolarizations, suggesting that DHPs have a higher affinity for inactivated channels. A number of photoaffinity labeling studies have identified a range of different binding sites on the α_1-subunit, but there are problems relating to the size of the photoreactive compounds and the size of the peptide fragments generated, such that the resolution of the method is not sufficient to allow the binding site to be determined very closely. Mutants called chimeras can be constructed, where parts of one subunit can be combined with parts of another to produce hybrid channels, with differing properties. Chimeras were used to study the involvement of parts of domains III and IV in the activity of DHPs. The chimeras were formed from combinations of parts of α_1-subunits which are DHP sensitive (cardiac α_1 C, see *Table 4.2*) and DHP insensitive (neuronal α_1 A). These experiments indicated that the S5–S6 linker region of domain IV, which contains the P5 loop, and the S6 segment are critical in determining DHP activity.

Benzothiazepines, like diltiazem, also bind to the extracellular portion of the α_1-subunit, but the exact nature of the binding site has been much less well defined, although it is thought to bind at a site which is also in domain IV.

4.5.2 Many other α_1-subunits have been cloned

Using probes derived from the skeletal muscle L channel sequence, six different genes encoding α_1-subunits have been found. Additionally, all of these genes can be spliced alternatively, giving a large number of alternative α_1-subunits. Identification of specific gene products with channel types is continuing, but a tentative scheme is shown in *Table 4.2*.

Table 4.2. Calcium channel α_1-subunits

Channel type	Gene[a]	Location
	α_1 S (2)	Skeletal muscle
L	α_1 C (3+)	Cardiac muscle, smooth muscle, brain
	α_1 D (4)	Endocrine, kidney, brain
P/Q?	α_1 A (5)	Neurons, kidney
N	α_1 B (2)	Neurons, endocrine
R/T?	α_1 E (3+)	Brain

[a]The number of currently known splice variants is shown in parentheses.

4.5.3 Other calcium channel subunits have also been cloned

As with the voltage-gated sodium channels, some of the calcium channel α_1-subunits can act as voltage-gated ion channels without the presence of other subunits, but these other subunits can modify the properties of the calcium channels.

The β-subunit is thought to be intracellular as it contains mainly hydrophilic residues and is not glycosylated. The presence of consensus sequences for phosphorylation by multiple protein kinases allows for the possibility of interactions with intracellular regulation mechanisms. So far, four genes have been discovered which code for the β-subunit, each with splice variants, giving a minimum of 13 variants and, as with the voltage-gated sodium channel, interactions between the α_1- and β-subunit have direct effects on channel gating.

The α_2- and δ-subunits are linked by disulfide bonds and are produced by a single gene with five known splice variants. The α_2-subunit is thought to lie on the extracellular face, whilst the δ-subunit forms a single TM segment. The α_2- and δ-subunits are thought to have act synergistically to increase current amplitude and activation rate.

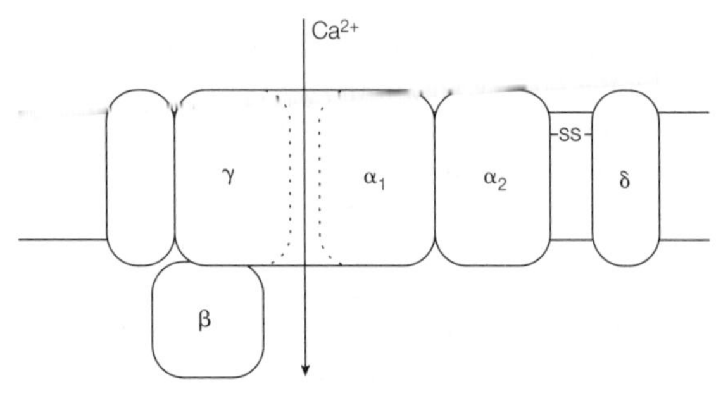

Figure 4.8. Arrangement of calcium channel subunits.

The role of the γ-subunit is not known; however, its structure consists of mainly hydrophobic residues and contains four predicted TM segments. It is the only one of the calcium channel subunits which has only been found as a single gene without splice variants, and is thought to be only expressed in skeletal muscle, although both L and N channels in brain are associated with a 100 kDa protein which may represent a novel brain-specific subunit (*Figure 4.8*).

With the existence of so many different variants and the enormous number of possible combinations, there are interesting questions that can be asked regarding the physiological significance of these variants. Firstly, do all combinations exist *in vivo* and are these distributed heterogeneously. Secondly, are there significant functional differences in the different subunit combinations. These questions of diversity and heterogeneity also apply to receptors, as shown in later chapters, and whether this actually matters is surely a central issue for contemporary neuroscience.

4.5.4 Calcium channels can interact with intracellular proteins

In skeletal muscle, there is a link between the L channels and calcium release channels in the sarcoplasmic reticulum (SR) which forms the basis of excitation–contraction coupling. Calcium entry through L channels is not large compared with that released from the SR, and the role of the L channels is mainly as a coupling device between the plasma membrane and the SR. A direct interaction between the α_1-subunit of the L channel and the SR calcium release channel (also known as the ryanodine receptor because it can be blocked by ryanodine) has been identified as involving the intracellular loop between domains II and III. In neurons, voltage-gated calcium channels allow calcium entry to stimulate exocytotic neurotransmitter release, a topic covered in Chapter 7. There are interactions between calcium channels and a number of intracellular proteins, and again the site of interaction seems to be the loop between domains II and III.

4.6 How similar are sodium and calcium channels?

There are clearly a number of similarities between sodium and calcium channels and it has been suggested that in the evolutionary development of ion channels, sodium and calcium channels are derived from a common ancestor. Sodium channels diverged from calcium channels with the appearance of the metazoa, only 1000 million years ago, relatively recently when compared with the evolution of mechanosensitive channels which pre-date the evolution of eukaryotes.

The close similarity between these two channel types has been shown in experiments involving site-directed mutagenesis of the SS2 segments of sodium channels. As previously described, the SS2 segment of the sodium channel contains a number of acidic amino acid residues (e.g. Glu387) which are involved in binding TTX and which form part of the external mouth and/or pore of the channel.

Comparisons of sequences between different sodium channels showed that at position 384 of the rat brain sodium channel II and at equivalent positions in other domains (*Table 4.3*), the amino acid residues are identical in sodium channels from brain, heart, skeletal muscle and the eel electroplax. This high degree of conservation suggests that these residues play an important role in channel function.

Table 4.3. Comparison of amino acid residues in SS2 segments of sodium and calcium channels

Position	Na^+ channels	Ca^{2+} channels
384	Asp	Glu
942	Glu	Glu
1422	Lys	Glu
1714	Ala	Glu

In calcium channels from brain, heart and skeletal muscle, at the same positions there are, again, identical residues, but this time they are all glutamate residues. Thus, at these positions, sodium and calcium channels only seem to vary by two amino acids, as aspartate is a conservative substitution for glutamate.

Mutants of the sodium channel were constructed with either the lysine at position 1422 converted to glutamate (K1422E) or the alanine (1714) converted to glutamate (A1714E) or both (K1422E:A1714E). Patch clamp currents were recorded in solutions containing Ba^{2+} as a Ca^{2+} substitute to record currents through calcium channels. Barium is used rather than calcium for two reasons. Firstly, many calcium channels are more permeable to barium than to calcium, so the measured current is larger. Secondly, the inactivation rate of many calcium channels is dependent on intracellular calcium, so using barium does not inactivate the channel being studied. In these studies, neither A1714E nor the wild-type channel showed any Ba^{2+} current, although the mutant A1714E had a reduced selectivity for sodium over potassium. However, the mutant K1422E had a significant Ba^{2+} current, and the double mutant K1422E:A1714 not only had a large Ba^{2+} current but was also selective for calcium over sodium at physiological concentrations. This demonstration of the importance of these residues in ion selectivity and conductance was confirmed by experiments on calcium channels where modification of the four glutamate residues to lysine reduced the high-affinity binding of calcium to the pore mouth, which is thought to be a prerequisite for calcium selective permeation. It is a demonstration of the close relationship between sodium and calcium channels that just a few amino acid substitutions can change the ion selectivity of the channel, and it is quite surprising that such a small number of changes can have such a significant effect on channel function.

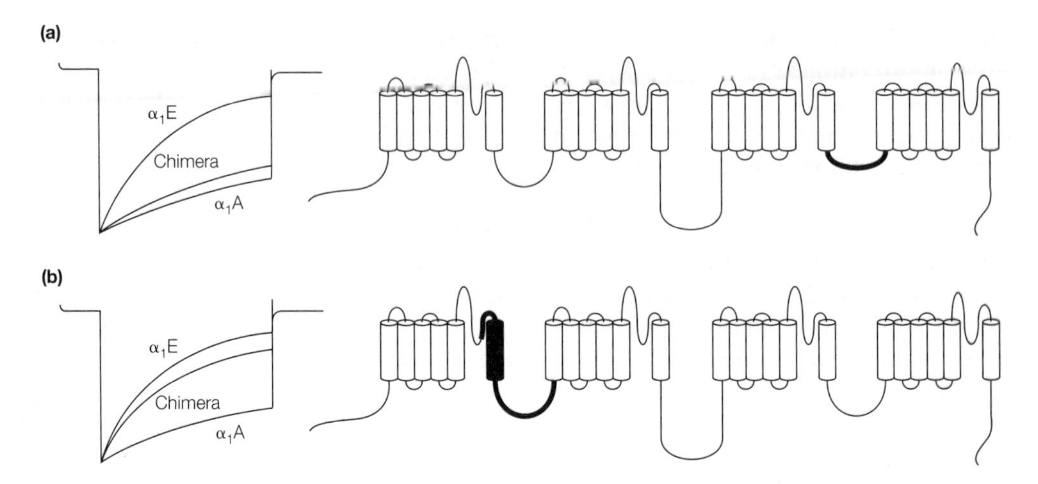

Figure 4.9. Chimeras of α_1-subunits produce different inactivation rates. The substituted section from the α_1E channel is shown in bold. Modified from Zhang, J.-F. *et al.* (1994) Molecular determinants of voltage-dependent inactivation in calcium channels, *Nature* **372**: 97–100.

However, a given function need not involve identical molecular determinants, even in closely related channels. Whereas sodium channel inactivation is thought to involve the cytoplasmic loop between domains III and IV, variations in the III–IV loop of the calcium channel have very little effect on channel inactivation. Two different calcium channel α_1-subunits (α_1A and α_1E) show different patterns of inactivation when expressed in oocytes (along with identical α_2- and β-subunits), with the α_1E subunit giving faster inactivation than α_1A. **Chimeras** in which the cytoplasmic loop III–IV of α_1E was put into the α_1A-subunit produced no significant increase in the inactivation rate (*Figure 4.9a*).

Other chimeras were then designed to discover the region of the channel responsible for the rate of inactivation. After testing many different combinations, it was shown that if the channel contained the region from the SS2 segment of domain I to the beginning of domain II then this determined the inactivation rate regardless of the rest of the channel (*Figure 4.9b*). Thus, channels made from α_1A containing this region from α_1E behaved like α_1E and vice versa. Chimeras with smaller and smaller insertions were then used to narrow down the region of interest until it was determined that the activation rate depends on nine or fewer amino acids between positions 322 and 375, all of which are in or very close to S6 of domain I.

In this way, calcium channel inactivation differs significantly from that of sodium channels although it has been suggested that sodium channel inactivation could consist of two components, one being the fast inactivation involving the III–IV loop which masks a slower type of inactivation involving the SS2 and S6 regions of domain I.

4.7 There is a large diversity of voltage-gated potassium channels

There are many types of potassium channel, each with its own distinctive electrophysiological and pharmacological properties; what they all have in common is that they tend to stabilize the membrane at the potassium **equilibrium potential**. This section will examine the role of just a few of these different kinds of potassium channel in the overall physiology of cells. It is important to appreciate the various functions of potassium channels since they are ubiquitously distributed – being present in most cells, including plants and even single-celled eukaryotes – and because it is probably fair to claim that the electrophysiological signatures of excitable cells is in large measure determined by the exact mix of potassium channel populations they possess. In order to classify the behavior of potassium channels, it is necessary first to examine the phenomenon of **rectification**. This is the property that membrane resistance varies with voltage, and can be demonstrated by performing voltage clamp experiments so that current–voltage (I–V) plots can be constructed (*Figure 4.10*).

Although some ion channels appear to obey Ohm's law, where current and voltage are linearly related, many do not, and it is this departure from Ohm's law that is rectification.

4.7.1 *Action potential repolarization requires K_V potassium channels*

A glance at the I–V plot of the potassium channel involved in action potential repolarization, which causes the delayed outward current in voltage clamp experiments (see Section 4.1), shows that it is an **outward rectifier** since it allows current to flow out but not in (*Figure 4.10b*). Hence this type of potassium channel is called the delayed outward rectifier (K_V) channel, and the current that it carries is depicted by $I_{K(V)}$. That K_V channels are heterogenous is shown by the fact that those in squid axon do not inactivate but those in frog myelinated nerve nodes of Ranvier do. Also, whilst the squid axon K_V is blocked by TEA only if the drug is perfused internally into the axon, the frog node K_V channel is blocked by low (0.4 mM) concentrations of TEA in the external bathing medium.

4.7.2 *Transient VDKCs control firing frequency of neurons*

A brief current flowing through a population of potassium channels which are activated by small depolarizations following a period of hyperpolarization can be seen in

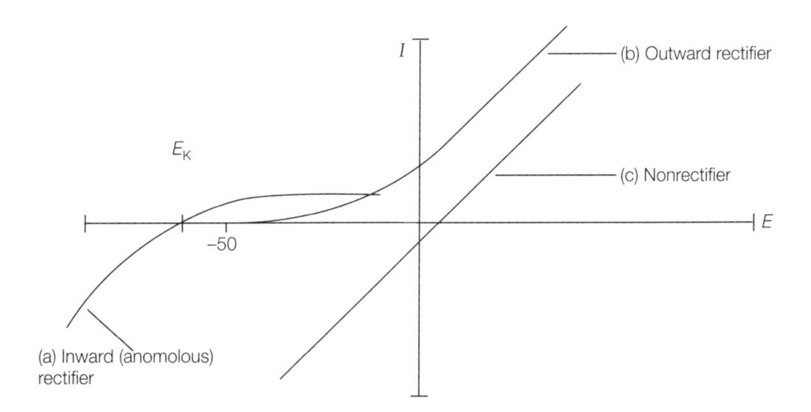

Figure 4.10. I–V plots of (a) cardiac inward rectifier; (b) neuron delayed outward rectifier; and (c) nonrectifier.

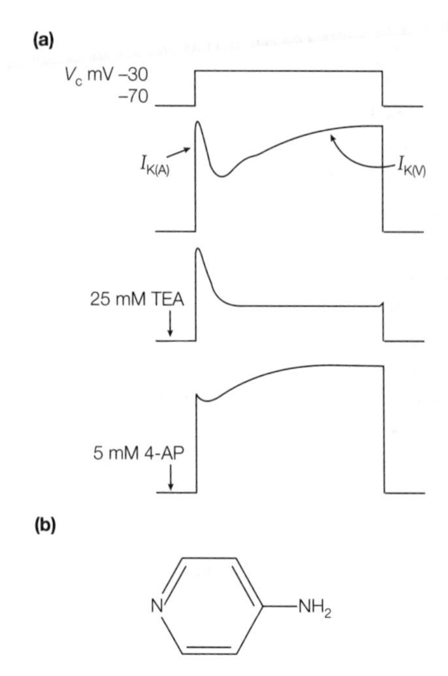

Figure 4.11. Pharmacologic separation of $I_{K(V)}$ and $I_{K(A)}$ currents in mollusc neuron: V_C is clamping command voltage. (b) Structure of 4-AP.

neurons. This current is termed the $I_{K(A)}$ current, and the channels that pass it K_A channels. K_A channels have pharmacological properties distinct from K_V channels, as shown below. $I_{K(V)}$ is blocked by external TEA, whilst $I_{K(A)}$ is unaffected. Conversely, 4-aminopyridine (4-AP) selectively blocks $I_{K(A)}$ (*Figure 4.11*).

The function of K_A channels was deduced by intracellular recording from stimulated molluscan neurons in the presence and absence of 4-AP. The experiments revealed that 4-AP reduces the interspike interval, the implication being that K_A channels normally act to increase the interspike interval, that is to reduce firing frequency. Because K_A channels activate at quite high potentials (i.e. >60 mV), $I_{K(A)}$ current will flow as soon as a neuron begins to depolarize, and the efflux of K^+ through the K_A channel will tend to oppose the depolarization. Essentially, this means that the $I_{K(A)}$ acts to prolong the afterhyperpolarization (AHP) following an action potential, and this increases the interspike interval. The end result is a slower firing frequency than would be seen in the absence of K_A channels. This allows neurons to encode a wide range of stimulus intensities since K_A channels prevent the neuron from reaching its maximum firing rate until stimulus intensity is higher than would be the case without them.

4.7.3 There are three types of Ca^{2+}-activated K^+ channel

So far all the channels discussed in this chapter have been activated by voltage. However, there are some voltage-activated ion channels that, in addition, are regulated by intracellular ligands. Furthermore, there are channels which are activated by intracellular ligands but are not voltage dependent. These variations on the ion channel theme are exemplified by Ca^{2+}-activated K^+ channels (K_{Ca}).

Table 4.4. Properties of Ca^{2+}-activated K^+ channels

Channel	Current	Conductance (pS)	Electrophysiology	Antagonists
B K_{Ca}	$I_{BK(Ca)}$	100–250	Voltage-dependent	TEA, charybdotoxin
I K_{Ca}	$I_{IK(Ca)}$	18–50	Voltage-independent	TEA, charybdotoxin
S K_{Ca}	$I_{SK(Ca)}$	6–14	Voltage-independent	Apamin

Potassium channels that are activated by intracellular Ca^{2+} over the physiological concentration range (10^{-8}–10^{-6} M) are ubiquitously distributed in excitable tissues and in some nonexcitable tissues such as salivary gland, and in red cells and macrophages. There are three types, distinguished by their conductances, called, rather imaginatively, big (B), intermediate (I) and small (S) K_{ca}. Their properties are summarized in *Table 4.4.*

Note that S K_{Ca} is blocked by picomolar concentrations of the bee venom peptide, apamin, which allows this channel to be distinguished from the other K_{Ca} channels which can be blocked by charybdotoxin (CrTx), a component of the venom of the scorpion *Leiurus quinquestriatus*. These toxins have proved to be useful tools for exploring the function of these potassium channels just as toxins have proved crucial in characterizing calcium channels.

The major role of K_{Ca} channels in excitable cells is to prolong the AHP following calcium action potentials that are seen in the dendrites of some neurons. Calcium entry during the action potential activates the potassium channels, so increasing K^+ efflux from the neuron. The action of K_{Ca} channels in some neurons is to produce long periods of hyperpolarization during which the cells are silent. Between these silent periods, the neurons fire with quite high frequencies. The overall effect is that the neurons fire bursts of action potentials.

Each burst terminates when the rise in intracellular Ca^{2+} reaches the point at which K_{Ca} channels open. Although K_{Ca} channels are involved in causing burst firing of neurons, most cells expressing K_{Ca} channels do not exhibit burst firing, indeed some are not even excitable (e.g. macrophages), suggesting that there are other functions for these channels. In some neurons, K_{Ca} channels may, like K_A channels, reduce the neuron firing frequency. Bursting is a property of some neurons in the hippocampus, a property of importance in mechanisms of plasticity and epilepsy.

4.8 Genes for potassium channels have been identified in naturally occurring mutant flies

Until recently, no high-affinity ligands for potassium channels were available that could act as tools for purifying the channel proteins in the way that TTX was important for VDSCs. Hence, the first genes coding for potassium channels were identified entirely by molecular genetic techniques.

The fruit fly *Drosophila melanogaster* is the workhorse (so to speak) of much experimental genetics. A number of mutant strains of fly have been identified which, when exposed to ether to anesthetize them, develop uncontrolled shakes in legs, wings and abdomen. Not surprisingly, these are called *Shaker* mutants.

The *Shaker* mutant has abnormal neuron and muscle electrophysiology. Nerve cell action potentials are prolonged in mutant flies because of a defect in repolarization. Voltage clamp studies of both neuron and muscle show a loss of the transient K^+ current, $I_{K(A)}$, which is known to be responsible for action potential repolarization in flies. The obvious inference that the *Shaker* locus was a gene encoding a mutant K_A channel prompted a search to locate the *Shaker* gene. This turned out to be a difficult task which took several years, but, eventually, using chromosome walking (***Box 4.4***, p. 75) and mapping techniques, the mutant gene was found. This enabled (in 1987) the DNA from the wild-type gene to be sequenced. RNA transcribed from the cDNA and injected into *Xenopus* oocytes resulted in the expression of channels that behaved exactly like K_A channels.

Actually, the *Shaker* locus is a huge gene containing several introns, and it encodes 12 closely related types of potassium channel by virtue of the fact that its primary transcript is subject to alternative splicing; different combinations of exons are selected during the processing into mature RNA.

Subsequently, oligonucleotides constructed to be complementary to short regions of the *Shaker* locus gene were used to probe genomic libraries, by hybridization under low-stringency conditions, to hunt for related genes. This revealed three other genes in *Drosophila*, called *Shal*, *Shaw* and *Shab*. Counterparts to all of the fly genes have been found in mammals which showed anything between 55 and 82% homology. Interestingly, in vertebrates, the *Shaker* gene homolog is a family of 12 intronless genes. So, whilst *Shaker* locus channel diversity in the fly is brought about by alternative splicing, in vertebrates *Shaker* channel subtypes are encoded by separate genes which presumably have arisen by gene duplication. The implication of this is that during early evolution two quite different strategies for generating potassium channel diversity evolved. Curiously, the method adopted by vertebrates (which evolved after invertebrates) is the most ancient strategy for achieving diversity; prokaryotic genes have no introns and so gene duplication is the only way of generating diversity.

The electrophysiological properties of the *Drosophila* gene products and their mammalian equivalents are quite diverse. *Shaker*, *Shal*, *Shab* and *Shaw* genes all produce channels that are outward rectifiers, and their mammalian homologs are termed K_V channels. So, for example, the 12 splice variants of the fly *Shaker* locus are called K_V 1.1 through K_V 1.12 in the mammal (*Table 4.5*).

Table 4.5. *Drosophila* potassium channel nomenclature and mammalian homologs

Drosophila gene	Mammalian homolog
Shaker	K_V 1.1–K_V 1.12
Shab	K_V 2.1–K_V 2.2
Shaw	K_V 3.1–K_V 3.4
Shal	K_V 4.1–K_V 4.2

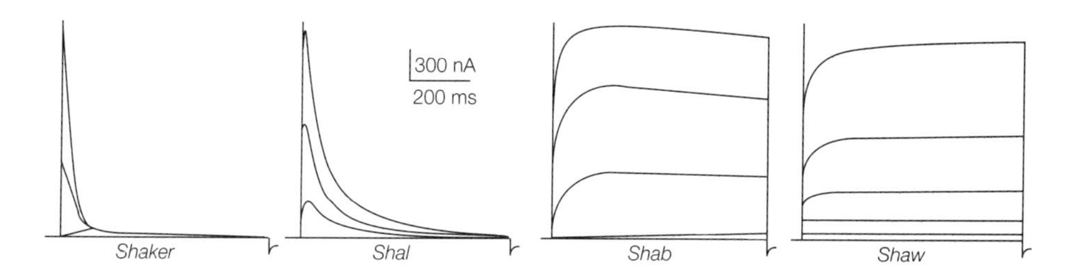

Figure 4.12. Voltage clamping of *Xenopus* oocytes expressing channels encoded by *Shaker*, *Shal*, *Shab* and *Shaw* transcripts. I_{out} was recorded in response to clamping between –80 and +20 mV in 10 mV steps from a holding potential of –90 mV.

When fly transcripts are expressed in *Xenopus* oocytes (*Figure 4.12*), both *Shaker* and *Shal* generate rapidly inactivating currents that resemble $I_{K(A)}$ in neurons of various species. *Shab* transcripts produce slowly inactivating currents, the mouse homolog of which closely resembles the delayed outward rectifier of rat hippocampal neurons. *Shaw* currents show no sign of inactivation.

4.8.1 Potassium channels are homologous to voltage-dependent Na^+channels

DNA sequencing reveals that the potassium channels encoded by fly and vertebrate *Shaker*, *Shal*, *Shaw* and *Shab* genes all resemble a single domain of the VSDC (*Figure 4.13*).

Each contains six putative TM hydrophobic segments – only some of which are α-helices – and has intracellular N- and C-terminal ends. A fifth gene, *Slo*, codes for the B K_{Ca} channel, and its sequence suggests that B K_{Ca} contains the usual potassium channel motif but also has a much longer C terminus and may have additional TM segments.

By analogy with VDSCs which clearly resemble four covalently linked *Shaker* proteins, VDKCs are assumed to be tetrameric **homo-oligomers**.

Recall that the *Drosophila Shaker* gene encodes 12 distinct channel proteins. Experiments in which *Xenopus* oocytes have been coinjected with mRNA coding for several *Shaker* channel proteins show that **hetero-oligomers** are formed, that is functional channels can be created by association of several distinct proteins of the *Shaker* gene family, though whether this happens *in vivo* is not known. Amazingly, hetero-oligomers are formed even if the mRNAs come from more than one species. In other words, it is possible to have potassium channels that are part fly and part mouse! By contrast, hetero-association of proteins from different gene families (e.g. *Shaker* + *Shal*, etc.) has never been seen to result in functional channels.

As with VDSCs, the S4 TM segment carries a cluster of positively charged residues and is thought to act as the voltage sensor for channel activation. It is noteworthy that the noninactivating *Drosophila Shaw* proteins have fewer positive charges on their S4 segments that the other channel families.

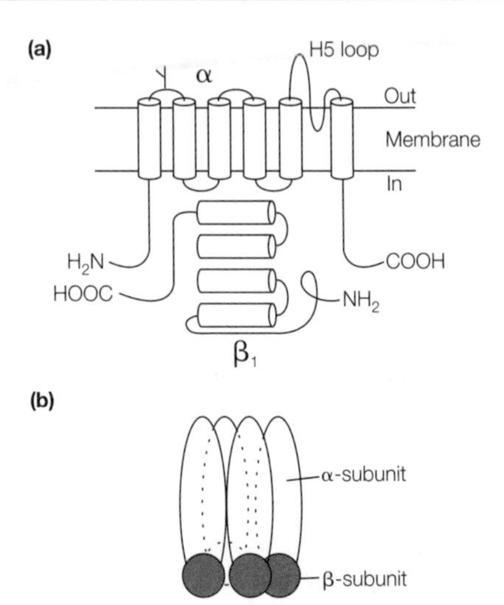

Figure 4.13. Postulated structure of voltage-dependent K^+ channel subunits. (a) The association of α- and β-subunits; (b) the octameric structure of the complete channel complex.

Site-directed mutagenesis studies, coupled with the use of selective toxins, have proved invaluable in unraveling which bits of the potassium channel are functionally important. In particular, attention has focused on identification of the pore-forming region, and on the mechanism of inactivation.

4.8.2 Pore-lining residues have been found

Charybdotoxin (CrTx) is an extracellular blocker of several types of potassium channel and is assumed, therefore, to bind a site close to the external mouth of the pore. Amino acid residues, mainly glutamate and aspartate, required for high-affinity CrTx binding are clustered on the extracellular ends of SS1 and SS2. Mutagenesis of these residues to positively charged residues reduces CrTx affinity 300-fold. The implication is that the SS1 and SS2 regions are part of the pore region.

This is supported by studies using TEA. Crucial residues for TEA blockade extracellularly are located on SS2. Thus, position 449 must be either a phenylalanine or a tyrosine residue to confer blockade. However, only eight residues away lies Thr441 which is needed for TEA to produce intracellular block; mutation of this residue results in a 10-fold reduction in affinity for TEA applied to the intracellular face of the channel. It was the discovery of the close proximity of the residues required for external and internal blockade by TEA which led to the inevitable conclusion that the SS2 region must span the membrane and so together the SS1 and SS2 regions formed a loop, the H5 loop which was associated with the ion pore. Actually it is by analogy with the potassium channel that, as we saw earlier, VDSCs are now thought to have H5 loops between segments S5 and S6 in each of the four domains. Further evidence that the H5 loop is part of the pore comes from the observations that:

(i) deletions of Tyr445 or Gly446 result in total loss of K^+ selectivity;
(ii) the creation of chimeric channels in which the H5 loop is swapped between different channel subtypes with distinctive conductances and affinities for extracellular and intracellular TEA binding shows that these pore properties swapped with the H5 loop.

4.8.3 *A potassium channel inactivation mechanism has been identified.*

Fast inactivation of VDKCs seems to occur by a quite remarkable process aptly described as the ball and chain model. The idea is that the N-terminal end of the molecule forms a ball or inactivation particle which occludes the intracellular mouth of the channel. Because of this mechanism, fast inactivation is also called N-type inactivation. Mutant channels in which a 50-amino acid residue sequence is inserted into the chain region of the N-terminal end show slower inactivation kinetics than wild-type channels. Presumably this is because the longer chain means that it is less likely that the inactivation particle finds its way into the internal mouth.

Mutants lacking the N-terminal 'ball' region fail to inactivate and, interestingly, when peptides with a sequence homologous to the ball region have been microinjected into *Xenopus* oocytes expressing *Shaker* potassium channel mutants with a deleted N terminus, inactivation is restored. The N-terminal ball is presumed to interact with the intracellular region of the H5 loop to produce fast inactivation since mutants with deletions in the H5 loop region do not fast-inactivate.

Despite the apparent differences in Na^+ and K^+ channel fast inactivation – the VDSC has no N-terminal inactivation particle – and the fact that there is no homology between the VDSC third cytoplasmic loop and the K^+ channel inactivation particle, there is evidence that inactivation in the two channel families is related.

Splicing of the VDSC third cytoplasmic loop sequence containing the amino acids -Ile–Phe–Met- onto the N terminus of a noninactivating VDKC (*Shaw*) generates N-type inactivation. Thus, the K^+ channel intracellular mouth recognizes the VDSC inactivation sequence. This implies an ancient common ancestral fast inactivation mechanism that evolved with the origin of the first voltage-dependent ion channels, probably some type of potassium channel in prokaryotes some two to three billion years ago. First VDKC then, later, VDSC inactivation mechanisms descended from this.

As it happens, N-terminal deleted *Shaker* mutants do inactivate slowly by a quite distinct mechanism which seems to depend on amino acids at the extracellular end of S6. This is termed slow or C-type inactivation and, although its physiological significance is currently a mystery in terms of kinetics and the part of the channel implicated, it may resemble the slow inactivation of the Ca^{2+} channel discussed earlier.

In summary, molecular biology has revealed a superfamily of K^+ channels, comprising *Shaker*, *Shal*, *Shab*, *Shaw* and *Slo* subfamilies, which between them encode populations of K_V and K_A channels and the B K_{Ca} channel.

4.8.4 *Potassium channels contain auxiliary proteins*

As for the voltage-dependent Na^+ and Ca^{2+} channels, associated with K^+ channels are auxiliary proteins. These have been discovered with the help of neurotoxins, such as α-dendrotoxin (α-DTx) in the venom of the black mamba *Dendroapsis polepsis*. α-DTx recognizes the same site as α-CrTx, and has been used to purify potassium channel complexes from bovine brain by affinity chromatography. The channels bind covalently to the toxin in the column by disulfide bonds and so, after eluting unwanted material with high salt buffer, the channel complexes are recovered by treating the column with reducing agents. Running the preparation on SDS–polyacrylamide gels shows two subunits, α and β. Treatment of the α-subunit with neuraminidase and endoglycosidase F reduced its relative molecular mass from 80 to 65 kDa, showing it to be a glycoprotein.

The primary sequence of the N-terminal end of the α-subunit is virtually identical to that predicted from the cDNA for the rat cerebral cortical *Shaker* homolog K_V 1.5. Subsequent work has shown that there are 20 α-subunit isoforms and they correspond to the K^+ channel superfamily. The 41-kDa β-subunits – two isoforms have thus far been identified – are auxiliary proteins closely associated with the α channel subunits; they cannot be dissociated even by high salt concentration. The β-subunits are thought to be cytoplasmic rather than TM proteins since they are largely hydrophilic, have no obvious hydrophobic amino acid TM sequences, have no N-terminal hydrophobic signal sequence, and are not glycosylated (*Figure 4.13a*). They contain consensus sequences for phosphorylation by PKA, PKC and casein kinase II, and indeed an endogenous kinase co-purifies with the channel complexes.

That the β-subunits are an essential component of mammalian K^+ channels *in vivo* is suggested by the fact that the biophysical properties of channels composed of α-subunits alone do not match those of native channels. The size of the purified channel complex (400 kDa) indicates an octamer stoichiometry, $\alpha_4\beta_4$ (*Figure 4.13b*).

Furthermore, only coexpression of both α- and β-subunits produces channels with native inactivation kinetics, as shown by experiments combining α-subunit from rat cortex K_V 1.1 (RCK1) with β in *Xenopus* oocytes. When RCK1 alone is expressed, the resulting current does not inactivate. However, when RCK1 and β-subunits are expressed together, the current inactivates as normal.

The β-subunits are thought to bind the N-terminal end of the α-subunits. The β-subunit has an extended N terminus which is thought to produce fast inactivation by forming a 'ball and chain' conformation which blocks the open internal mouth of the channel formed by the α-subunits. Hence, in mammals, K^+ channel inactivation seems to be the same sort of mechanism as in the fly, except that additional (β) proteins are involved. Presumably the β-subunits provide another vehicle for generating diversity of mammalian K^+ channels.

Although there are some structural similarities, the β-subunits associated with K^+ channels have little amino acid sequence homology with the β-subunits associated with Ca^{2+} channels. Moreover, whilst the β-subunits associated with both channel types regulate gating kinetics, there is no indication that they use a similar mechanism. Indeed, this is unlikely since the predominant modes of inactivation of the two channel types appear quite distinct; fast N-type inactivation for K^+ channels and slow C-type inactivation for Ca^{2+} channels. What is more, the β-subunits required for native Na^+ channel gating behavior are structurally quite distinct from the calcium and potassium β-subunits. The evolution of ion channel auxiliary proteins was clearly a messy business.

4.9 Not all K^+ channels belong to the same superfamily

Recently, K^+ channels have been sequenced that bear little relationship to the superfamily considered so far (*Table 4.6*). Amongst these is the **inward rectifier** (see *Figure 4.10a*) isolated from heart muscle, K_{IR} (*Figure 4.14*). It has only two postulated TM sequences, between which lies an H5 loop, and it possesses a long C terminus which has phosphorylation consensus sequences. It is thought to associate into a tetramer to form a functional channel. Now K_{IR} channels are voltage-dependent, they activate and inactivate, yet they have none of the motifs that we might expect. There is no positively charged TM segment homologous to S4, neither is there an apparent inactivation particle. Hence, molecular mechanisms of gating in these channels are not yet understood.

The importance of K_{IR} for neuroscience is twofold. Firstly, they, or channels resembling them, are likely to be responsible for the inward K^+ currents activated by hyperpolarization seen widely throughout the central nervous system (CNS) such as olfactory cortex, neocortex, hippocampus, forebrain cholinergic cells, lateral geniculate nucleus, locus ceruleus and raphé nuclei. The role of inward rectifiers in the nervous system is not immediately obvious, but one suggestion is that in periods of intense activity the external potassium concentration will rise, so activating the Na^+/K^+ ATPase. As this transporter allows influx of $2K^+$ for an efflux of $3Na^+$ it is electrogenic, and its activity generates a modest hyperpolarization. This hyperpolarization will result in I_{in} through inwardly rectifying K^+ channels in both neurons and glia. In neurons, this will tend to stabilize the resting membrane potential around the potassium equilibrium potential, E_K. In other words, the extent of the hyperpolarization will be limited. In this way, other voltage-dependent ion channels in the neuron membrane are kept at a potential where they can display appropriate gating behavior and so maintain excitable properties of the neuron. In glia, activation of inward rectifiers allows a large K^+ influx, so lowering K^+ concentration in the extracellular space. In this way, glial cells help maintain ion homeostasis in the extracellular space.

Secondly, a species of K^+ channel that is modulated by nucleotides, the K_{ATP} channel, has been cloned and sequenced from cardiac muscle. The sequence turned out to show high homology with previously sequenced inward rectifiers (K_{IR}). Interestingly, mRNA coding for K_{ATP} channels has been identified throughout the brain. This was achieved by extracting total RNA from brain areas and reverse transcribing it using random

Table 4.6. Nomenclature of K_{IR} channels

Mammalian gene	Channel clone	Designation	Notes
Kir 1.1a,1.1b	ROMK1/ROMK2	K_{IR}	Rat outer medulla, splice variants
Kir 2.1–2.3	IRK1–3	K_{IR}	
Kir 3.1–3.3	GIRK1–3[a]	K_{IR}	G protein-linked
Kir 3.4	CIR[a]	K_{IR}	Cardiac
Kir 4.1	BIR10	K_{IR}	Brain
Kir 5.1	BIR9	K_{IR}	Brain
Kir 6.1	uK_{ATP}	K_{ATP}	

[a] GIR3 and CIR are thought to form the acetylcholine-activated K^+ channel, K_{ACh}.

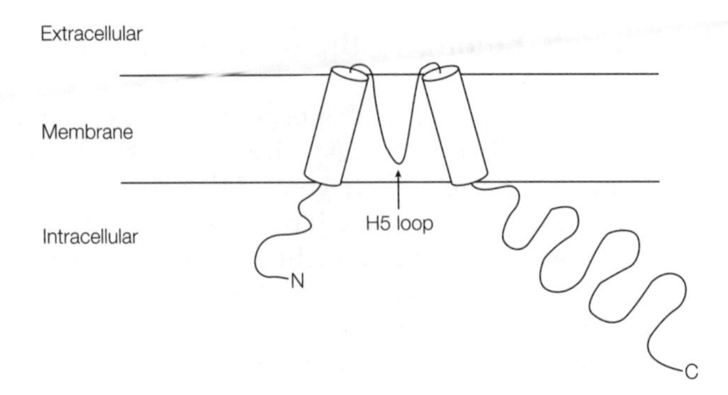

Figure 4.14. Putative structure of a K_{IR} channel.

primers. The cDNA of interest was then amplified by PCR using unique oligonucleotides directed against known 5′-untranslated and N-terminal coding sequences. The reaction products were probed by a specific radiolabeled oligonucleotide directed to a sequence between the amplification primers using **Southern blotting**. Whilst this procedure may seem rather more complicated than the **Northern blotting** technique commonly employed to localize tissue-specific expression of mRNA, it has the advantage that low-abundance mRNAs can be detected more readily. Unfortunately, in neuroscience, it is frequently the low-abundance message that is of most interest.

K_{ATP} channels are closed when the ratio of ATP:ADP is high. Under conditions of very intense cellular activity or hypoxia, there is a fall in the ATP:ADP ratio and K_{ATP} channels open, allowing K^+ efflux and membrane hyperpolarization. This increase in membrane potential will tend to reduce excitability and hence neural activity, so curtailing the oxygen requirements of the cell. In this way, K_{ATP} channels are thought to match membrane potential to metabolic status, so minimizing oxidative stress. While this role of K_{ATP} channels is well established in heart muscle, it is less clear in the nervous system, but important to understand. If K_{ATP} channel activity is neuroprotective in brain, as it is cardioprotective in heart, this opens up therapeutic possibilities for treatment of severe epilepsy, strokes and other disorders in which hypoxia plays a part. There is much work still to be done in this area. The notion that K_{ATP} channels are inward rectifiers, electrophysiologically, is by no means universally accepted, and it has been suggested that K_{ATP} channels are a modified type of K_V channel.

Box 4.1. Voltage clamping

Voltage clamping is a technique which makes it possible to measure the currents that flow across a cell membrane at a given potential. Although intracellular recording can measure the potential difference across the cell membrane, it is not possible to derive the current directly. From Ohm's law, $I = V/R$, so, in order to calculate the current, the resistance must also be known and this is not easily measured. Voltage clamping circumvents this problem by using the intracellular microelectrode to record the potential difference established by the current flow, and a feedback amplifier to inject a current, the short-circuit current (I_{SC}), across the membrane to maintain the potential difference at the initial value. Since the I_{SC} is just sufficient to counteract the change in potential difference, it must be the same size (although opposite in direction) as the current we wish to measure. Essentially, voltage clamping fixes the membrane potential at a given value and measures the current needed to maintain this potential. It is because the voltage across the membrane is held fixed that the technique is called voltage clamping. A cell membrane can be clamped at any appropriate potential, which is called the command potential.

There are two reasons why it is not possible to measure the current flow directly with an ammeter. Firstly, there is insufficient time to measure the current flow across a membrane if the potential is allowed to change along with the current. Peak rate of depolarization during the action potential may be as high as 500 V sec^{-1}, and a system to monitor the underlying currents must therefore have a response time of microseconds and there are no ammeters fast enough to do this. Secondly, many of the ion channels which control the flow of currents across the cell membrane are voltage dependent, that is the channels may open and close in response to the membrane potential. Hence, if the membrane potential is allowed to change, the currents will also change continuously. Voltage clamping overcomes this problem by fixing the voltage.

Practically, the cell is impaled with two microelectrodes, one to measure the voltage and another to inject the current (*Figure 1*). The voltage electrode measures the membrane potential (V_m). A variable voltage source generates a command voltage (V_c) which is fed into the positive input of the voltage clamp amplifier and V_m is fed into the negative input. Whenever V_m differs from V_c, the amplifier generates a current, I_{SC}, proportional to the difference in voltage. This I_{SC} is of appropriate amplitude and sign to drive V_m back to V_c within microseconds, and is

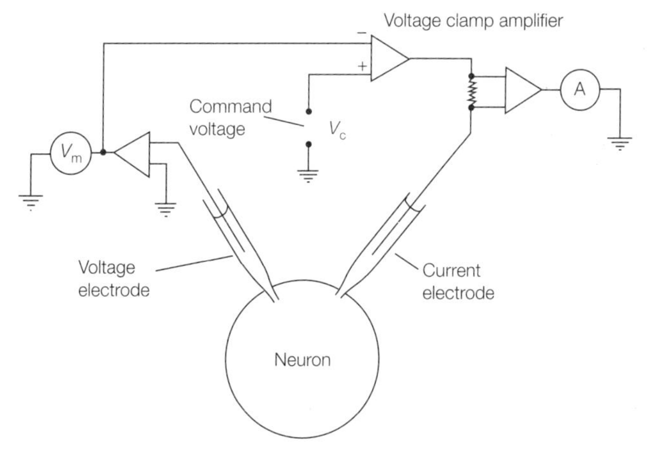

Figure 1. Voltage clamp circuitry.

Continued

measured by an ammeter (A). The voltage clamp is thus a negative feedback device designed to hold V_m equal to V_c in which the error signal provides an almost instantaneous measure of the events which produced the error. A variation of this technique, called single electrode voltage clamping uses the same method except that only one intracellular electrode is used and there is rapid switching between measuring voltage and current injection by the same electrode. This avoids having to insert two electrodes and can be used in smaller cells than traditional voltage clamping.

Box 4.2. Patch clamping

The technique of patch clamping, which is a development from voltage clamping, was first described in 1976 by Bert Sakmann and Erwin Neher as a method for measuring the currents flowing through single acetylcholine-activated channels in frog skeletal muscle. Over the next 3 years, the method was refined and extended into a range of techniques designed to measure currents through single ion channels, and has proved to be an immensely powerful tool in the understanding of the physiology, pharmacology and molecular biology of ion channels at the level of individual molecules. Patch clamping can be used to investigate any type of ion channel, and has revolutionized our knowledge of ion channels. In recognition of the importance of these methods, Sakmann and Neher were awarded the 1991 Nobel Prize in Physiology or Medicine.

Patch clamping works in essence by isolating a tiny patch of membrane beneath the microelectrode so that all extraneous noise is eliminated and only the signal from the patch is recorded. Microelectrodes which have a tip diameter of about 1 μm are fire-polished to smooth the ends and coated with hydrophobic resin to reduce capacitance. The microelectrode is placed gently on the surface of the cell, and a small amount of suction applied which greatly increases the seal between the cell membrane and the tip of the microelectrode. Once the seal, which is of the order of gigaohms, is established, it is possible to see current records as in *Figure 1* where the channel switches between closed and open states. In the ideal situation, the small patch of membrane contains only a few channels, and each square wave represents the opening of a single channel.

This type of patch clamping is called **single-channel recording** in the cell-attached mode and is one of a number of configurations that can be used. In cell-attached mode, voltage-activated channels can be opened by applying a voltage step to the pipette. This mode can also be used to study the effect of ligands applied to the remainder of the cell on second messenger-activated channels that may be in the patch.

Other types of single-channel recording use outside-out or inside-out patches (*Figure 2*). These modes allow the application of putative ligands or second messengers to either the intracellular or extracellular faces of the membrane in order to activate or modulate the channel activity. Most commonly, the outside-out mode is used to study

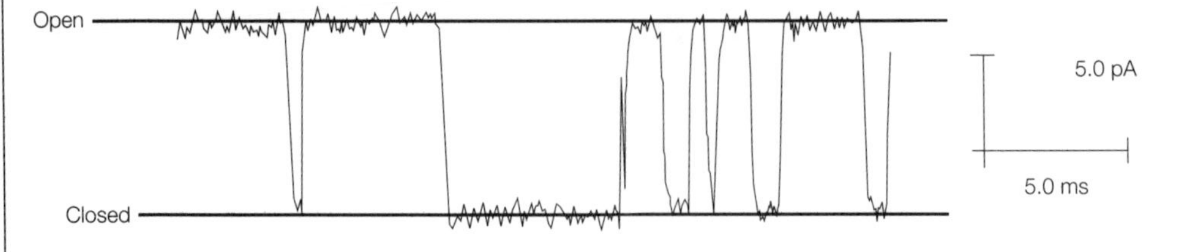

Figure 1. An example of single-channel records.

Continued

the effect of neurotransmitters as by adding these to the bath solution they can be rapidly and easily exchanged, which has advantages for performing complicated experiments such as examining dose–response relationships. Inside-out mode is usually used for the detailed examination of second messengers since these can be applied to the bath.

The other mode of patch clamping is called whole-cell mode and is essentially a form of intracellular recording, but one which can be used on the very cells which are too small to be impaled with conventional microelectrodes. After a gigaohm seal has been formed, the patch of membrane is ruptured by applying either a transient pulse of suction or a high-voltage pulse. This enables the measurement of the macroscopic currents that flow through the channels in the whole-cell membrane.

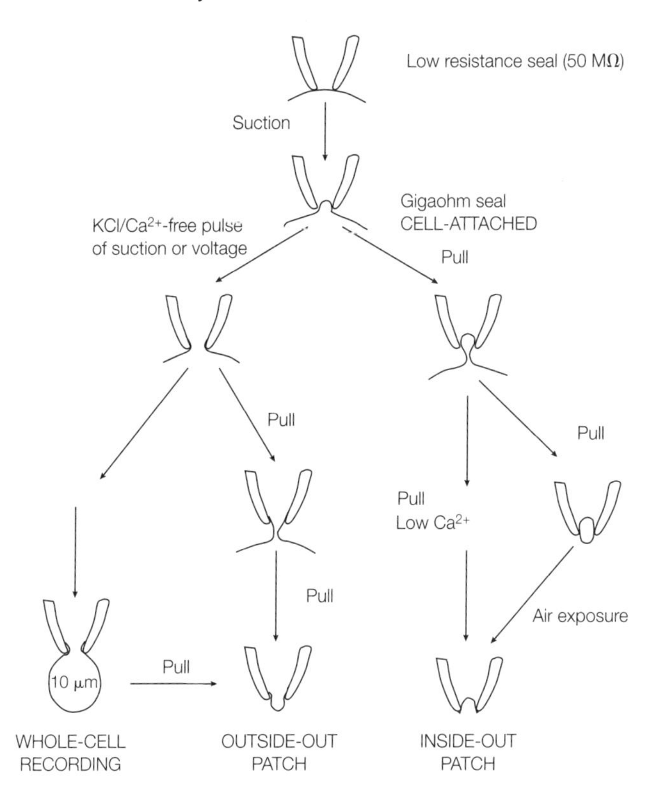

Figure 2. Patch clamp configurations.

Box 4.3. Site-directed mutagenesis

This technique allows alteration to be made in the DNA sequence in order to introduce specific changes in the amino acids of the protein. For example, changing a single base of the codon AAG to CAG will cause glutamate to be translated rather than lysine. This would remove a positively charged residue and replace it with a negatively charged one, an overall change of two charges. In this way, the role of a specific amino acid residue or residues can be examined.

The earliest versions of this technique involved cutting the DNA with specific restriction enzymes and then modifying the DNA of the cut ends to produce an insertion or a deletion at the restriction site. Refinements of this technique used multiple restriction sites to cut the DNA in the region of interest and to replace parts of the normal, 'wild-type', DNA with DNA containing new, synthetically generated, sequences.

More recent developments in the methods of site-directed mutagenesis use synthetic oligonucleotides to introduce the mutation at the desired position. An oligonucleotide of about 15–30 nucleotides, which is complementary to the section of the gene required but differing by one or two nucleotides near the center of the oligonucleotide, is annealed to the single-stranded DNA of the gene under conditions of low stringency. This means that despite the mismatches at the mutation site, the flanking sequences can confer sufficient stability to the hybrid. The rest of the single-stranded DNA is now copied using DNA polymerase to give a double-stranded DNA molecule with one wild-type strand and one mutant strand. When transformed in *E. coli*, this produces daughter plasmids, some of which will carry the mutant strands. The mutant colonies can then be identified by hybridization with the same synthetic oligonucleotide (now ^{32}P labeled), but at high stringency, when only clones containing the mutant plasmid are labeled.

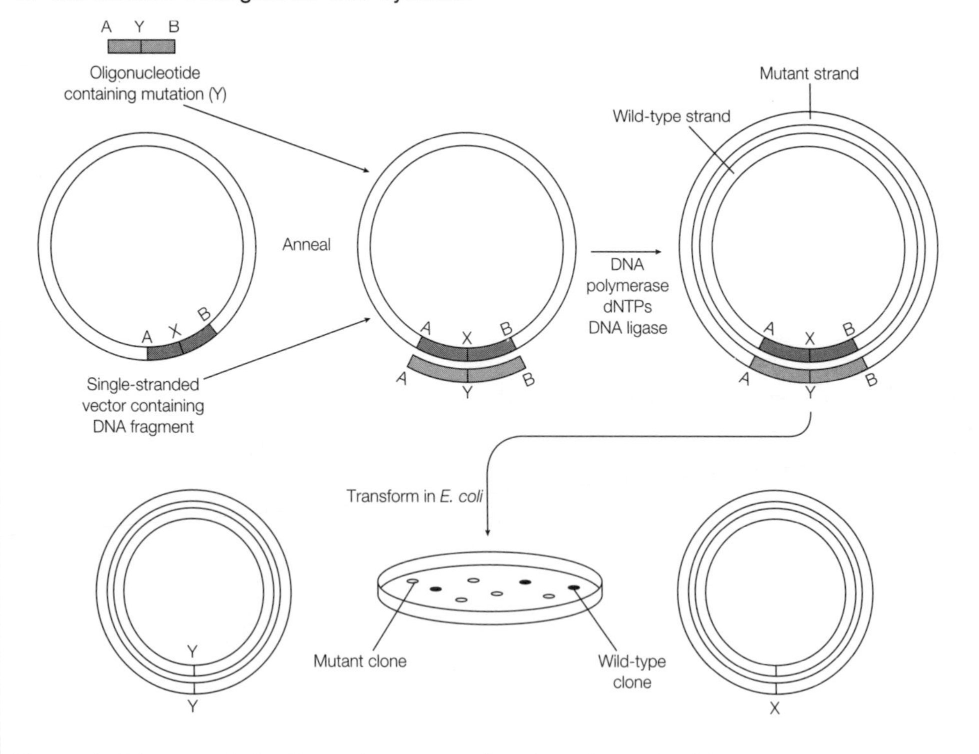

Figure 1. Diagram of site-directed mutagenesis using oligonucleotides.

Box 4.4. Chromosome walking

The gene responsible for a particular phenotype may be associated with a region of a chromosome if it tends to be inherited with a marker gene whose position is known. Using the method of **linkage analysis**, it is possible to estimate how far away from the known gene the unknown gene is found. Using chromosome walking, it is possible to explore the region of interest to locate and sequence the gene responsible for the phenotype by moving systematically along the chromosome from the marker gene to the unknown gene.

The DNA containing both the marker and the unknown gene is cut into overlapping fragments and used to construct a DNA library. The largest clone containing the marker gene is identified, by probing the library with an oligonucleotide probe for the marker, and the clone sequenced. A probe is then made from the part of the sequence which is furthest from the marker in the direction of the unknown gene, and this is used to probe the library for clones which contain this new sequence (*Figure 1*, Probe A). This new clone will contain a stretch of DNA which is closer to the unknown gene than the original clone containing the marker. This new clone is then sequenced and again a probe is derived from the furthest end (Probe B). This is then used to probe for sequences which are even closer to the unknown gene. This is repeated until a clone is obtained which is the correct distance from the marker gene and thus contains the unknown gene. The gene can then be sequenced and the first steps made to identify the structure and function of the gene product.

In the case of the *Drosophila Shaker* gene, chromosome walking was initiated in a region of the X chromosome called 16F/17 where the *Shaker* gene locus previously had been identified.

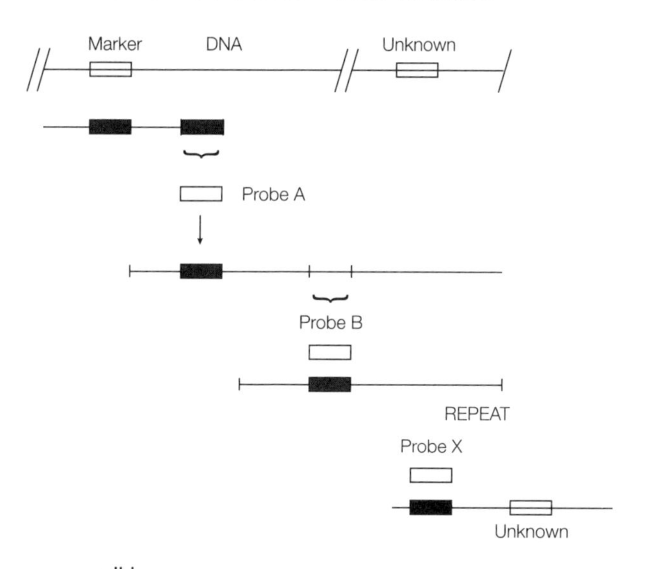

Figure 1. Chromosome walking.

Chapter 5

Ionotropic receptors

5.1 Introduction

Chapter 2 explained how the nAChR was cloned and sequenced and how deductions about how it might be assembled *in situ* were made. This chapter shows that the powerful techniques of genetic engineering can be used to confirm guesses about relationships between the receptor and cell membrane, reveal the regions of the molecule important in ligand binding, in ion selectivity and in gating or intracellular modulation, and moreover that nicotinic receptors are just one member of a large superfamily of related receptors. Glutamate and γ-aminobutyric acid (GABA), quantitatively the most important transmitters in the mammalian brain, mediate fast transmission via **ionotropic receptors**. These receptors are implicated in a number of pathologies and are targets for clinically important drugs.

5.2 Nicotinic acetylcholine receptors have been studied extensively

5.2.1 *A 4TM topology for nAChR is most likely*

For some years, the prevailing model of nAChR subunits was that they have five TM (5TM) segments which included an amphipathic transmembrane domain (MA) between M3 and M4. However, currently this model is not widely accepted for two reasons. One is because most recent work suggests the M2 not the MA segment is crucial to the ion pore. Indeed, the MA segment is poorly conserved whereas the M2 segment is the most highly conserved region in the vertebrate nicotinic receptors, hinting that it is of vital importance. The other reason is that the C terminus appears to be extracellular, consistent with a 4TM topology (see *Figure 2.19*). Several independent approaches have led to this conclusion. Firstly, *Torpedo* (though not vertebrate) nAChRs aggregate into dimers *in vivo* via disulfide bonds between cysteines at the C-terminal ends of the δ-subunits. These disulfide bonds are reduced by application of 2-mercaptoethanesulfonate and glutathione to sealed outside-out vesicles just as well in the presence and absence of detergents (*Figure 5.1a*). By contrast, detergents enhanced reduction of the receptor dimer in inside-out vesicles. (*Figure 5.1b*). This shows that the C termini of the δ-subunits, at least, face into the synaptic cleft.

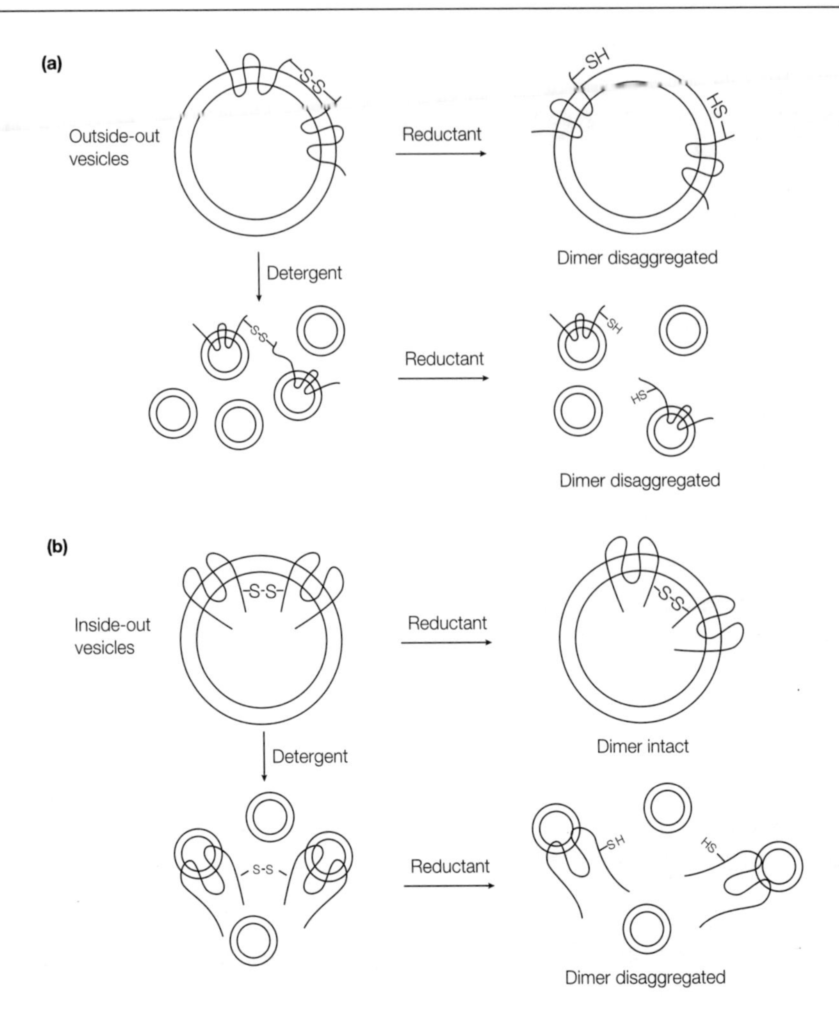

Figure 5.1. Only reducing agents on the extracellular face of the nAChR disrupt the disulfide links between the C termini of δ-subunits. (a) Receptor in outside-out vesicles; (b) receptor in inside-out vesicles.

Secondly, producing mutant α-subunits in which glycosylation consensus sites were engineered in various parts of the molecule and then determining which mutants could in fact be glycosylated, strongly supports the notion of extracellular N and C termini, with the rest of the protein either transmembrane or intracellular.

Thirdly, and perhaps most convincingly, the 4TM topology is most consistent with the results of **fusion protein** experiments. These involve inserting foreign DNA into a bacterial gene in such a way that the open reading frames of both foreign DNA and gene are preserved. Thus the hybrid gene has both reading frames perfectly fused, and when transcribed and translated results in a fusion protein. It is usual to insert the foreign DNA near the start of the bacterial gene. In general, fusion proteins are a useful way of encouraging bacteria to express foreign genes efficiently. Expression can be under the control of the bacterial promoter of choice (the lac promoter is popular) and so regulated by the experimenter. Fusion proteins tend to be relatively stable to degradation by bacterial proteases and – expressed in appropriate bacterial genes – will be exported out of the cell into the periplasmic space from which they readily can be harvested.

In the case of nAChR, fusion proteins were constructed in which a C-terminal 142 amino acid fragment of prolactin (PRL), used as an antigenic marker, was placed after each of the putative TM segments of the α-subunits. Transcripts coding for each of the fusion proteins were incubated with rabbit reticulocyte lysates and dog pancreatic microsomes. This cell-free system contains everything needed for translation and insertion of the fusion proteins into the microsomes. The microsomes were now treated with proteinase K on the basis that digestion of the PRL fragment would only occur if it were outside the vesicle, equivalent to the cytoplasmic site *in vivo*. The presence of the PRL fragment is reported by prolactin antiserum.

Figure 5.2. Deduced topology of α-subunit–PRL fusion proteins confirm the 4TM model for nAChR subunits.

Figure 5.2 illustrates the findings obtained using the five α-subunit fusion proteins αPM1, αPM2, αPM3, αPMA and αPM4. PRL-immunoreactive peptides were recovered from αPM2 and αPM4, showing that their PRL fragments had not been exposed to the protease and so must have been inside the microsome, equivalent to the extracellular face *in situ*. αPM1, αPM3 and αPMA failed to react to PRL antiserum after digestion, showing that their PRL fragments were outside the microsome. Results with δ-subunit fusion proteins were identical. These data provide convincing evidence for the 4TM model.

Finally, the 4TM model places the loop between M3 and M4 segments in the cytoplasm, a notion supported by the presence here of phosphorylation consensus sequences.

5.2.2 nAChR ligand-binding sites occur at the α–δ subunit and α–γ subunit interfaces

ACh, potent agonists and competitive antagonists of nAChR all contain a positively charged quaternary nitrogen and, as we saw in Chapter 2, this was assumed to bind to an anionic site comprised of aspartate or glutamate residues. **Photoaffinity labeling** of the *Torpedo* α-subunit bound only the cysteines in positions 192 and 193 and a cluster of four aromatic residues of the α-subunit. All these residues are fully conserved in α-subunits across species, implying a crucial functional role. Site-directed mutagenesis of these residues causes a dramatic decrease in the affinity for ACh. However, acidic amino acid residues are also implicated by elegant experiments using a 0.9 nm cross-linker, one end of which reacts with sulfhydryl groups and the other end with carboxyl groups. This is predicated on the notion that the subsite binding the quaternary ammonium group of ACh is about 1 nm away from the cysteine residues in positions 192 and 193 in *Torpedo*. In both *Torpedo* and mouse, the cross-linker located acidic amino acid residues within 1 nm of the conserved cysteines, but on the δ- not the α-subunit. Mutation of these acidic residue results in large decreases in affinity for ACh. This indicates that the δ-subunit contributes together with an α-subunit to one of the two ACh-binding sites in the receptor. Moreover, homologous acidic amino acid residues are located in the γ- and ε-subunits and so may contribute, with the other α-subunit, to the second ACh-binding site in fetal and adult receptor respectively. All of the above considerations allow a model to be constructed of how nAChR binds ACh (*Figure 5.3*)

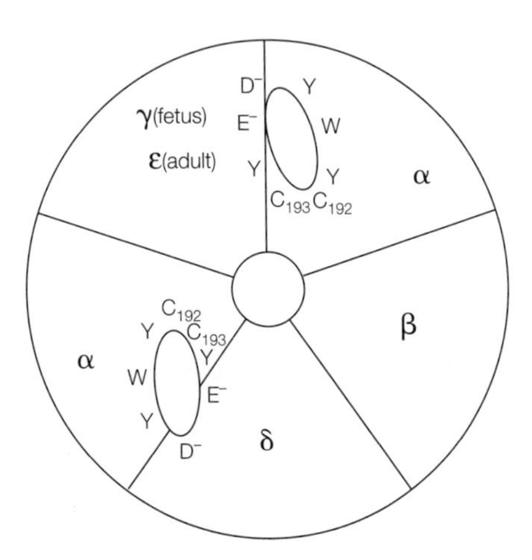

Figure 5.3. Cartoon depicting the putative location of two ACh-binding sites, formed at the interface of α–δ- and α–γ (ε)-subunits. ε is expressed in the adult and replaces the fetal γ-subunit.

Interestingly the β-subunits have no acidic residues corresponding to those in the α-, δ- or ε-subunits, and coexpression of α- plus β-subunits fails to form proper ACh-binding sites. Occupation of the binding site with ACh contracts the site and is postulated to cause a shift of one subunit relative to the other at the binding site. When this occurs simultaneously at both binding sites, the conformational changes necessary to open the channel are triggered.

5.3 The M2 segments contribute to the ion pore

Considerable evidence implicates the M2 segments as part of the channel.

(i) By constructing chimeric receptors in *Xenopus* oocytes, it had been shown that the δ-subunit was an important determinant of the unitary conductance of the nicotinic receptor. The native *Torpedo* receptor has a large conductance, but the calf receptor has a small conductance. **Chimeras** in which the δ-subunits had been swapped were found to have the conductance of the δ-subunit species.

To localize the part of the δ-subunit responsible for the species difference in conductance, cDNA constructs encoding 11 chimeric δ-subunits in which portions of the *Torpedo* δ-subunits were replaced systematically by corresponding regions of the calf δ-subunit were prepared by cutting and ligating plasmids containing either *Torpedo* or calf δ-subunits. The constructs were transcribed *in vitro*, and resulting chimeric δ-subunit mRNA injected with *Torpedo* α-, β- and γ-subunit mRNA into *Xenopus* oocytes. Single-channel current measurements of voltage-clamped cell-attached patches allowed I–V plots to be constructed for ACh-activated chimeras from which conductances were calculated. The results were clear. Conductances were either *Torpedo*-like or calf-like, and were determined by the provenance of the M2 segment only.

(ii) A second strategy has been to use **open channel blockers** as probes to locate the ion pore. These compounds, which include chlorpromazine (CPZ), triphenylmethylphosphonium (TPP), phencyclidine (PCP) and some local anesthetics, are noncompetitive antagonists which stabilize the desensitized state of the nAChR. That open channel blockers are orders of magnitude more potent in the presence of agonist implies that their target is within the ion channel. For example, [^{3}H]PCP binds to *Torpedo* nAChR-rich membranes 10 000-fold faster if the cholinergic agonist carbamylcholine is also present. CPZ, TPP and other compounds have been used to photolabel the *Torpedo* receptor, since they will react covalently with amino acid residues when activated by a brief pulse of UV light. Proteolysis and sequencing of the labeled peptides shows binding to a serine conserved in all subunits, and labeling of a few additional residues within M2 (*Figure 5.4*). The most obvious inference to draw from these data is that the M2 segment lines the channel.

(iii) Site-directed mutagenesis within M2 provides robust evidence that it is part of the channel. Mutant channels in which alanine residues had been substituted for the serines photolabeled in M2 altered the binding affinity for the open channel blocker QX-222, showing that these serine residues are targets for these compounds. Surprisingly, however, these mutations had little of no effect on ion conductance. However, mutating other serine residues in the M2 segment of the *Torpedo* α-subunit did succeed in reducing ion conductance. Moreover, mutations in all subunits at one position (263) produced decreases in conduction proportional to the volume or hydrophobicity of the substituted side chain. It is thought that residues at position 263 form a narrow ring deep in the channel important in ion selectivity. It was not surprising, in view of the *Torpedo*–calf chimera studies, that mutations in the δ-subunit at position 263 had by far the biggest effect.

(iv) Site-directed mutagenesis of residues flanking the M2 segment can have dramatic effects on ion fluxes. In a massive study, *Xenopus* oocytes were injected with mutant and wild-type *Torpedo* α-, β-, γ- and δ-subunit mRNAs in a 2:1:1:1 molar ratio. After 3–5 days, single-channel currents were measured in inside-out patches, with ACh in the pipette, to generate I–V plots for each mutant receptor. The mutations

Cytoplasmic ←| |→ Extracellular

α M T L S I *S+ V L L S L T V F

β M S L S I *S+ A L* L A V T V F

γ C* T L S I *S V L* L A Q T I F

δ M S T A I *S+ V L L A Q A V F

Figure 5.4. Aligned sequences of *Torpedo* nAChR subunit M2 segments showing photoaffinity labeling by CPZ (*) and TPP(+).

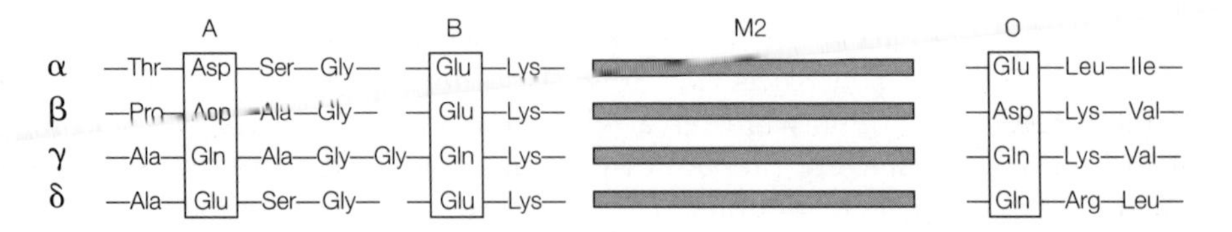

Figure 5.5. Identification of the conduction pore of the AChR by point mutations in the vicinity of M2.

were made in conserved negatively charged and glutamine residues arranged in three clusters which flank M2. (*Figure 5.5*)

The wild-type nAChR is **nonrectifying**; it allows current to pass just as readily in or out and so has a linear I–V plot. By contrast, mutations of the anionic clusters A and C, which reduce the negative charge, cause inward and outward rectification of ion transport respectively. In other words, alterations in A reduce outward current while those in C reduce inward current. That the same mutation of electrical polarity (negative to positive) in the two anionic clusters produced a different direction of rectification implies that the two rings are on opposite sides of the membrane, i.e. M2 is membrane spanning. Moreover, an approximately inverse relationship was apparent between channel conductance and total net negative charge, that is the more negative charges that were mutated, the smaller the conductance. Mutations of the intermediate cluster (B) had the largest effect on conductance and variable effects on rectification. Further evidence for the membrane topology of clusters A and C was provided by the effects of Mg^{2+}. When added to the extracellular side of wild-type or cluster A mutant channels, Mg^{2+} (0.5 mM) decreases inward but not outward current. In contrast, addition of Mg^{2+} to the intracellular side decreased the outward but not the inward current in wild-type and cluster C mutant channels.

Remarkably just three single point mutations in the α-subunit in the region of M2 result in a channel selective for anions rather than cations. Hence ion selectivity seems to be governed by very few residues in ligand-gated ion channels (LGICs) just as in voltage-dependent ion channels. Perhaps the time of the 'designer' ion channel is not so far away.

(v) In a particularly ingenious use of site-directed mutagenesis, nine consecutive residues which lie in the middle of the mouse M2 α-subunit were mutated one at a time to cysteines and the mutants expressed together with native β-, γ- and δ-subunits in *Xenopus* oocytes. The accessibility of the substituted cysteines to the aqueous compartment was tested by examining whether the highly polar molecule methanethiosulfonate ethylammonium (MEA), which reacts with sulfhydryl groups, would inhibit ACh-evoked currents when applied to the outside of oocytes. It was reasoned that only cysteines either on the extracellular surface of the receptor or lining the channel would be available to react with the MEA, by virtue of their aqueous environment. The experiment showed that every alternate residue mutant was irreversibly blocked by MEA, even when the channel was closed. Several deductions can be made from this result. Firstly, the gate which closes the channel must lie closer to the cytoplasmic side than the deepest mutated residue (equivalent to *Torpedo* aligned α267). Secondly, that alternate residues are accessible to the reagent suggests that M2 is not an α-helix, as TM segments are often assumed to be, but a β-sheet. What follows from this is that as five parallel β-strands cannot enclose a cylinder with a 0.7-nm diameter lumen, as demanded by X-ray diffraction and other studies, regions other than M2 must contribute to the channel lining. At present there is no consensus about what these additional regions might be. A model of the channel based on experiments examining M2 is given in *Figure 5.6*.

As with all such models, it should be regarded with healthy scepticism. At the very least it is incomplete, at worst it may be wrong. It assumes that the relationship between mutant structure and mutant function is a simple one. It is far from obvious that this is always the case. Mutating a residue might cause conformational changes in apparently quite remote parts of a molecule and/or have knock-on consequences for quaternary structure and function. This would greatly complicate the interpretation of site-directed mutagenesis experiments.

5.3.1 nAChR subunit TM segments may be β-sheet conformation

Structural considerations suggest that all TM regions in the nicotinic receptor may be β-sheet rather than the hitherto assumed α-helix. One powerful argument is that β-sheet invested subunits interact via hydrogen bonding between the polypeptide

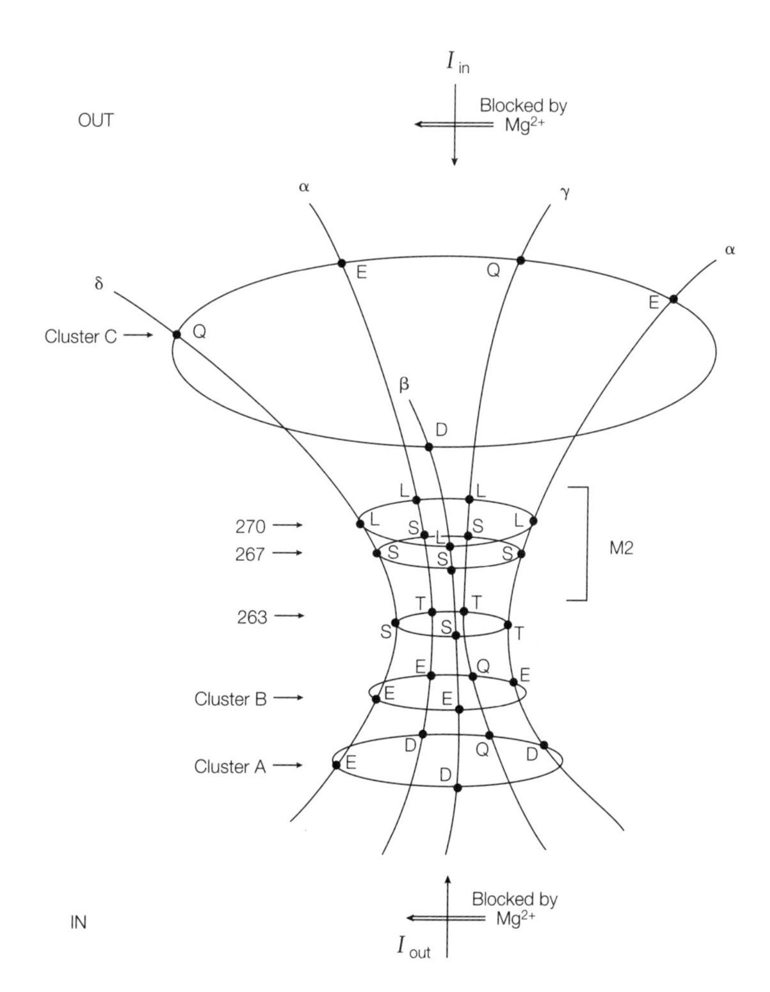

Figure 5.6. A model of the nicotinic receptor ion channel. Numbering of residues is for the *Torpedo* α-subunit.

backbone, whereas with α-helices the interaction is via the side chains. Hence, higher order structure is much less dependent on specific amino acid sequences in the case of β-sheet secondary structure, and so the basic organization of the receptor in the membrane would be fixed for a wide range of channel subtypes. This would ensure that despite the diversity of nicotinic receptors, crucial regions of the pore gating mechanism would be aligned correctly with respect to the binding site. Moreover, because β-sheets have exposed hydrogen bonding groups, in the hydrophobic interior of the lipid bilayer β-sheet invested subunits would aggregate spontaneously. Thus, we have an explanation for the assembly of the quaternary structure of the receptor.

5.3.2 *Channel gating currently is a mystery*

Although the precise sequence of molecular events that causes the channel to open when ACh binds is not yet known, elegant experiments by Nigel Unwin in Cambridge have revealed structural differences between closed and open nicotinic receptors that at least highlight which regions of the receptor are involved in the closed to open transition. The experiments used electron microscopy to study nicotinic receptors arranged in a particularly symmetrical array, comparing closed receptors with those frozen in the open state after brief exposure to ACh. Membrane isolated from *Torpedo* electric organ can be made to crystallize into long tubular vesicles when resuspended in low salt solution. These contain receptors incorporated in an outside-out configuration. To capture nAChRs in the open state, the vesicles were sprayed with tiny (~1 μm diameter) droplets containing 100 mM ACh, 5 msec before being frozen in ethane cooled to –160°C. They were analyzed by electron diffraction down to a resolution of about 0.9 nm. Small but reproducible differences could be measured between ACh-sprayed vesicles and nonactivated ones. The study revealed that activation by ACh caused a cavity about half way down the extracellular region of the α_δ-subunit, which is presumed to be one of the two ACh-binding sites, to disappear. Moreover, three apparent α-helical rods placed around the original cavity are displaced anti-clockwise by approximately 30° on exposure to ACh. Similar but smaller changes occur in the α_γ-subunit. At the level of the TM region of the receptor, activation by ACh caused displacements of five electron-dense rods assumed to be the M2 α-helices, resulting in opening of the central pore. The opening has two remarkable consequences. Firstly, it exposes hydroxyl groups on each α-helix which come to lie virtually parallel to the pore axis. It is tempting to speculate that these are important in ion permeation. Secondly, in the open state, the narrowest part of the channel (~1 nm across) shifts to lie about 1 nm deeper than in the closed state. The open channel constriction appears to correspond to several threonine residues. That mutation of one of these threonine residues has already been found to enhance or reduce ion flow depending on the volume of the substituted residue supports the model.

It is worth noting that the model developed as a result of the work contradicts the interpretation placed on the cysteine mutation experiments reviewed earlier in two respects. Firstly, the model requires that the M2 TM regions are α-helices. Secondly, it requires that the leucine residues do indeed form the narrowest part of the closed channel, that is the gate. This is contradicted by the cysteine substitution mutations which suggested that the gate lies deeper than the leucine ring. In agreement with the model, however, is the finding that cysteines substituted for the α-subunit leucines proved to be available to the sulfhydryl reagent MEA when the channel was open, since in the open state the narrowest part of the channel *does* lie deeper than the leucine ring.

5.4 GABA is the major CNS inhibitory transmitter

It was the high abundance of nicotinic receptor in electric organs and in mammalian muscle which made them obvious candidates for molecular study. In the CNS, however, cholinergic neurons make up only a few percent of the total number of neurons, and cholinoceptors are correspondingly rare. Central nicotinic receptors are even more sparse, since estimates suggest that most cholinergic receptors in the CNS are muscarinic. In quantitative terms, by far the most important of the 100 or so molecules postulated to be neurotransmitters in the mammalian brain are the simple amino acids, γ-aminobutyric acid (GABA) and glutamate. The great majority of all brain synapses use one or other of these molecules, which are responsible for most of the fast transmission in the CNS via LGICs. In general – though not invariably – GABA is inhibitory whereas glutamate is excitatory on postsynaptic neurons.

Traditionally, pharmacologists have distinguished between fast responses, which can be mimicked by muscimol and blocked by bicuculline, attributed to $GABA_A$ receptors, and slower bicuculline-insensitive responses that can be elicited by the agonist baclofen at $GABA_B$ receptors. Currently three populations of GABA receptor have been defined; $GABA_A$, $GABA_B$ and $GABA_C$. $GABA_B$ receptors are **metabotropic receptors** (see Chapter 6), whilst the others are ionotropic.

5.4.1 $GABA_A$ receptors mediate fast inhibitory responses

$GABA_A$ receptors are found on neurons, glial cells and adrenal medulla cells. They are responsible for most of the fast inhibitory responses in the CNS. Activation of the receptor by GABA opens an intrinsic chloride selective ion conductance which tends to clamp the membrane potential at the chloride reversal potential, E_{Cl}, approximately –60 mV. The chloride current shunts depolarizing currents and so reduces the probability that excitatory inputs will drive the cell across the threshold for firing action potentials. In cortical pyramidal cells, the axosomatic location of the 50 or so GABAergic synapses gives them a very high weighting. Thus, activation of only a few GABA synapses is sufficient to offset activation of many hundreds of excitatory synaptic inputs into the dendrite tree.

5.5 $GABA_A$ receptors bind many different ligands

One of the reasons why $GABA_A$ receptors have attracted much attention is that they have recognition sites not only for their endogenous ligand GABA but also for several major classes of drugs which, on binding, allosterically modulate the kinetics of binding of GABA and the kinetics of the ion channel. Many of these drugs are in clinical use, but understanding in molecular detail how they work has obvious implications for developing novel therapeutic agents.

5.5.1 Benzodiazepine-binding sites

1,4-Benzodiazepines (BZs) such as diazepam increase the affinity of the GABA site for GABA and increase the frequency of opening of the Cl^- channel, resulting in potentiation of the effects of GABA without prolonging its effects. GABA enhances the binding of BZs. BZs like diazepam are acting as allosteric agonists at the $GABA_A$ receptor, and this accounts for their usefulness as minor tranquillizers and anticonvulsants. There is a good correlation between potency of BZs and their affinity for binding to the $GABA_A$ BZ-binding site.

Several compounds, both BZs and some members of the β-carboline series of esters, bind competitively to the BZ site and reduce the frequency of opening of the Cl^- channels. Their effects are the opposite of what is expected of a typical BZ and so they are described as inverse agonists. They are either convulsant or proconvulsant (i.e. they potentiate seizures triggered by other means) and anxiogenic (anxiety-promoting).

The binding and effects of both agonists and inverse agonists are blocked by BZ antagonists, such as flumazenil. BZ antagonists prevent enhancement of GABA effects but do not reduce basal conductance of Cl^-. GABA antagonists prevent gating of Cl^- channels in spite of the presence of BZs. *Table 5.1* summarizes the effects of ligands at the BZ-binding site.

Autoradiography of brain slices with radioligands such as [^{3}H]muscimol to measure $GABA_A$ receptors and [^{3}H]flunitrazepam or [^{3}H]flumazenil to measure BZ-binding sites consistently shows that there are more GABA than BZ sites. The implication is that some $GABA_A$ receptors lack BZ-binding sites and hence that there is $GABA_A$ receptor heterogeneity.

5.5.2 Barbiturate-binding site

Barbiturates increase the affinity of the $GABA_A$ receptor for both GABA and BZs in a manner which correlates with their potency as anesthetics and hypnotics. They increase the mean open time (τ) of the Cl^- ion channel and so prolong the actions of GABA.

Table 5.1. Ligands acting at the BZ-binding site of the $GABA_A$ receptor and their effects

Action	Drug	Effect
Agonists	Flunitrazepam Zolpidem Abecarnil (partial) β-CCP	Increase Cl^- current Anticonvulsant Anxiolytic Amnestic
Inverse agonists	DMCM β-CCE Ro194603 β-CCM (partial)	Reduce Cl^- current Convulsant (or proconvulsant) Anxiogenic Increases avoidance learning
Antagonist	Flumazenil ZK93426	No effect on Cl^- current Inhibits agonist and inverse agonist binding

β-CCE, ethyl-β-carboline-3-carboxylate; β-CCM, methyl-β-carboline-3-carboxylate; β-CCP, *n*-propyl-β-carboline-3-carboxylate; DMCM, methyl-6,7-dimethoxy-4-ethyl-β-carboline-3-carboxylate; ZK93426, ethyl-6-benzyloxy-4-methoxymethyl-β-carboline-3-carboxylate.

At high concentration, barbiturates open the Cl^- channel even in the absence of GABA. The barbiturate-binding site has not been well characterized because of its low affinity; [^{3}H]phenobarbitone binds with a K_D of 100 μm. There is a close correlation between octanol:water partition coefficients, ability to enhance GABA binding and ability to displace [^{3}H]phenobarbitone binding for a whole series of barbiturates and, furthermore, barbiturate binding is abolished by detergents. All this might signify that the binding site has a lipid component.

5.5.3 Anesthetic steroid-binding site

A number of 5α-reduced metabolites of progesterone are short-acting anesthetics at high dose and anxiolytic at low dose. They act to increase the affinity of the $GABA_A$ receptor for GABA and BZs and to increase the mean channel open time of the Cl^- ion channel, but via a binding site distinct from that of barbiturates. These steroids have a wider therapeutic window than the short-acting barbiturate anesthetics.

5.5.4 The chloride ion channel is a target for convulsants

Picrotoxin, a potent convulsant, reduces the frequency of opening of the Cl^- channel without an effect on GABA or BZ binding, and is thought to bind close to the ion channel, sterically hindering chloride flux. *T*-Butylbicyclophorothionate (TBPS), a second convulsant, may bind to the same site, but then exerts its effects rather differently. BZs, barbiturates and anesthetic steroids inhibit, while convulsant β-carbolines increase TBPS binding, suggesting that TBPS binds to the closed state of the channel. Pentylenetetrazole (PZT) and penicillin are convulsant and may also act on the ion channel directly.

5.5.5 Additional GABA receptor interactions

Apart from the well-defined binding sites, described above, interactions with other molecules occur. These include volatile general anesthetics and Zn^{2+}, which binds to the extracellular face of the receptor and may be an endogenous modulator of neurotransmission. Interestingly, ethanol probably alters CNS function in part via $GABA_A$

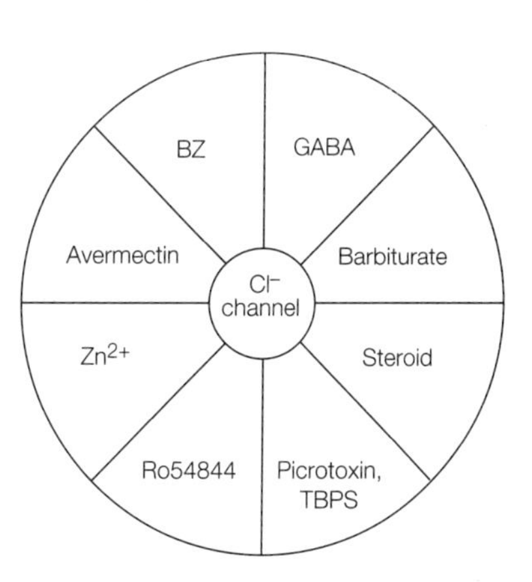

Figure 5.7. There are many independent binding sites on the $GABA_A$ receptor complex.

receptors. It enhances GABA-mediated $^{36}Cl^-$ influx into spinal neurons and brain synaptosomes. *Figure 5.7* summarizes the binding sites on the $GABA_A$ receptor.

5.6 $GABA_A$ receptor heterogeneity is evident from pharmacology

It is clear that the $GABA_A$ receptor is a remarkably complex macromolecule. However, it turns out that the complexity is much greater than originally suspected, since molecular biology studies have revealed a potentially huge diversity of $GABA_A$ receptor subtypes. In fact this heterogeneity is hinted at by conventional pharmacology.

Strains of mice have been bred with large differences in sensitivity to convulsants and anticonvulsants, or to the ataxic effects of ethanol and diazepam, which reflect functional heterogeneity in the $GABA_A$ receptor they express.

At least two distinct BZ-binding sites have been distinguished by their different affinities for BZs, the triazolopyridazine CL218872 (3-methyl-6-[(3-trifluoromethyl)phenyl]-1,2,4-triazolo-[4,3β]pyridazine, β-carbolines and clonazepam (see *Table 5.2*). The predominant type I (BZ_1) binding site which has the highest affinity for CL218872 is particularly dense in the cerebellum. The type II (BZ_2) site is preferentially enriched in spinal cord, striatum and hippocampus.

Table 5.2. Benzodiazepine receptor heterogeneity

Characteristic	BZ_1	BZ_2
Binding affinity for BZs	nM	μM
Affinity for CL218872 and β-CCM	High	Low
Clonazepam binding	+	–
Location	Cerebellum	Spinal cord, striatum, hippocampus

5.6.1 *$GABA_A$ receptor subunits were first cloned from bovine brain*

In 1987, Eric Barnard and 11 other authors reported the sequence of two subunits, α and β, of a $GABA_A$ receptor and, noting their high homology with subunits of the nAChR, argued for the existence of a superfamily of LGICs, a view now universally accepted.

The strategy for cloning the $GABA_A$ subunits was not very different in principle to that used for nicotinic receptor subunits. The $GABA_A$ receptor was purified from bovine cerebral cortex by BZ **affinity chromatography**. On SDS–polyacrylamide gels under reducing conditions two subunits could be resolved. Oligonucleotide probes were prepared on the basis of microsequencing of four peptides generated either by cleavage of the receptor protein with cyanogen bromide and separation using high-performance liquid chromatography (HPLC), or from a trypsin digest of the α-subunit peptide eluted from SDS–polyacrylamide gels. The probes were labeled with ^{32}P and used to screen a cDNA library, derived from bovine brain poly(A)$^+$ RNA, and constructed in λ phage using *Eco*RI (and other) linkers. The identified clones were sequenced by the chain termination method of Sanger to give the sequences of the α- and β-subunits. To confirm the authenticity of the cloned cDNAs, they were spliced separately into plasmids containing a bacteriophage SP6 promoter and transcribed *in vitro*. The resulting two pure RNA species were injected either alone or together into *Xenopus* oocytes which were voltage clamped. Only when both α- and β-subunit RNAs were coinjected into the same oocyte did GABA produce a current, suggesting that both subunits are needed to form a functional receptor. The effect of GABA was dose-dependent, antagonized by bicuculline, blocked by picrotoxin and potentiated by chlorazepate (a BZ) and phenobarbitone. All of this strongly suggested that the native $GABA_A$ receptor was a hetero-oligomer of α- and β-subunits; an $\alpha_2\beta_2$ stoichiometry was postulated. Moreover, photoaffinity labeling studies suggested that the β-subunit contained the GABA-binding site whilst the α-subunit had the BZ-binding site. Hence there was every reason to suspect that the $\alpha_2\beta_2$ structure was complete.

5.6.2 *Multiple subunits exist for $GABA_A$ receptors*

Within 4 years of the cloning of the first two subunits, numerous others were cloned and sequenced from several species by homology screening. Currently, five subunits are known, α, β, γ, δ and ρ, which share 15–25% homology with each other. However, each subunit comes in a variety of forms, so called **isoforms**, which share about 70% sequence homology with each other. At present, six α, four β, four γ, one δ and two ρ have been identified. Isoforms arise from distinct genes, but alternative splicing produces additional variants of some isoforms so that a truly baroque diversity of $GABA_A$ receptors is theoretically possible.

All subunits contain about 450 amino acids and are strongly conserved across species. Each subunit appears to conform to the same structural motif in having four putative TM segments, two to four N-terminal glycosylation sites and a large, poorly conserved, cytoplasmic loop between M3 and M4, which often has consensus sequences for phosphorylation by either A kinases or C kinases.

Barnard's early studies on the cloned $GABA_A$ receptors suffered from a lack of knowledge of the properties of the native receptor. However, as these became known, it was observed that the cloned receptor differed from the native one, in particular the cloned receptor bound BZs inconsistently and too weakly. By 1989, however, it became obvious that normal BZ pharmacology is conferred by coexpressing the γ_2-isoform together with the α_1 and β_1 subtypes. In retrospect, it now appears that the SDS–polyacrylamide gel electrophoresis of purified native receptor had failed to resolve γ-subunits, possibly because they comigrate along with α-subunits.

Coexpression studies have been extremely important in elucidating the roles of particular subunits and their isoforms in $GABA_A$ receptors and in showing how $GABA_A$ receptor diversity arises. It is now clear that while any β-isoform will do, it is the α-isoform which determines the nature of the BZ-binding site; α_1 gives BZ_1 whereas α_2 and α_3 both give BZ_2 sites. The α-subunit also influences the affinity of the receptor for GABA, with receptors containing α_5 having a far greater affinity (ED_{50} 14 μM) than those containing α_1 (ED_{50} ~1 mM). Positive cooperativity of channel gating is seen in native receptors, and this is best reproduced in receptors engineered to contain the α_5-isoform.

There are considerable differences in the distribution of α isoforms throughout the brain. The α_1-isoform is by far the most abundant and widely distributed throughout the brain. The α_3-isoform is largely confined to the forebrain cortex, but at low density. The α_6-isoform is localized exclusively in the granule cells of the cerebellar cortex.

$GABA_A$ receptors isolated from rat cerebellum by immunoprecipitation using antiserum raised against the C terminus of the α_6-isoform have an unusual BZ pharmacology. These cerebellar $GABA_A$ receptors have low affinity for BZs, except for the inverse agonist sapmazenil. This can be replicated by coexpression of α_6, β_2, γ_2 subunits. Sapmazenil is an antagonist of the motor incoordination and ataxia produced by ethanol intoxication. Hence, the α_6-containing $GABA_A$ receptors uniquely localized in the cerebellum are presumably responsible for ethanol-induced impairment in motor function.

β-subunits are thought to contribute to the GABA-binding site. However, channels assembled exclusively from α-, β- or δ-subunits can be gated by GABA, which rules out a unique role for the β-subunit in agonist binding.

It appears that the γ-subunit enables BZ binding. Whilst the γ_2-isoform coexpressed with α_1, and β_1 produces an authentic BZ-binding site, replacing γ_2 with γ_1 produces a receptor for which flumazenil (normally an antagonist) and the β-carboline, DMCM (usually an inverse agonist), both acquire agonist behavior, potentiating the effects of GABA. Since this pharmacology is not seen *in vivo*, it has been suggested that the α_1, β_1, γ_1-containing variant does not exist naturally. However, some combination of subtypes with γ_1 presumably occurs since γ_1 mRNA is predominant in neurons of the septum, amygdala and hypothalamus. Receptors in which γ_2 are replaced by the δ-subunit fail to bind BZs. Such receptors do appear to be expressed *in vivo* and may explain the earlier autoradiography experiments which implied that not all $GABA_A$ receptors bound radiolabeled flunitrazepam.

Ethanol has been shown to potentiate $GABA_A$-mediated chloride conductance. Now, the γ_2-isoform comes in two alternatively spliced variants which vary in length and so are termed γ2L and γ2S. Coexpression shows that ethanol interactions are seen in receptors with the γ2L rather than the γ2S variant.

The exact subunit composition of any native $GABA_A$ receptor is not yet known, but mRNAs encoding α_1, β_2 and γ_2 are often colocalized in the CNS. Moreover, the $GABA_A$ receptor, like the nAChR, appears to have a pentameric quaternary structure as revealed by image analysis of electron micrographs of uranyl acetate-stained, purified, pig $GABA_A$ receptors. Hence the best guess for the cerebral cortical receptor with BZ_1 pharmacology – the most common species – is $\alpha_1\alpha_X\beta_1\beta_2\gamma_{2L}$, where X denotes an uncertain isoform.

If every possible permutation of isoform now known, including the spliced variants, were expressed in brain as a pentameric quaternary structure this would result in over a quarter of a million distinct $GABA_A$ receptors! However, it is assumed that only a small subset (perhaps 100 or so) are actually assembled, since it is known from coexpression studies that some combinations either cannot assemble to form functional receptors or else do so with such low efficiency as to be unlikely candidates for a native receptor.

While there are clearly quite a number of distinct $GABA_A$ receptors occurring naturally that differ in their pharmacologies, localization in the CNS, cellular distribution, developmental expression and the like which can be attributed to their subunit make-up, the extent of the diversity seems unnecessarily large. Is it simply a tolerated accident of genetics; repeated gene duplication which spawns many isoforms that persist because one may often substitute for another with no appreciable change in function? Is it that nonfunctional combinations are assembled which have a physiological role other than as GABA receptors?

5.6.3 *Binding sites are being clarified by site-directed mutagenesis*

The GABA-binding site bears some similarities to the ACh-binding site on the nicotinic receptor, and also with the glycine receptor, another member of the superfamily which is an intrinsic chloride channel. Like the nAChR, the ligand-binding domain of the $GABA_A$ receptor involves two adjacent subunits. Single point mutation of a phenylalanine residue to leucine in the N-terminal extracellular domain of rat α_1 produced dramatic decreases in agonist and antagonist affinities. In addition, point wise mutagenesis of two short regions in the N-terminal extracellular domain of the β_2-subunit reduced activation by GABA. The two sequences, each of only four amino acids, lie only 41 residues apart, and are identical except that they are mirror images of each other.

It is known that the $GABA_A$ receptor has two binding sites for GABA which display cooperativity. What is not known is whether each four-residue binding domain contributes to the binding of separate GABA molecules or whether both domains bind a single GABA. In the latter case, if it is assumed that the receptor complex has an $\alpha_2\beta_2\gamma$ stoichiometry then presumably the two GABA-binding sites are provided by the two interfaces between α- and β-subunits. There are obvious parallels with the binding sites in the nAChR. Site-directed mutagenesis indicates that both α- and γ-subunits contribute to the BZ-binding site. A glycine residue upstream of the first TM segment in the α_1-subunit has proved to be an important determinant of BZ binding. Its counterpart in all other α-isoforms is glutamate. Mutation of these glutamate residues to glycine alters the pharmacology of the receptor from BZ_2 to BZ_1.

The importance of the γ_2-subunit in BZ binding is confirmed by site-directed mutagenesis of a single N-terminal threonine residue to serine, which produced dramatic changes in BZ pharmacology.

Some homology has been noted between the BZ-binding site, the GABA-binding site and even the ACh-binding site in the nicotinic receptor. Hence, even though BZ is now an allosteric modulator of GABA binding, the BZ site probably evolved from an agonist-binding site.

5.6.4 *$GABA_A$ receptor phosphorylation*

LGICs are modulated by phosphorylation. $GABA_A$ receptor subunits are targets for a variety of protein kinases such as PKA, PKC and calcium/calmodulin-dependent protein kinase II (CaMKII). PKA causes a decrease in the peak amplitude of the chloride current in cells coexpressing α_1- and β_1-subunits, and it also decreases the rate of rapid desensitization. Interestingly, the effect of ethanol on $GABA_A$ receptors containing the γ_{2L}-subunit may be mediated by phosphorylation since only the long variant contains the serine which may be phosphorylated by PKC or CaMKII, and mutation of this serine abolishes ethanol sensitivity.

5.7 $GABA_C$ receptors

mRNA from bovine retina expressed in *Xenopus* oocytes provided the first evidence, in 1991, for a novel GABA receptor that is insensitive to both bicuculline and baclofen. Recording from rat retinal slices reveals a residual inward current carried by Cl^- in bipolar cells dosed with GABA in the presence of bicuculline. $GABA_C$ receptors have a higher affinity for GABA than do $GABA_A$ receptors, and their Cl^- channel is clearly distinct (see *Table 5.3*). Responses of $GABA_C$ receptors are therefore likely to be longer lasting and generated by lower GABA concentrations. $GABA_C$ receptors are activated selectively by folded GABA analogs such as *cis*-4-aminocrotonic acid (CACA) though with low potency.

$GABA_C$ receptors appear to be composed of ρ-subunits the first of which, ρ1, was cloned from a human cDNA library in 1991. When expressed in *Xenopus* oocytes, ρ-subunits form homo-oligomeric Cl^- channels with very similar pharmacology to oocytes expressing retinal mRNA and to $GABA_C$ receptors in retinal slices. It has been suggested that ρ-subunits are $GABA_C$ subunits. However, ρ-subunits do not coassemble with either α- or β-subunits. Use of single-cell PCR and *in situ* hybridization has shown that ρ-subunits are highly expressed in individual rod bipolar cells.

Although ρ1 is confined to retina, ρ2 is expressed in other brain regions (e.g. hippocampus and neocortex), which presumably accounts for the $GABA_C$-like responses that have been reported in the brain.

There is good evidence that $GABA_C$ receptors can be modulated intracellularly. The cytoplasmic loop between M3 and M4 of rat ρ1 and ρ2 and human ρ1 contains consensus sequences for PKC-mediated phosphorylation. $GABA_C$ chloride currents in retina are reduced by activation of 5-hydroxytryptamine (5-HT_2) receptors and those metabotropic glutamate receptors (mGluRs) that are coupled to the polyphosphoinositide second messenger system.

Table 5.3. A comparison of ionotropic GABA receptors

Characteristic	$GABA_A$	$GABA_C$
EC_{50} for GABA	10–100 μM	1–4 μM
Pore diameter	0.56 nM	0.51 nM
Main conductance	27–30 pS	7 pS
Mean channel open time	30 msec	150 msec

5.8 Multiple glutamate receptors exist

The first hints that glutamate might act via more than one receptor came from regional differences in potency of excitatory amino acids and analogs. Neurons in the dorsal horn of the spinal cord were shown to be more sensitive to glutamate than to aspartate, whilst the reverse was the case for ventral horn cells. The analogs *N*-methyl-D-aspartate (NMDA) and kainate both clearly acted at glutamate receptors, but responses to NMDA had longer latency and were more protracted than those to kainate. Moreover, NMDA responses were more prevalent in ventral than dorsal horn. In a crucial experiment, it was shown that low concentrations of Mg^{2+} in the bathing medium depressed depolarizing responses of cells in isolated frog spinal cord to some excitatory amino acids, particularly aspartate and NMDA, but not to others such as kainate or quisqualate. The implication of these studies was that there are at least two types of glutamate receptor. Further progress in differentiating multiple glutamate receptors was made with the discovery of several antagonists of NMDA receptors, the most selective of which was D(–)-2-amino-5-phosphonovalerate (D-AP5, D-APV). A number of experiments, at all levels of analysis from behavioral through to biochemical, suggested the existence of three classes of glutamate receptor, named for their preferential agonists, NMDA, quisqualate and kainate receptors. A further advance was made by the synthesis, in 1980, of the putative quisqualate agonist, α-amino-3-hydroxy-5-methyl-4-isoxazoleproprionic acid (AMPA). It became clear subsequently that quisqualate and AMPA did not label identical binding sites and that quisqualate was a far less selective ligand than originally presumed, so a more appropriate classification of NMDA, AMPA and kainate receptors was adopted. At about the same time, quisqualate and ibotenate were found to mediate responses quite unlike those seen previously and this led to the discovery of mGluRs (see Chapter 6).

It had long been assumed that NMDA, AMPA and kainate receptors are ionotropic because the time course of responses generated at these receptors is rapid and Na^+-dependent. Voltage clamp studies of locust muscle, where glutamate is the transmitter at the neuromuscular junction, showed that the glutamate receptor gates a cationic (Na, K, Ca) conductance.

5.8.1 *The first glutamate receptor was cloned by fractionating message*

The lack of ligands with sufficiently high affinity resulted in several failures of standard protein purification biochemistry to isolate glutamate receptors. This meant that a quite different strategy was needed from that used to clone nACh, glycine and GABA receptor subunits. The method, employed by Stephen Heinmann's lab at the Salk Institute in California in 1989, used an iterative procedure to screen mRNA fractions for their ability to cause expression of kainate-evoked depolarization in *Xenopus* oocytes. A flow chart of the procedure is shown in *Figure 5.8.*

Initially, rat forebrain poly(A)$^+$ RNA was prepared and 75 ng injected into a *Xenopus* oocyte. Intracellular recording from the oocyte 2 days later revealed depolarizing responses to glutamate, NMDA, quisqualate and kainate. The remaining poly(A)$^+$ RNA was then used to construct a cDNA library which consisted of 850 000 clones! To make the assay more manageable, DNA was pooled into 18 sublibraries, linearized and

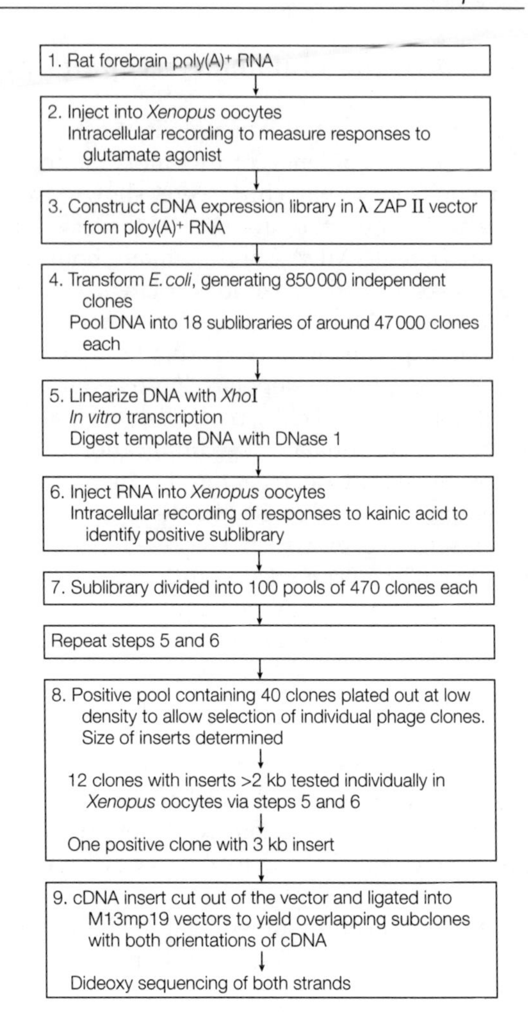

Figure 5.8. Cloning and sequencing of the first glutamate receptor.

transcribed *in vitro* to produce RNA for injection into oocytes. RNA derived from one sublibrary resulted in reproducible and rapid depolarizations to kainate. Further subdivisions of clones comprising this library eventually allowed identification of a single clone carrying a 3-kb insert which could elicit kainate responses in oocytes.

5.8.2 *Ionotropic AMPA/kainate receptors are only weakly homologous to other ligand-gated ion channels*

Several additional members of the same family of ionotropic AMPA/kainate receptors, were cloned by homology screening of cDNA libraries with probes based on the sequence of the first glutamate receptor, under low stringency conditions. To date, seven genes encoding AMPA/kainate receptors have been discovered (GluR1–GluR7).

By analogy with ACh, $GABA_A$ and glycine receptors and other members of the LGIC superfamily, all of the GluRs originally were postulated to have 4TM segments. The N terminus is assumed to be extracellular because it bears glycosylation consensus sequences, and the loop between TMII and TMIII was thought to be cytoplasmic. Apparently inconsistent with this, subsequent site-directed mutagenesis showed that TMII must form part of the pore.

There are considerable differences between GluRs and other LGICs which makes it sensible to regard GluRs as comprising a separate superfamily.

(i) The overall sequence homology between GluRs and the conventional LGICs is only about 20%, though homology is higher for the TM regions.
(ii) GluRs are about twice the size of other LGICs because they possess longer N-terminal extracellular domains.
(iii) All N-terminal regions of the LGIC superfamily contain a 15-amino acid cysteine-rich loop. This is missing in GluRs.
(iv) It is now known that GluRs are subject to RNA editing, a post-translational modification not seen in LGICs.

Once it became clear that GluRs were different from other LGICs, it became possible to postulate other models, such as one in which GluRs have only 3TM segments (*Figure 5.9*).

In this model, the loop between TMIII and TMIV is extracellular, and the presence of functional glycosylation sites in this TMIII–TMIV loop strongly suggests that the C terminus is cytoplasmic, as required by an odd number of TM segments. Phosphorylation of GluR1–R4 by CaMKII and PKC has been shown *in vitro* and *in vivo*, and the C termini of these receptors bear **consensus sequences** for these kinases. Remarkably, the 3TM model shows striking homologies between GluRs and K_V channels in that the hydropathicity profiles of the TMI–TMIII region in GluRs is strikingly similar to the S5–H5–S6 region of the K_V channel, although inserted in the opposite sense into the membrane, that is the H5 loop in GluRs in inserted from the intracellular face. Furthermore, a number of functionally important amino acids are highly conserved between GluRs with the 3TM conformation and potassium channels. The obvious idea is that the TMI–TMIII region of the GluRs is derived from some ancient K^+ channel.

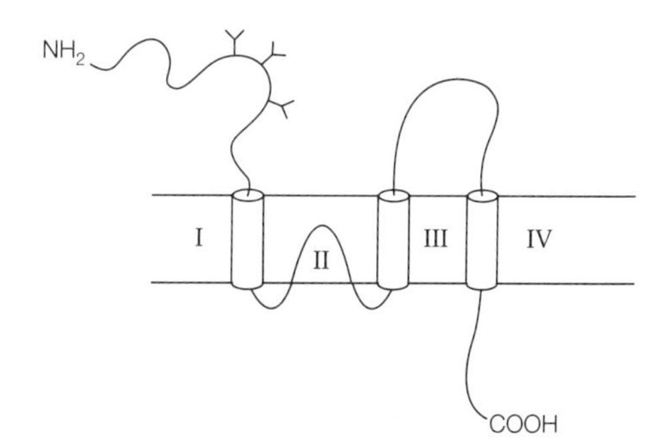

Figure 5.9. A 3TM model for GluRs. Here TMII in the 4TM model becomes a pore-forming loop.

5.9 Separate AMPA and kainate receptors can be distinguished

Sequence homologies indicate the existence of two subfamilies of nonNMDA GluRs. The first consists of GluR1–GluR4, which share about 70% sequence homology, whilst the second comprises GluR5 and GluR6, which share 81% homology. The two subfamilies have distinctive pharmacology. AMPA is more potent than kainate at GluR1–R4. By contrast, GluR6 is readily activated by kainate and is completely insensitive to AMPA. Both glutamate and AMPA but not kainate cause desensitization of the GluR1–R4 subfamily, whereas kainate causes rapid desensitization of GluR6. *In situ* hybridization in brain sections of adult mice shows a good correlation between density of GluR6 transcripts and [^{3}H]kainate-binding sites. GluR5 is only weakly activated by glutamate, and is unaffected by the other agonists. In summary, it seems sensible currently to regard GluR1–R4 as AMPA receptors and GluR6 as a kainate receptor.

5.9.1 *The native AMPA receptor is probably a hetero-oligomer*

A key question is whether the native receptor is composed of one or multiple subtypes. One approach to answering this is to compare the electrophysiological responses found in coexpression experiments with those of native receptors. Firstly, the failure of GluR6 to form channels when coexpressed with any of GluR1–R4 reinforces the notion of two quite distinct subfamilies. Next, GluR1 or GluR3 alone and GluR1/GluR3 coexpression gives rise to inwardly rectifying Ca^{2+} (and Mg^{2+}) permeable channels. However, coexpression of GluR1 or GluR3 with GluR2 results in a nearly linear I–V plot and loss of divalent cation permeability, properties that match the traditional AMPA/kainate receptor identified in pyramidal neurons, or seen when hippocampal poly(A)$^+$ RNA is expressed in *Xenopus* oocytes.

5.9.2 *Mutant channels allow the pore-forming region to be defined*

The molecular basis of the differences in rectification and Ca^{2+} permeation has been shown to be determined by a single amino acid presumed to lie in the pore. Subunit chimeras between GluR2 and GluR1 or GluR3 have shown that a region from position 350 to position 750 determines the shape of the I–V curve. In particular, the TMII (H5) region contains a conserved glutamine (Q) in all but one of the GluRs, but is arginine (R) in GluR2. Replacing Q by R in GluR3 or GluR4, by site-directed mutagenesis, causes the mutant channels to become outward rectifiers. Conversely, replacing R for Q in GluR2 turns this channel into an inward rectifier. Furthermore, the mutant GluR2 channel became Ca^{2+} conducting whereas the mutant GluR3 and GluR4 lost their Ca^{2+} conductance (gCa) (*Table 5.4*).

Table 5.4. Effects of mutations on glutamate receptor rectification and calcium conductance

Receptor	Wild-type	Mutant
GluR2	R_{586} Outward rectifying No gCa	Q_{586} Inward rectifying gCa
GluR3 GluR4	Q_{590} Q_{587} Inward rectifying gCa	R_{590} R_{587} Outward rectifying no gCa

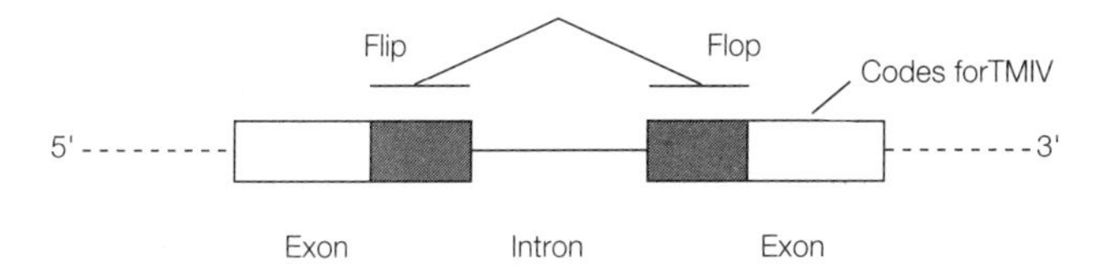

Figure 5.10. Alternative splicing of GluR1–GluR4 transcripts.

5.9.3 *RNA processing adds to functional diversity of AMPA receptors*

GluR1–GluR4 subunits come in two splice variants, termed flip and flop, which have different properties. Two 38-amino acid residue stretches are encoded by adjacent exons, flip and flop (*Figure 5.10*). Excision of the intron leaves either the flip or the flop sequence. Glutamate is five times more potent in activating flip than flop, and the expression of these variants appears to be developmentally regulated.

In addition to alternative splicing, GluRs, alone among the LGICs, are subject to post-transcriptional modification, **RNA editing**. This involves oxidative deamination of adenosine to inosine in specific pre-mRNA codons, which changes the amino acid coded for. RNA editing occurs at a number of sites in GluRs; most noteworthy is the modification of a CAG codon to CIG in exon 11 of GluR2, GluR5 and GluR6. This results in the substitution of R for Q at the crucial residue (GluR2 R_{586}) in TMII that is implicated in determining the rectification and ion permeation properties of the channels. In the case of GluR2, only the edited version is expressed. However, both GluR5 and GluR6 can exist in both edited and unedited versions with their respective different electrophysiological properties (*Table 5.4*). Other types of RNA editing also occur. R/G editing in GluR2 results in edited subunits that recover more rapidly from desensitization, and both I for V and Y for C occur in GluR6 with discernible influences on channel function. The enzyme responsible for the editing has not been identified unambiguously but, more importantly, it is necessary to understand how RNA editing is controlled so that the functional differences it generates are regulated.

5.10 NMDA receptors

The study of NMDA receptors by classical techniques initially proceeded more rapidly than that of AMPA/kainate receptors because of the earlier development of selective NMDA antagonists, such as D-AP5. Investigation was motivated further by the realization that NMDA receptors are involved in exciting aspects of brain function such as neuronal plasticity – including learning and memory – and in important pathologies, stroke and epilepsy, with obvious implications for drug discovery.

5.10.1 *NMDA receptors harbor binding sites for several classes of molecule*

Like the $GABA_A$ receptor, NMDA receptors have multiple binding sites. The recognition site for glutamate also binds the other agonists, aspartate and NMDA, and competitive antagonists such as D-AP5.

The discovery in 1987 by Johnson and Ascher that glycine dramatically potentiated the effects of NMDA was surprising. Glycine, a well-documented inhibitory transmitter, acts as a **co-agonist** of NMDA receptors, binding with high affinity to a site distinct from that which binds glutamate. Since the concentration of glycine required to maximally 'prime' the NMDA receptor for subsequent gating by glutamate is of the same order as the presumed extracellular glycine concentration, about 1 μM, it is difficult to envisage that any regulation of this co-agonist site is possible *in vivo*. However, both D-serine and D-alanine are agonists, and kynurenate is an antagonist at this glycine site. Kynurenate is synthesized and released from glial cells, and may thus competitively block the effect of glycine.

A third binding site exists for a number of drugs which act as open channel blockers, including phencyclidine (PCP), more colloquially known as angel dust, the closely related dissociative anesthetic, ketamine, and dizocilpine (MK801). It seems likely that the hallucinogenic effects of PCP, ketamine and related compounds are mediated via NMDA receptors.

Further binding sites exist on NMDA receptors for polyamines such as spermine and for Zn^{2+}, and these ligands may modulate the receptor in complex ways, the functional significance of which is far from clear.

5.10.2 NMDA gates a cation channel

The ion channel gated by NMDA is permeable to Ca^{2+}, Na^+ and K^+, but most of the current is carried by Ca^{2+}. A striking feature of the NMDA receptor is that the ion channel is blocked by extracellular Mg^{2+} ions in a voltage-dependent manner (recall Section 5.4). This is illustrated by voltage clamping of NMDA receptors expressed in *Xenopus* oocytes (*Figure 5.11*).

In the absence of external Mg^{2+}, the NMDA receptor has an I–V plot which is virtually linear; the channel is nonrectifying. However, with physiological concentrations of external Mg^{2+}, strong rectification is seen. Mg^{2+} blocks the inward current at potentials close to the resting potential, and this blockade is lifted by depolarization. Hence, at the resting potential, the binding of agonists to the NMDA receptor does not enable the permeation of ions. Activation of NMDA receptors requires both agonist and a

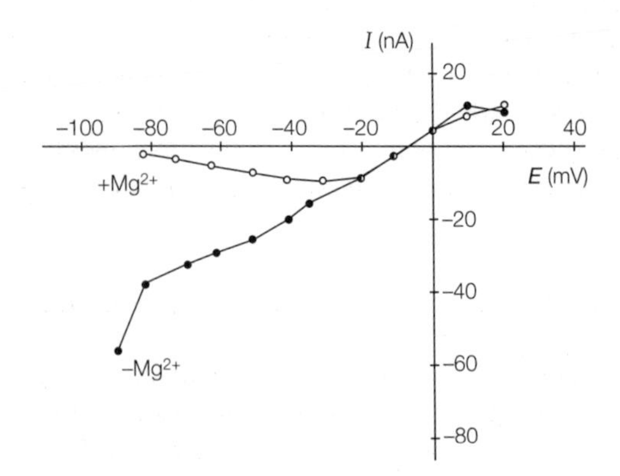

Figure 5.11. I–V plot for the NMDA receptor, challenged with 100 μM NMDA both in the presence (○) and absence (●) of Mg^{2+}.

reduction in membrane potential. This property is crucial to the role of NMDA receptors in certain types of plasticity that will be discussed more fully in Chapter 8.

5.10.3 *NMDA receptors probably have a topology resembling other ionotropic glutamate receptors*

The first NMDA subunit was cloned by fractionating brain mRNA and assaying it for its capacity to generate NMDA-mediated responses in *Xenopus* oocytes, rather like GluR1. Attempts at homology cloning based on probes for nonNMDA GluRs had failed, not surprising when it transpired that NMDA receptor subunits have rather low homologies with other glutamate receptors. However, cloning of the first NMDA receptor opened the flood gates, and numerous others were soon sequenced.

Two gene families are responsible for encoding five distinct subunits. NMDAR1 is ubiquitously distributed in brain, and assembles into homomeric channels with the pharmacology and electrophysiological properties of native NMDA receptors. At least seven splice variants of NMDAR1 exist. NMDAR2 (A–D) are coded for by distinct genes and do not form homomeric channels but, as heteromers with NMDAR1, they help to give rise to the diversity of NMDARs displayed *in vivo*.

As with GluRs, the current view favors the 3TM model for all the ionotropic glutamate receptors (iGluRs) including NMDARs (*Figure 5.9*), and evidence for the 3TM topology for NMDARs is fairly convincing.

Firstly, it means that the NMDAR subunits contain the inverted potassium channel motif – like the GluRs – made up from TMI, TMIII and the intervening loop analogous to the H5 or P loop of VDKCs. Indeed, this motif precisely resembles the structure of inward rectifying potassium channels. Now, interestingly, the inward rectification of K_{IR} channels is produced by internal Mg^{2+} blockade at positive potential, whereas NMDARs (with their inverted K^+ channel-like domain) are blocked by external Mg^{2+} at negative potentials. Moreover, an asparagine residue, Asn598, of NMDAR1 has been identified as crucial for Mg^{2+} blockade, and the 3TM model places the residue close to the tip of the H5 loop, in a position exactly homologous to Thr441 of the *Shaker* potassium channels that is essential for internal TEA block (see Section 4.8).

Secondly, the C terminus of NMDAR subunits has been shown to be phosphorylated by PKC, showing that is intracellular. This phosphorylation potentiates NMDA responses by increasing the probability of channel opening and decreasing the voltage-dependent Mg^{2+} block.

Finally, the 3TM model requires that the loop between TMIII and TMIV be extracellular. The discovery that residues in this region directly affect the binding of glycine supports this view. In fact, in all iGluRs, there are close homologies between **periplasmic** amino acid-binding proteins of gram-negative bacteria and the TMIII–TMIV loop, and the part of the N terminus adjacent to TMI comprising the glutamate-binding site. The crystal structures of some of these bacterial proteins have been determined to high resolution, providing insights into secondary and tertiary structure of the extracellular regions of all glutamate receptors.

In evolutionary terms, it appears that iGluRs are mosaics, constructed from several ancient proteins, including potassium channels and amino acid-binding proteins.

5.10.4 The NMDAR1 subunit has several isoforms derived by alternative splicing

The NMDAR1 gene is large. It has 22 exons, three of which (5, 21 and 22) generate a transcript that can undergo alternative splicing, in theory producing eight NR1 isoforms. In fact, seven of these possible alternatives have been identified in cDNA libraries. The eighth has been genetically engineered. When expressed in *Xenopus* oocytes, each splice variant forms channels with distinct electrophysiological and pharmacological properties. Moreover, hybridization studies with probes that recognize specific isoforms show that they are localized to particular brain regions and expressed differentially during development. There is no doubt then that alternative splicing contributes to the functional diversity of NMDARs.

The splice variants are classified according to whether they have or lack each of three **splice cassettes** termed N1, C1 and C2 (see *Figure 5.12*). Exon 5 encodes a 21-amino acid residue splice cassette (N1) which can be inserted into the N terminus of the protein. At the C terminus, variants occur by deletion of C1 or C2 or both cassettes (*Figure 5.12a*, options i, ii and iii respectively). Splicing out C2 removes a first stop codon, resulting in translation of an unrelated sequence C2′ further downstream. Hence, all isoforms lacking C2 have C2′, and vice versa (*Table 5.5*).

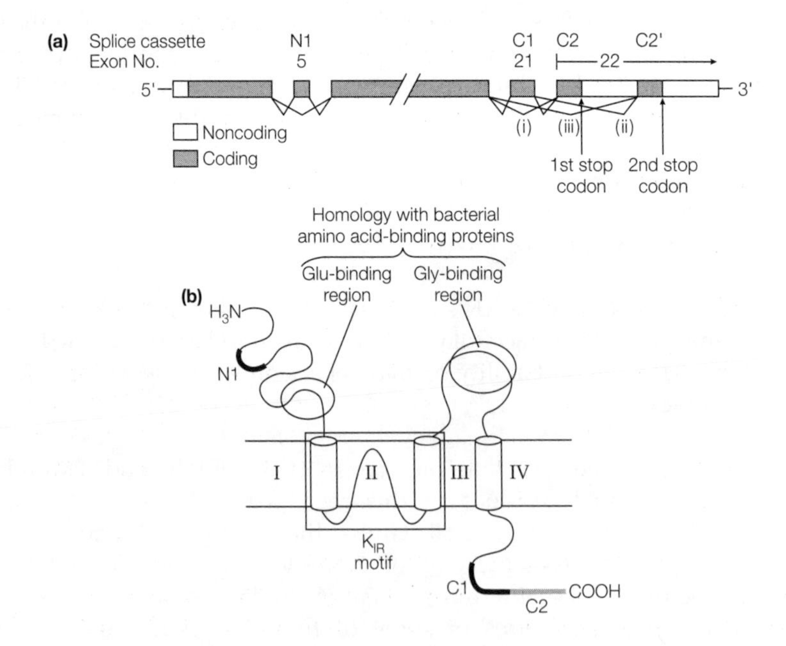

Figure 5.12. (a) Alternative splicing in NMDAR1 mRNA. (b) Postulated structure of the NMDAR1 showing the location of the splice cassettes (see text for details).

Table 5.5. NMDAR1 splice variants classified according to the absence (–) or presence (+) of three splice cassettes

Variant	Cassette			
	N1	C1	C2	C2′
$NR1_{000}$	–	–	–	+
$NR1_{001}$	–	–	+	–
$NR1_{010}$	–	+	–	+
$NR1_{011}$[a]	–	+	+	–
$NR1_{100}$	+	–	–	+
$NR1_{101}$	+	–	+	–
$NR1_{110}$[b]	+	+	–	+
$NR1_{111}$	+	+	+	–

[a]First variant to be cloned; the most abundant isoform in rat forebrain.
[b]Has been genetically engineered; may not occur naturally.
Modified from Zukin and Bennett (1995) *Trends Neurosci.* **18**: 306–313.

5.11 Properties of NMDARs were studied by coexpression

NMDAR2 subunits (NR2A–D) are larger and have only 20% homology with NMDAR1 subunits. Coexpression studies show that they cannot form functional channels alone. However, immunoprecipitation experiments show that NMDAR1 and NMDAR2 subunits in rat embryo cortical membranes coprecipitate, implying that some native NMDARs are heteromers of both types of subunit.

Functional NMDARs are thought to be pentameric. The properties of any given receptor will be determined by the NMDAR1 splice variant(s) and which, if any, NMDAR2 subunit(s) it contains. Investigations of NMDAR1 homomeric receptors expressed in *Xenopus* oocytes show them to have most of the properties of native neuronal NMDARs and have revealed the effects that alternative spicing confers. Pair-wise comparisons of splice variants that differ only in a single splice cassette can indicate the function of individual cassettes.

The presence on the N1 cassette increases agonist affinity and generates larger macroscopic currents. At low glycine concentrations, all splice variants show potentiation by spermine. At saturating glycine concentrations (10 μM), no potentiation by spermine is seen in those variants possessing the N1 cassette. Splice variants lacking N1 show potentiation with 1 μM Zn^{2+}, but those with N1 show no potentiation at this concentration. Interestingly, potentiation by both spermine and Zn^{2+} can be recovered in variants containing N1 when six positively charged amino acids are mutated to alanines.

The relationships between alternative splicing and phosphorylation by PKC is paradoxical. Although the C1 cassette contains most of the PKC consensus sequences, variants lacking C1 exhibit PKC-evoked potentiation indistinguishable from those with C1. Furthermore, variants with N1 but lacking C2 are markedly more potentiated by PKC than those without N1 but with C2. It currently is a mystery how an N-terminal

LIVERPOOL JOHN MOORES UNIVERSITY
LEARNING SERVICES

extracellular region can have such a profound influence on a process mediated by phosphorylation of the intracellular C terminus of the molecule.

Heteromeric NMDARs display a remarkable variety of pharmacologies depending upon both the NR1 splice variant and any NR2 subunit present. No heteromeric NR1–NR2 receptors show potentiation by 1 μM Zn^{2+}, but, like neuronal receptors, they are inhibited at high concentrations. Potentiation by spermine at saturating glycine concentrations only occurs in NR1–NR2 receptors in which the NR1 variant lacks the N1 insert and the NR2 is the B isoform. With the appropriate NR1 variants, NR2A and NR2B are permissive in allowing potentiation by PKC, but the presence of the NR2C isoform prevents PKC modulation of the channel.

5.11.1 *NMDAR expression is region specific*

Probes directed against the N1, C1 and C2 regions have been used in *in situ* hybridization studies to explore the localization of the NR1 variants. Those variants lacking N1 are distributed throughout the hippocampus, including the dentate gyrus, whereas N1-containing variants are restricted to CA3, for example. In general, rostral structures of the brain (cerebral cortex and caudate) have C1- and C2-containing variants, whereas more caudal structures (thalamus, tectum and cerebellum) express variants lacking C1 and C2. Interestingly, in the hippocampus, N1-containing variants which can be potentiated by PKC are restricted to the CA3 region.

5.11.2 *NMDARs appear to be developmentally regulated*

The earliest NMDAR1 variants to be expressed in the rat embryo tend to be those lacking N1 inserts, so presumably phosphorylation of NMDA receptors assumes greater importance later in development. NMDAR2 subunits are also expressed at different times in development. So, for example, while NR2B is expressed in the hippocampus soon after birth in the rat, NR2A and NR2C are not expressed until after postnatal day 12. Furthermore, expression of NMDAR2 subunits is cell specific. Matching this differential expression to the physiological properties of receptors assembled from these subunits should provide clues to how NMDA-mediated transmission is regulated in the brain in time and space. Though this approach is not far advanced presently, it can suggest testable hypotheses. It is tempting to speculate, for instance, that the early expression of receptors lacking N1 but with NR2B subunits might imply a role for polyamine modulation of NMDARs in early postnatal development.

Chapter

Metabotropic receptors and signal transduction mechanisms

6.1 Introduction

When the responses of postsynaptic cells are measured after stimulation of the presynaptic cell, there may be a fast excitatory postsynaptic potential (epsp) or inhibitory postsynaptic potential (ipsp) which are generated by the action of neurotransmitters on ionotropic receptors. However, in many cells, the response of the postsynaptic cell is either a slow change in potential or no observable electrical change at all. In these cases, the receptors belong to a large number of receptors which act by modifying the intracellular biochemistry of the postsynaptic cell through the alteration of second messengers and the activation of intracellular enzymes. These slow-acting receptors, which because of their actions on intracellular metabolism are termed **metabotropic receptors**, fall into a number of families, the largest of which are the G protein-linked receptors, but there are also receptors with intrinsic enzyme activity, such as the tyrosine kinases and the guanylate cyclases.

6.2 G protein-linked receptors activate a cascade of protein interactions

In many cases, the receptors on the postsynaptic cells belong to a large family called G protein-linked receptors. The actions of these receptors occurs via a sequence involving a group of molecules called G proteins, so called because they bind GTP and have GTPase activity. These G proteins in turn interact with other cellular elements to produce a response, which could be, amongst others, a change in the levels of a **second messenger** or the activation of an ion channel.

At its simplest, this cascade of reactions works as follows (*Figure 6.1*). Following binding of the neurotransmitter to the receptor, the receptor undergoes a conformational change which allows it to bind a G protein complex. This complex, which consists of three proteins (α, β and γ) plus GDP, is found freely diffusing in the cytoplasmic face of the plasma membrane, and is inactive when the α-subunit has GDP bound (*Figure 6.1a*). The binding of the G protein complex to the receptor allows the α-subunit to exchange GDP for GTP (*Figure 6.1b*). The G protein complex then dissociates into α-GTP and $\beta\gamma$-subunits. These can then diffuse away from the receptor, which, whilst it still has agonist bound, can activate other G protein complexes. The α-subunit can then interact

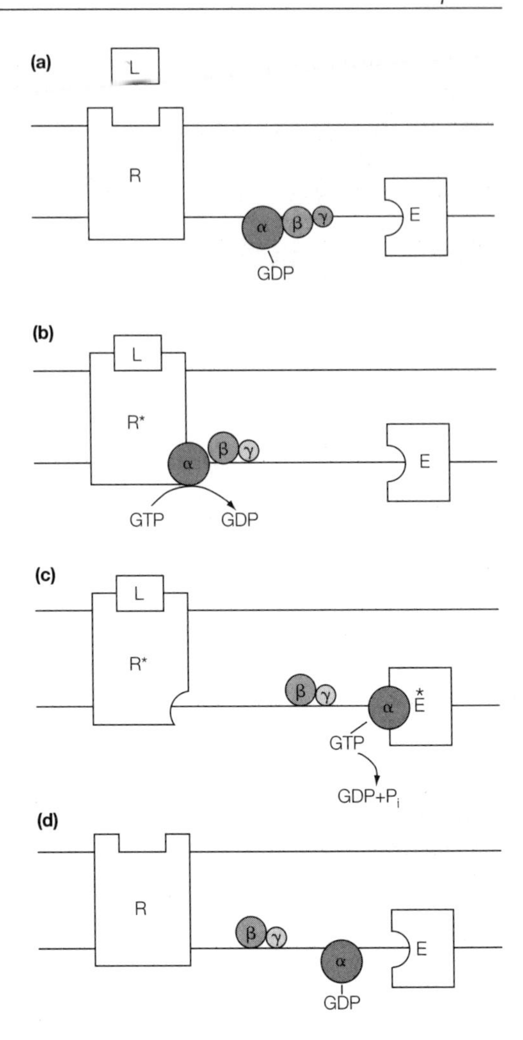

Figure 6.1. Summary of G protein-linked receptor, G protein and effector cascade. L = ligand, R = receptor, E = effector.

with its target protein (*Figure 6.1c*), which is usually, but not always, a membrane-bound enzyme, such as adenylate cyclase (AC), which then either increases or decreases its rate of production of its second messenger [i.e. cyclic AMP (cAMP)]. This continues until the GTP bound to the α-subunit is hydrolyzed to GDP by the intrinsic GTPase activity of the α-subunit. The α-subunit is then released from the target protein (*Figure 6.1d*) and recombines with the βγ-subunits, thus returning the G protein complex to its inactive state. In some cases, an active role for the βγ-subunits has also been observed. Because of the amplification available at each step, the binding of a single agonist molecule produce hundreds or even thousands of second messenger molecules.

6.3 G protein-linked receptors form a superfamily of receptors

The first of these receptors to be cloned and sequenced was the photoreceptor protein, rhodopsin, which acts as a 'light receptor'. The subsequent analysis of other

G protein-linked receptors such as the β-adrenergic receptor and the muscarinic ACh receptor, showed that many of these proteins share a high degree of sequence homology and are presumed to have a similar structure. Some of the more commonly known G protein-linked receptors are listed in *Table 6.1* but the full list is very long and is growing continuously. Indeed there is at least one book which is almost entirely devoted to listing these.

Many of the substances listed in *Table 6.1* interact with multiple receptor subtypes, for example, the α-adrenoceptor has two divisions, α1 and α2. α1 receptors have at least three subtypes, whilst three subtypes of α2 receptors have so far been identified, giving a total of six different receptors.

Table 6.1. Some examples of G protein-linked receptors

Receptor	Abbreviation	Subtypes
Adenosine	A	1,2A,2B,3
α-Adrenergic	α-AR	α1 (A, B, D), α2 (A, B, C)
β-Adrenergic	β-AR	β1, β2, β3
Muscarininc cholinergic	mAChR	M1–M5
Dopamine	D	1–5
GABAergic	$GABA_B$	
Glutamatergic	mGluR	1–8
Histamine	H	1–3
5-HT (serotonin)	5-HT	1(A, B, D, E, F), 2(A–C), 4, 6, 7
Purinergic	P	2U, 2Y
Serum calcium		
Rhodopsin		
Olfaction		
Taste		
Angiotensin	AT	1, 2
Bradykinin	B	1, 2
Calcitonin		
Cannabanoid	CB	1, 2
Endothelin	ET	A, B
Glucagon		
Opiates	μ, δ and κ	
Oxytocin	OT	
Parathyroid hormone		
Prostanoid	DP	
	FP	
	IP	
	TP	
	EP	1–4
Somatostatin	SST	1–5
Substance P		
Thyrotropin		
Vasopressin		

As well as the receptors for neurotransmitters and hormones, members of the G protein-linked family are sensory transducers. They mediate the response to light by the rod and cone cells, using the molecules rhodopsin and color opsins. Receptors for the chemical senses, taste and smell, are also G protein-linked receptors, and it is thought that the odorant receptors may have the largest number of receptor subtypes yet found, with estimates ranging from 400 to 1000 distinct receptors. Interestingly, these different receptors are not produced by alternative splicing of a small number of genes but it seems that each receptor is coded by a distinct gene.

In order to give an idea of the diversity of G protein-linked receptors, the subtypes for some of the receptors are listed in *Table 6.1*. However, this list is not exhaustive. Many of these receptors have been first identified from their genes using low-stringency probes based on sequences from other G protein-linked receptors.

6.4 G protein-linked receptors share a common structure

The significant homologies between many of these receptors is mirrored in their common predicted structure. All of these receptors are thought to have seven transmembrane-spanning segments which gives rise to an alternative name for these receptors as seven transmembrane (7TM) receptors (*Figure 6.2*).

They all have an extracellular N terminus of widely varying size, and a cytoplasmic C terminus. Many of these receptors also have an extended cytoplasmic loop between the S5 and S6 segments.

The proposed arrangement of the TM segments in the membrane was based originally on **X-ray diffraction** studies on bacteriorhodopsin (bRH). This molecule is a photon-coupled proton pump with 7TM segments, and although it shares little sequence homology with G protein-coupled receptors, it was, at the time, the only similar protein which could be crystallized successfully. However, more recently, it has proved possible to crystallize a true G protein-linked receptor, rhodopsin.

The original bRH studies showed an arrangement of seven TM segments and, although this is also the case for rhodopsin, the perpendicular arrangement of the TM segments is quite different (*Figure 6.3a*).

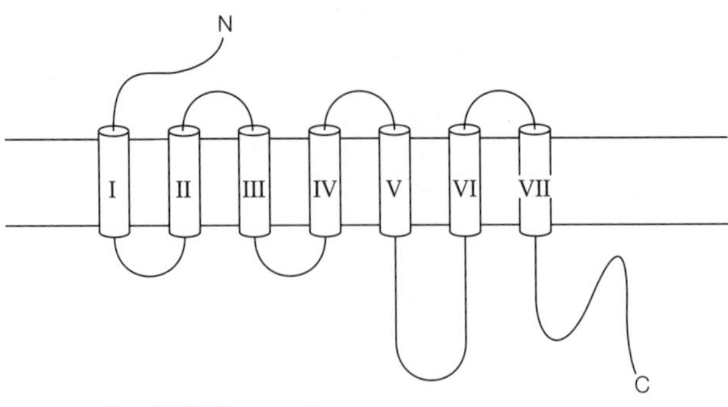

Figure 6.2. Structure of a generalized 7TM receptor.

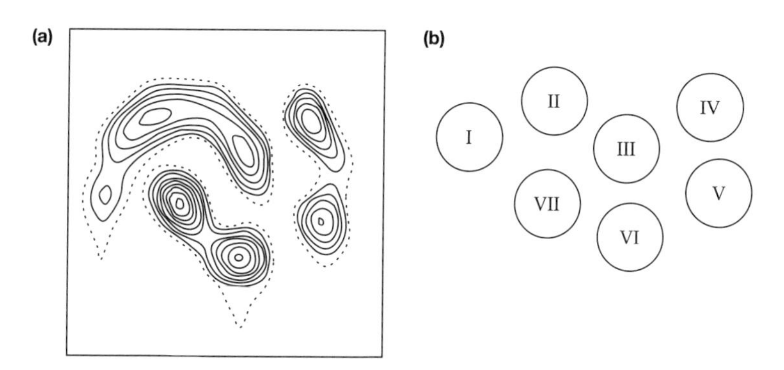

Figure 6.3. (a) X-ray diffraction pattern of rhodopsin. Reprinted with permission from *Nature*, Schertler, G.F.X., Villa, C. and Henderson, R., Projection structure of rhodopsin **362**, pp. 770–2. Copyright 1993 Macmillan Magazines Limited. (b) Predicted arrangement of the seven helices.

It is very reassuring that the diverse studies of hydropathy analysis and X-ray diffraction have come to the same conclusions regarding the predicted structures.

The arrangement of the individual helices has been deduced from the features of a large number of receptors (*Figure 6.3b*). It was proposed that the lipophilic surface of each TM segment faces outwards.

6.4.1 *Metabotropic glutamate receptors have evolved separately*

The superfamily of 7TM receptors can be arranged in subfamilies based on their sequence homologies. However, not all 7TM receptors have sequence homologies with the main group of receptors as exemplified by the mAChR and the β-AR. The metabotropic glutamate receptors (mGluRs) have no significant sequence homology with the other G protein-linked receptors, and are considerably larger (~1200 amino acids) with a very large extracellular N terminus and a relatively short loop between S5 and S6. This lack of similarity has led to the suggestion that these glutamate receptors have evolved separately from the other 7TM receptors.

In the absence of specific high-affinity ligands and because of their lack of sequence homologies with other 7TM receptors, mGluRs were not discovered by probing gene libraries but by using a technique involving the expression of mRNA in oocytes.

Previous work had already shown that when *Xenopus* oocytes are injected with poly(A)$^{+}$ RNA derived from whole rat brain (thus containing mRNA coding for all expressed proteins), they express many proteins including glutamate receptors. These glutamate receptors are diverse, and include ionotropic receptors activated by kainate (see Chapter 5), but receptors activated by quisqualate are also present which, via intracellular enzymes, cause the activation of a long-lasting chloride current. Thus the activation of a chloride current upon application of metabotropic receptor agonists, quisqualate and ±amino-cyclopentane dicarboxylate (ACPD), could act as an assay for the mGluRs in *Xenopus* oocytes.

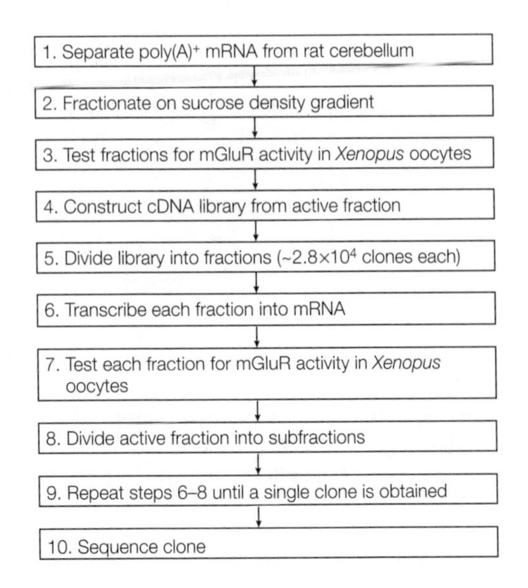

Figure 6.4. Method for cloning mGluRs.

The technique employed to isolate the first mGluR used poly(A)$^+$ mRNA from rat cerebellum (*Figure 6.4*). This was centrifuged on a sucrose density gradient which will sort mRNAs by size, and different fractions were injected into *Xenopus* oocytes which were tested for the presence of agonist-activated chloride currents. The poly(A)$^+$ RNA fraction giving a large response was then used to produce a cDNA library. The library was itself divided into fractions and mRNA produced from each fraction. These fractions were again tested for mGluR activity. The active fraction was again subdivided, new mRNA fractions produced and tested and so on until an individual clone (pmGR1) was identified which could then be sequenced. The identified clone gave a predicted polypeptide of 1199 amino acids.

Following this, a further seven mGluRs have been discovered (*Table 6.2*), using probes derived from the sequence of the first clone. These mGluRs can be divided into three groups on the basis of sequence similarities and common signal transduction pathways. For some of the receptors, a number of splice variants have also been identified, for example mGluR1 which has five variants named a, b, c, d and e.

Table 6.2. mGlu receptors

Group	Subtype	Variants	Actions
I	mGluR1	a, b, c, d, e	↑PLC activity
	mGluR5	a, b	↓I_K
II	mGluR2		↓AC activity
	mGluR3		↓I_{Ca}
			↑I_K
III	mGluR4	a, b	↓AC activity
	mGluR6		↓I_{Ca}
	mGluR7	a, b	
	mGluR8		

In situ hybridization has been used to examine the distribution of the mRNA coding for different mGluRs in brain, and they have been found to be highly localized, with mGluR1 mRNA concentrated in all neurons in the dentate gyrus and the CA2–CA4 regions of the hippocampus, whilst mGluR5 mRNA is found only in pyramidal cells in hippocampal regions CA1–CA4 and in granule cells of the dentate gyrus. The roles of mGluRs are examined in later chapters.

6.4.2 *Ligand-binding domains vary between different G protein-linked receptors*

The different ligands which activate G protein-linked receptors are very varied in both size and structure, from the minute photons, which through the interactions between the light-sensitive pigment, retinal and the opsin protein make up rhodopsin, to the large peptide hormones such as follicle-stimulating hormone (FSH). There are many different mechanisms by which the ligands bind to the receptors. For a number of small ligands, such as noradrenaline (NA), the binding site is deep in the membrane and involves the binding to critical residues in the TM domains. Other ligands are thought to bind to varying degrees to the external face of the TM segments and the N-terminal domain (*Figure 6.5*).

The large extracellular domain of the mGluR has sequences which resemble a family of bacterial proteins called periplasmic amino acid-binding proteins (PBPs). Interestingly, these are the same proteins which have been compared with the glutamate- and glycine-binding sites of ionotropic glutamate receptors (see 5.10.3). These molecules

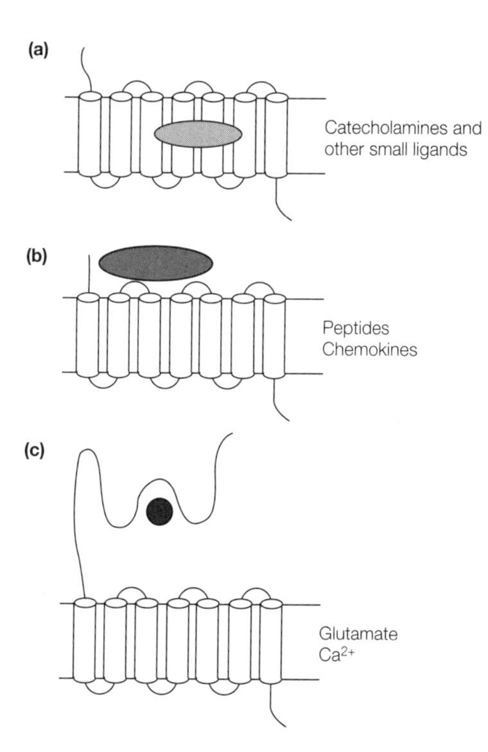

Figure 6.5. Ligand-binding sites for different G protein-linked receptor ligands. (a) Catecholamines; (b) peptides; (c) glutamate and Ca^{2+}.

have been shown to undergo a large conformational change when binding their ligands which has been compared with the action of a Venus Flytrap. This has led to the suggestion that mGluRs bind glutamate to the N-terminal domain which then 'closes' over the protein surface, trapping the ligand between the N-terminal domain and the loops of the TM segments.

6.4.3 G protein-linked receptors interact with G proteins via common intracellular domains

As different receptors interact with different G proteins, attention has focused on the sections of the receptors which exhibit divergent sequences. The regions thought to be of particular importance are the third cytoplasmic loop between S5 and S6 and the intracellular C terminus (*Figure 6.2*).

Evidence for the involvement of the S5 and S6 segments and the connecting intracellular loop in binding to specific G proteins comes from experiments where chimeras were constructed from α1-and β2-adrenergic receptors. A construct derived from a β2-AR which contained the S5–S6 loop from α1-AR showed the ligand specificity of the normal β2-AR but, instead of coupling a G protein which then mediated the stimulation of adenylate cyclase, it was coupled to increases in inositol 1,4,5-trisphosphate (IP_3). This experiment thus showed that the determining factor in coupling to specific G proteins was contained within the substituted region.

It has been shown that individual ligands acting through different receptor subtypes can activate different G proteins and thus affect more than one second messenger system. In at least one case this can be achieved through alternative splicing. The activation of the receptor subtype EP3 for prostaglandin E_2 (PGE_2) can produce increases or decreases in cAMP levels. Four isoforms of the EP3 receptor are produced by the **alternative splicing** of the same gene. These different isoforms A–D differ only in the portion which starts 10 amino acids after the S7 segment and ends at the C terminus. When cDNAs encoding each of the isoforms were expressed in Chinese hamster ovary (CHO) cells, the response to an EP3 receptor agonist, MB-28767, depended on the isoform expressed. Some isoforms were coupled to G proteins, which stimulated adenylate cyclase, whilst others inhibited. Other isoforms were coupled to G proteins which stimulated the production of IP_3 and raised intracellular calcium. Thus, in this case, it is clearly the C-terminal domain which determines G protein interactions.

The evidence thus suggests that either the S5–S6 loop or the C-terminal domain can determine which G protein can be bound to the activated receptor. Whether this varies with the type of receptor or whether all 7TM receptors have a common mode of G protein interaction has not been elucidated; however, clearly both regions are important.

6.4.4 Desensitization of the receptors is associated with phosphorylation

In order to prevent excessive stimulation, a mechanism exists which switches off the receptor so that even in the continued presence of agonist there is a reduction in the response. This desensitization is associated with the phosphorylation of the receptor by a specific receptor kinase which recognizes the agonist-bound form of the receptor. This is termed homologous desensitization, since only the specific receptor is desensitized and the cell's response to other agonists acting through G proteins is unimpaired. An example of a specific receptor kinase is βARK which phosphorylates the β-AR at serine residues in the intracellular C terminus.

By contrast, heterologous desensitization occurs when a stimulated receptor activates, via second messengers, kinases with multiple targets such as PKA or PKC. These will phosphorylate, and so desensitize many different receptor types in the same cell. Heterologous desensitization may be a negative feedback mechanism to curtail excessive stimulation of common second messenger pathways.

Although it has been shown that phosphorylation of receptors is necessary for desensitization to occur, there is a suggestion that the receptor only becomes fully desensitized after the binding of arrestin to the phosphorylated receptor. This inhibits the binding of the G proteins to the receptor and thus prevents G protein activation. There is some evidence that different arrestin molecules bind specifically to different receptors. Visual arrestin which normally binds to phosphorylated rhodopsin has a low affinity for both adrenergic and muscarinic receptors, and β-arrestin binds preferentially to adrenergic receptors.

G protein-linked receptors also show agonist-induced internalization of receptors, where receptors are endocytosed by clathrin-coated vesicles. The receptors may then be recycled back to the membrane or degraded. The ratio of recycled/degraded receptors may depend on the continuing presence of agonist. The permanent removal of receptors from the membrane is an important response to chronic drug treatment.

6.5 There are many distinct G proteins

G proteins are not integral membrane proteins but are anchored to the cytoplasmic face of the plasma membrane and, at rest, consist of a trimer of proteins called α, β and γ. Each protein is coded for by multiple genes, giving a wide repertoire of possible combinations.

6.5.1 *The are four families of α-subunits*

The largest group are the α-subunits. These are between 39 and 52 kDa and are coded for by at least 17 distinct genes, with alternative splicing of these genes generating even more diversity. The α-subunits have been grouped into four families, G_s, G_i, G_q and G_{12}. Initially this was based on their effect on adenylate cyclase, since members of the G_s family stimulated the enzyme activity and those of G_i inhibited. Another family, G_q, was found to have effects on phospholipase C enzymes (*Table 6.3*).

The subsequent sequencing of the genes coding for the G proteins has revealed sequence similarities between the α-subunits which mirrors their functions, and so a family tree based on amino acid identity can still maintain and extend the original classification, although not all the members of each family have identical actions (*Figure 6.6*).

Table 6.3. Properties of G protein α-subunits

Family	Subclass	Action	Distribution
s	s	↑AC, ↑I_{Ca}	Ubiquitous
	olf	↑AC	Olfactory epithelium
i	i1	↓AC, ↑I_K, ↓I_{Ca}	Brain
	i2	↓AC, ↑I_K, ↓I_{Ca}	Ubiquitous
	i3	↓AC, ↑I_K, ↓I_{Ca}	Ubiquitous
	o1	↓AC, ↑I_K, ↓I_{Ca}	Brain
	o2	↓AC, ↑I_K, ↓I_{Ca}	Brain
	t1	↑cGMP PDE	Retinal rods
	t2	↑cGMP PDE	Retinal cones
	gust	↑cGMP PDE	Taste buds
	z	?	Brain, platelets
q	q	↑PLC-β	Ubiquitous
	11	↑PLC-β	Ubiquitous
	14	↑PLC-β	Stroma cells
	15	↑PLC-β	B lymphocytes
	16	↑PLC-β	Monocytes and T cells
12	12	?	Ubiquitous
	13	?	Ubiquitous

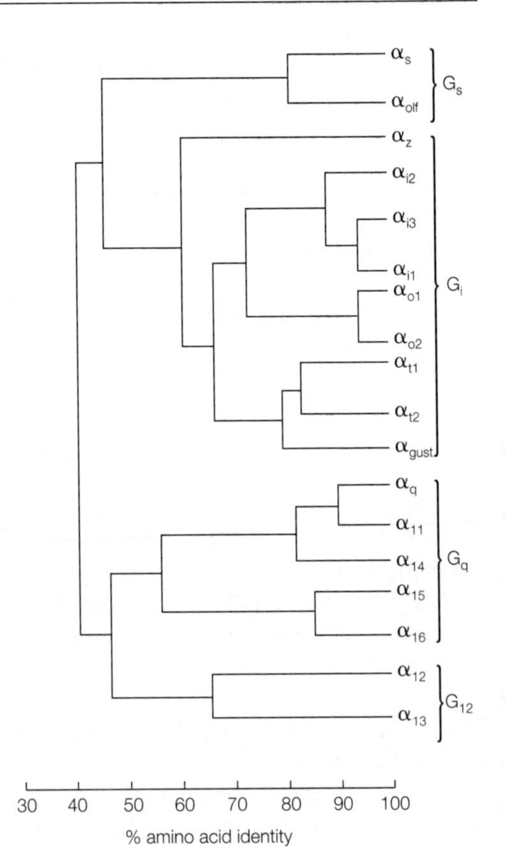

Figure 6.6. Relationships between G protein α-subunits based on amino acid identity.

In the inactive state, the α-subunit has GDP bound, and it is the interaction between the G protein trimer and the activated receptor which reduces the affinity of GDP allowing exchange for GTP, which is usually present in the cytoplasm in higher concentrations than GDP. The α-subunit remains active until the intrinsic GTPase activity of the α-subunit converts GTP to GDP, thus inactivating the α-subunit.

The α-subunits, although predominantly hydrophilic, are localized mainly in membrane fractions. The anchoring of the α-subunits to the membrane may be due to their coupling to βγ-subunits which are themselves membrane associated. Some, but not all, α-subunits have a 14-carbon myristic acid added to their N termini at Gly2 which could provide a means for hydrophobic interactions between this fatty acid and membrane phospholipids. Some of the α-subunits, notably α_s, are also modified, this time by the addition of palmitic acid to cysteine residues within the N-terminal region.

6.5.2 *The β- and γ-subunits are less diverse*

Under normal conditions, the β- and γ-subunits are purified as a complex and are tightly, but noncovalently bound. So far, there have been five different β- (35–39 kDa) and seven different γ- (7 kDa) subunits cloned. Many of them are ubiquitous, although there are some types whose distribution is more restricted. For example, both β_3 and γ_1 are restricted to photoreceptors. Although there are 35 possible combinations of βγ-subunits, not all combinations occur, and in any case there are no known significant differences in the actions of different complexes. This leaves open the question as to why there are so many different β- and γ-subunits.

A region in the C terminus of the γ-subunits consisting of a CAAX motif (C = cysteine, A = aliphatic, X = leucine or serine) indicates a consensus sequence for isoprenylation of the cysteine residue. This **post-translational modification** of the subunit does not affect the interaction between the β- and γ-subunits but is required for anchoring the complex in the membrane, for efficient coupling to the α-subunit and for effector regulation. It has been suggested that one of the major roles of the β- and γ-subunits is to increase the membrane-bound concentration of α-subunit and thus promote coupling between the α-subunit and the receptor.

6.6 The actions of G proteins can be studied using a variety of techniques

A number of compounds have been important in the study of G protein-mediated effects and have helped to elucidate both the second messenger systems activated by a given receptor and associated G protein and the downstream consequences in the cell's response associated with this.

6.6.1 *Pertussis and cholera toxins can affect G proteins*

Two of these compounds are the bacterial toxins, pertussis toxin (PTX) from *Bordatella pertussis*, which causes whooping cough, and cholera toxin (ChTX) from *Vibrio cholerae*.

Both these toxins act on specific subclasses of the G protein α-subunits and can be used to assign, at least provisionally, a novel G protein to a range of subclasses. PTX affects the α-subunits of the i, o, t and gust subclasses, whilst ChTX affects the subclasses s, olf, t and possibly gust. Both toxins act by catalyzing the ADP ribosylation of the α-subunit. However, the effect of these modifications is very different. PTX modifies a cysteine residue located four residues in from the C terminus, and this stabilizes the G protein in the GDP-bound state and thus prevents G protein activation. In the case of G_i, for example, this causes persistent activation of the effector. ChTX modifies a conserved arginine residue located at approximately the midpoint of the amino acid chain. This modification suppresses the activity of the intrinsic GTPase. This locks the G protein in the active GTP-bound form.

6.6.2 Guanine nucleotide analogs can be used to study the effects of G proteins

A number of compounds based on the structure of guanine nucleotides have been used in the study of G protein action. Two nonhydrolyzable analogs of GTP, GTPγS and Gpp(NH)p, irreversibly activate G proteins and, because of this, cause continual activation of the linked effector, even in the absence of the receptor ligand. A GDP analog, GDPβS, binds to the same site as GDP but with a much higher affinity, which means that it does not exchange for GTP, preventing activation of the G protein.

6.7 G protein α-subunits are linked to a range of effectors

After the dissociation of the α-subunit from the βγ-subunits, it can diffuse along the cytoplasmic face of the bilayer until it reaches its effector. The first effectors which were found to be linked to receptors via G proteins were the enzymes adenylate cyclase and the retinal cGMP phosphodiesterase. Further effectors have since been identified, and the list is still growing. Many of the effectors change the intracellular levels of second messengers which activate intracellular enzymes, such as protein kinases, phospholipases and proteases (see Appendix 3). However, second messengers also affect the opening and closing of ion channels and can thus influence cell excitability.

6.7.1 Adenylate cyclase can be dually regulated by G proteins

The enzyme, adenylate cyclase, which is found in most cells where it catalyzes the production of the second messenger, cAMP, can be both stimulated and inhibited by different G proteins. Members of the G_s family stimulate and some of the G_i family inhibit the enzyme's activity.

Adenylate cyclase is a 100-kDa integral membrane glycoprotein with two repeats of six TM segments separated by a large cytoplasmic loop (C1) and with a large C-terminal domain (C2). Six distinct isoforms (ACI–ACVI) have been identified and vary in their similarity to one another from 37 to 68%. This diversity in structure may underlie the fact that some forms of adenylate cyclase may not only be controlled by α-subunits but can also be regulated by calcium, calmodulin (CaM) and/or βγ-subunits (see Section 6.8.1).

A major action of cAMP is to activate the cAMP-dependent protein kinase (PKA) which phosphorylates many intracellular proteins. This phosphorylation, which is usually of either serine or threonine residues, alters the activity of the target protein. This phosphorylation may be reversed by the action of cytosolic phosphatase enzymes.

6.7.2 Retinal G proteins activate a phosphodiesterase

In the retina, the G protein-linked receptor, rhodopsin, is activated via retinal by light. The activated G protein α-subunit, transducin (G_t), itself increases the activity of a cyclic GMP phosphodiesterase (cGMP PDE). This reduces the intracellular concentration of cyclic guanosine monophosphate (cGMP) and this reduces the current, called the dark current, which flows through a cGMP-activated sodium channel. The cGMP PDE consists of a tetramer of $\alpha\beta\gamma_2$ subunits (not to be confused with the α-, β- and γ-subunits of G proteins). The catalytic activity resides in the α- and β-subunits, but in the dark is inhibited by the two γ-subunits. After its activation by light, G_t binds to the γ-subunits, which removes the inhibition and allows the α- and β-subunits to hydrolyze cGMP to 5′-GMP. This system has a very high sensitivity as a single activated rhodopsin molecule may activate over 30 000 molecules of transducin. However, the system can also be switched off rapidly through the action of rhodopsin kinase which, by phosphorylation of rhodopsin and the subsequent binding of visual arrestin, desensitizes the receptor by inhibiting the coupling of the receptor and the G protein.

The inherited visual defect, retinitis pigmentosa, starts with night blindness and can then develop into complete loss of vision. It is a heterogeneous disorder which may be inherited as an autosomal dominant (adRP), an autosomal recessive (arRP) or an X-linked (xlRP) disease, depending on which mutations are involved. In some cases of adRP and arRP, the defect lies in mutations in the rhodopsin molecule (Section 6.12). Other recessive forms (arRP) of the disease have been linked to defects in the cGMP PDE.

6.7.3 G protein activation of phospholipase C produces multiple second messengers

The enzyme phospholipase C (PLC) is activated by α-subunits of the G_q and possibly the G_i families although all of them stimulate the enzyme's activity, even those of the G_i family. PLC catalyzes the cleavage of the membrane phospholipid, phosphatidylinositol-4,5-bisphosphate (PIP_2) into two compounds, inositol-1,4,5-trisphosphate (IP_3) and diacylglycerol (DAG), both of which can act as second messengers (*Figure 6.7*).

As it is water soluble, IP_3 can diffuse into the cytoplasm where it binds to specific IP_3 receptors, which are located on the membranes of intracellular calcium stores. This causes the release of calcium, another second messenger, which can have further effects on cell biochemistry through activation of molecules such as CaM and thus Ca/CAM-dependent protein kinases. Calcium levels can also change cell excitability through binding to calcium-dependent ion channels such as the calcium-dependent K^+ channels described in Chapter 4. The lipid moiety DAG remains in the membrane, but can

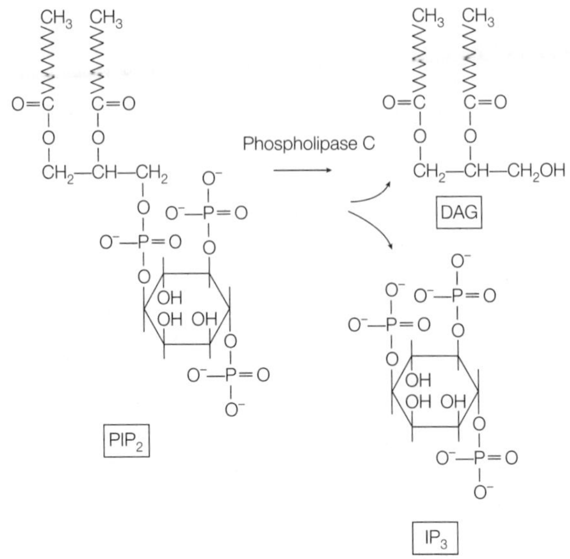

Figure 6.7. Cleavage of PIP_2 produces IP_3 and DAG.

diffuse laterally and, in the presence of calcium and another membrane lipid, phosphatidylserine, can activate PKC.

There are three classes of PLC called PLC-β, PLC-γ and PLC-δ, although, at present, it is thought that only the four members of the PLC-β class (PLC-β 1–4) are activated via G_q proteins. The two types of PLC-γ are activated by phosphorylation by tyrosine kinase-linked receptors (see Section 6.10), whilst the method of activation of PLC-δ (four isoforms) is unknown. Although the overall sequence homology between the different PLC isoforms in not particularly high (21–56%), there are two regions called X and Y which are more highly conserved, with up to 80% sequence homology. Experiments involving mutations in these highly conserved regions suggest that these are the regions responsible for the catalytic activity of the enzymes.

6.7.4 The G protein linking receptors and the enzyme phospholipase A_2 is unknown

The activation of the enzyme phospholipase A_2 (PLA_2) has been linked to a number of G protein-linked receptors. The link via specific G proteins is still under investigation, although there is some evidence that it may be via G_o. Three groups of PLA_2 have been identified, cytoplasmic $cPLA_2$ which is activated by micromolar levels of Ca^{2+}, secreted $sPLA_2$ which requires millimolar Ca^{2+}, and a cytosolic Ca^{2+}-independent $iPLA_2$ which is activated by ATP. PLA_2 catalyzes the hydrolysis of membrane phospholipids, and one of the major products of this hydrolysis is arachidonic acid (AA), the precursor of many eicosanoids. These molecules have many functions in autocrine and paracrine regulation and have been suggested as possible retrograde messengers in synaptic plasticity (see 8.7.2).

6.8 G protein βγ-subunits can alter the activity of α-subunits

Initially, there were thought to be two major roles for βγ-subunits. The first of these was, as previously mentioned, the anchorage of α-subunits in the membrane. The second was as a regulator of the activity of the α-subunits. The βγ-subunits stabilize the binding of GDP to the α-subunit, which at the same time inhibits the binding of GTP and thus prevents activation. This ensures that in the absence of stimulation there is a low level of basal activation of α-subunits. Also, because the βγ-subunits seem to be interchangeable, this means that when a given G protein-linked receptor is stimulated and the appropriate α-subunit is activated, this releases βγ-subunits which could go on to suppress the activation of other α-subunits. This process would allow the receptor-mediated inhibition of G protein activation. The size of such an effect would, of course, depend of the number of βγ-subunits released and their affinity for the α-subunits. In the light of this, the only βγ-mediated inhibition commonly seen is the inhibition of adenylate cyclase, as the affinity of α_s for βγ is relatively high compared with that of other α-subunits.

6.8.1 *βγ-subunits can also affect the activity of intracellular enzymes*

The direct role of βγ-subunits has been observed in the stimulation of particular PLC-β isoforms. The four isoforms of PLC-β can be stimulated independently by both the α- and βγ-subunits (see *Table 6.4*). However, whilst PLC-β1 and PLC-β3 isoforms are more responsive than PLC-β2 and PLC-β4 to stimulation by the α-subunits of the G_q family, PLC-β2 and PLC-β3 are more sensitive than PLC-β1 to activation by βγ-subunits. PLC-β4 is not activated at all by βγ-subunits. Each of the subunits can act independently to stimulate the PLC, which means that whilst the stimulation of PLC-β1 will occur mainly through the activation of receptors linked to G proteins of the q family, PLC-β2 and PLC-β3 can be stimulated by the production of free βγ-subunits through the activation of other G proteins.

In the case of adenylate cyclase, all α_s-subunits stimulate all of the isoforms (ACI–ACVI) and α_i inhibits the activity of ACI, ACV and ACVI. The actions of βγ-subunits are isoform selective, with ACI being inhibited by βγ and ACII stimulated. Unlike the independent effect of βγ seen with PLC, the action of βγ on ACs is dependent on the enzyme being stimulated by α_s The inhibition of ACI is not due solely to the reversal of the α_s stimulation by mass action binding of βγ to α_s-subunits, but is through a direct pathway, as shown by the inhibition of purified ACI by application of βγ. As both G_i and G_o proteins are present in the brain in much higher concentrations than G_s, the activity of ACII could be greatly enhanced by the simultaneous stimulation of

Table 6.4. PLC-β isoforms are differentially sensitive to stimulation by α- and βγ-subunits

	PLC-β1	PLC-β2	PLC-β3	PLC-β4
α_q	+++	+	+++	+
βγ	+	+++	+++	0

+++, highly stimulated; +, weakly stimulated; 0, not stimulated.

both the G_s and G_o (or G_i) pathways. This situation allows the levels of cAMP to reflect the simultaneous activation of two separate pathways.

Because βγ-subunits from any source seem to be able to influence the activity of PLC and AC, it is important that in the absence of stimulation the levels of free βγ-subunits is kept low. The easiest way for this to occur is by the production of equimolar amounts of α- and βγ-subunits. In this way, the binding of α-GDP and βγ results in inactive trimers. However, as the cell may produce many different α-, β- and γ-subunits, this method of regulation seems difficult to coordinate.

Alternative methods of regulation have been suggested which involve the association of βγ-subunits with other intracellular proteins which may act to sequester the βγ-subunits in an inactive pool. In the retina, the soluble phosphoprotein, phosducin, is thought to regulate signaling by the tethering of βγ-subunits. In the dark, phosducin is phosphorylated by cyclic nucleotide-dependent PKA. After a period of exposure to light, cyclic nucleotide levels decrease and phosducin is dephosphorylated by protein phosphatase 2A. Dephosphorylated phosducin competes more effectively with α_t (transducin) for binding to βγ and thus reduces the amount of trimeric G_t. This would desensitize the response to light by preventing G protein activation.

A further action of βγ in the G protein signaling cascade is its ability to stimulate the activity of the receptor kinases which phosphorylate G protein-linked receptors.

6.9 Many ion channels are modulated by G proteins

The activity of a number of ion channels has been shown to be affected by the activation of specific G protein-linked receptors. In many cases, this may be a second messenger-mediated effect. For example, the activity of calcium-dependent K^+ channels can be affected by the activation of PLC by G_q and hence the production of IP_3 and the subsequent release of Ca^{2+}. Similarly, the activation of cAMP-dependent protein kinases and the subsequent phosphorylation of ion channels can be linked to the G protein-mediated stimulation of adenylate cyclase. An example of this is the sympathetic stimulation of the heart via the β-AR, which through activation of G_s and the stimulation of adenylate cyclase, causes the activation of PKA. This produces phosphorylation of a number of proteins, including the L-type calcium channel. This channel then remains open for longer, allowing greater calcium influx which increases the force of contraction.

The levels of cAMP also affect the gating of a channel which is found in the cilia of olfactory epithelial cells. As mentioned in Section 6.3, odorant receptors are 7TM G protein-linked receptors. The G proteins involved in the signal transduction are G_{olf}, which are members of the α_S family and thus act by stimulating the activity of adenylate cyclase. The cAMP binds to a cyclic nucleotide-gated channel, increasing the cation conductance and thus depolarizing the cell. If the depolarization is sufficient then an action potential is produced and is propagated to the olfactory bulb. This is similar, at least in the use of a cyclic nucleotide-gated channel, to the transduction of light in the retina and the closure of cGMP-gated channels.

6.9.1 Direct membrane-delimited regulation by α-subunits has been shown

There are a number of cases in which the regulation of an ion channel has been shown to occur, not through the production of a second messenger but by a direct interaction in the membrane between the α-subunit and the channel. In many neurons, presynaptic inhibition of calcium channels is regulated by G proteins. A number of experiments have shown that, in certain cases, this regulation is via the PTX-sensitive G_o and that specific subtypes of the α-, β- and γ-subunits are required for the transmitter-specific inhibition to occur. There are a number of ways in which the involvement of a particular G protein can be inferred. As described in Section 6.6, the sensitivity to PTX or ChTX can ascribe the response to a family of G proteins. The second messenger linkages can also be assessed by the use of the irreversible inhibitors or activators. However, in order to discover the specific isoform used, other methods must be employed. Reconstitution experiments using bacterially expressed cloned G protein α-subunits have been used, although these subunits are less potent than native subunits. Another approach uses α-subunit-specific antibodies to immunoprecipitate the receptor–G protein complexes, although there are limitations in the ability of antibodies to differentiate between closely homologous α-subunits.

A novel approach has been the use of **antisense knockouts** (***Box 6.1***, p. 126) selectively to remove individual subtypes of G proteins. In one experiment, the inhibition of calcium currents by somatostatin and the muscarinic agonist, carbachol, was examined electrophysiologically, in pituitary GH_3 cells which had been injected with **antisense oligonucleotides** directed against specific α-, β- and γ-subunits. Oligonucleotides specific for α_o2, β1 or γ3 selectively blocked somatostatin-induced inhibition whilst inhibition by carbachol was blocked by antisense probes to α_o1, β3 or γ4. There was no effect of α_i sense or antisense oligonucleotides. Although this experiment showed that specific β- and γ-subunits were also involved, the receptor-mediated inhibition also occurred with disruption of the α-subunit alone, thus clearly indicating the effect of α-subunits on calcium current inhibition.

There is as yet no indication of exactly how interactions between G proteins and channels act to change the channel activity, but there are clearly membrane-delimited effects due to α-subunits.

6.9.2 Ion channels can also be activated by βγ-subunits

In cardiac myocytes, there is a population of K^+ channels which are activated by acetylcholine acting on muscarinic (M2) receptors, which are G protein linked. Their activation hyperpolarizes the cells, which increases the time between contractions and thus slows the heart. In patch clamp experiments, it was shown that in cell-attached mode, they could not be activated by ACh applied to the cell, indicating that a diffusible second messenger was not involved. However, the channels could be activated by ACh in the pipette, indicating a membrane-delimited effect. In whole-cell mode, the channel could only be activated by ACh in the presence of intracellular GTP, and the response could be blocked by PTX, suggesting the involvement of a G protein. Although it was thought initially that the channel was opened directly by activated α-subunits, it has since been clearly shown that it is the βγ-subunits which gate the

K^+ current. This has been shown in two ways. Firstly, it has been shown that βγ-subunits associate directly with the channel, and secondly, that the effect of the βγ-subunits can be inhibited by the addition of GDP-bound, and hence inactive, α-subunits, which will bind the βγ-subunits into the inactive trimer. Hence, just as the level of free βγ-subunits can suppress the activation of α-subunits, the levels of inactive α-subunits could influence the activity of βγ-subunits. With the cloning of the mACh-regulated K^+ channel from heart, called GIRK1 (which is probably the K_{IR} channel described in Chapter 4.7.6), it has been shown that coexpression of GIRK1 with the genes for β and γ resulted in cells which showed constitutive channel activity. This was not seen in cells coexpressing GIRK1 and α genes, indicating that the previously observed α-mediated activation may have been artifactual. GIRK1 is the first member of a family of G protein-coupled K^+ channels, and mRNA for GIRK1 has been detected in the forebrain and the cerebellum.

6.10 Many growth factor receptors activate an intrinsic tyrosine kinase

The protein kinases, such as PKA and PKC, which are activated by cAMP and other second messengers, phosphorylate serine and threonine residues on many intracellular targets. Another large group of protein kinases can be distinguished from these, as they only phosphorylate tyrosine residues. These tyrosine kinases fall into two groups. Firstly, there are the cytosolic, nonreceptor protein tyrosine kinases (PTKs). These are activated via a number of receptors, including those for cytokines, and the antigen receptors. The second type of tyrosine kinases are the receptor tyrosine kinases (RTKs), which are a family of more than 50 different transmembrane receptors which have a PTK domain in their intracellular domains. The ligands for these receptors are growth factors, such as epidermal growth factor (EGF), insulin and platelet-derived growth factor (PDGF). Activation of these receptors produces a clustering of receptors and subsequent autophosphorylation of the catalytic domains. This is thought to induce a conformational change which allows the binding and phosphorylation of other substrates.

6.10.1 Receptor tyrosine kinases can activate PLC-γ

One of the first pathways which has been shown to be linked to activation of RTKs is the PLC-γ class. As mentioned in Section 6.7.3, the two isoforms of PLC-γ are activated via tyrosine kinase phosphorylation, producing events which parallel those seen in the activation of PLC-β by G proteins. The activation of PLC-γ occurs by the binding of the PLC-γ to the autophosphorylated receptor through Src homology 2 (SH2) domains. These domains, which are present on a large number of intracellular proteins, bind to specific phosphotyrosines and can mediate protein–protein interactions. Similar binding domains, called PTB and SH3, have been found which also mediate protein interactions.

6.10.2 RTKs are involved in long-term intracellular events

Other important intracellular targets of RTKs are the mitogen-activated protein kinases (MAPKs), although MAPK is not phosphorylated directly by the RTK but via a series of other proteins including the small membrane-bound guanine nucleotide-binding protein, Ras (see Chapter 8). MAPK is involved in the pathways associated with long-term transcriptional events such as cell differentiation and proliferation, but is also thought to be involved in shorter term events, such as neuro-transmitter secretion. Studies have shown that MAPK can also be activated by G protein-linked receptors via the activation of intracellular tyrosine kinases, such as PKY2, which itself is activated by intracellular calcium levels and PKC. This and other interactions allows the integration of both short-term and long-term signaling mechanisms.

6.11 Cyclic GMP levels are controlled by two separate pathways

Despite the close structural similarities between cAMP and cGMP, the structure and regulation of guanylate cyclase (GC) are very different from those of adenylate cyclase (AC). GCs are found in two forms, one is membrane bound like AC, but the other is cytoplasmic.

The membrane-bound GCs are transmembrane proteins with an extracellular N terminus which binds to peptides, such as atrial natriuretic peptide (ANP). There is a single TM segment and an intracellular C terminus which contains the catalytic domain, which catalyzes the conversion of GTP to cGMP. In the kidney, the activation of these receptors by ANP causes diuresis and natriuresis, and they are important in the control of electrolytes and blood volume. However, ANP is also synthesized in the brain, although its role here is unknown, and at least one brain-specific ANP receptor/guanylate cyclase has been described.

The cytoplasmic form of GC was first discovered in the smooth muscle cells of blood vessels, where it is involved in the control of vascular tone. Muscarinic activation of G protein-linked receptors on endothelial cells causes the activation of PLC and the eventual release of Ca^{2+} from intracellular stores. This activates a Ca^{2+}/calmodulin-dependent nitric oxide synthetase (NOS) which produces nitric oxide (NO) from arginine (*Figure 6.8*). NO is a small gaseous molecule with a short half-life which can diffuse easily through membranes. NO produced by endothelial cells then diffuses to the adjacent smooth muscle cells where its target is the soluble GC. This enzyme contains a catalytic domain which resembles that of the membrane-bound GC, and a heme group, which acts as a receptor for the NO, thus activating the cyclase. Increased cGMP levels activate a cGMP-dependent protein kinase, and the subsequent phosphorylation of muscle proteins leads to relaxation of the blood vessel.

However, NOS is not restricted to endothelial cells; there are a number of different forms, both constitutive (cNOS) and inducible (iNOS), and specifically there is a form of cNOS which is restricted to neurons (ncNOS). The possible action of NO as a diffusible retrograde messenger involved in brain plasticity is a subject that will be addressed in Chapter 8, and its possible role in neurodegenerative disease will be discussed in Chapter 9.

Calcium may also be involved in the control of GC. As already described in Section 6.7.2, in the retina, cGMP levels are reduced by a G protein-activated cGMP PDE, and the reduction in cGMP reduces the dark current flowing though cGMP-activated

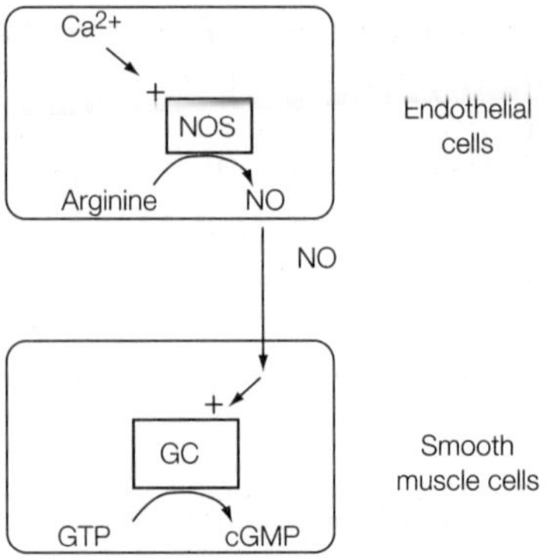

Ca2+ + NOS Arginine NO NO Endothelial cells + GC GTP cGMP Smooth muscle cells

Figure 6.8. Soluble guanylate cyclase in smooth muscle is stimulated by NO derived from endothelial cells.

channels. These channels are permeable to sodium but are also permeable to calcium and, in the light, intracellular Ca^{2+} levels are reduced at the same time as the hyperpolarization due to the reduced influx of Na^{+}/Ca^{2+}. In photoreceptors, there is a Ca^{2+}-sensitive protein, recoverin, which is activated by *reduced* levels of Ca^{2+}. Thus, levels of active recoverin increase in the light. Activated recoverin stimulates GC and so, in continuing illumination, the levels of cGMP increase by the action of recoverin on GC, contributing to the adaptive response of the retina to sustained light.

6.12 Mutations in 7TM receptors underlie a variety of diseases

A number of diseases have already been linked to defects in 7TM receptors, and with the techniques now available to isolate and sequence genes many more will surely be found. Mutations in 7TM receptors can be grouped according to whether they cause a loss of function or, conversely, cause **constitutive activation** of the receptors in the absence of the appropriate stimulus.

In Section 6.7.2, it was noted that some forms of retinitis pigmentosa have been associated with defects in rhodopsin. In one **autosomal recessive** form (arRP), a mutation which adds a stop codon to the third TM segment produces a rhodopsin molecule with no function. However, as this is a recessive form, the lack of function can be compensated in heterozygotes by the presence of a normal gene product. This is a clear demonstration of how a loss-of-function mutation may only be apparent in the homozygotes. This is not always the case. In another loss-of-function mutation of rhodopsin, the improper folding of the mutated rhodopsin protein produces abnormal trafficking of the mutated protein and eventual death of the rod cell. This mutation is **autosomal dominant** and cannot be compensated for by the presence of normal rhodopsin.

Other loss-of-function mutations that have been identified include X-linked mutations in the cone opsins leading to color blindness, X-linked mutations in the vasopressin 2 receptor causing some forms of congenital nephrogenic diabetes insipidus, and an autosomal dominant mutation in the calcium sensor, leading to familial hypocalciuric hypercalcemia.

Constitutively activating mutations are interesting in the sense that despite the many feedback mechanisms which exist to turn off the effects of continued signaling, this can, in some cases, be overridden by the constitutively activated receptor. From the study of the exact amino acid substitutions of these mutants, it is clear that it is possible to have a constitutively activated receptor with mutations occurring in a wide variety of different positions in the receptor (*Figure 6.9*).

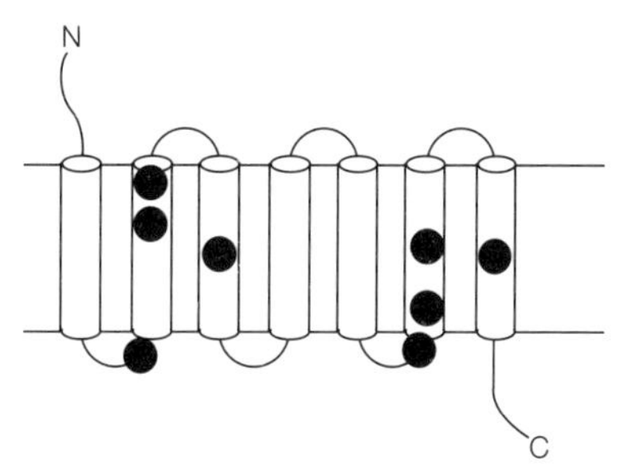

Figure 6.9. Positions of mutations of diverse constitutively activating receptors (●).

This has led to the suggestion that, in the absence of ligand, the receptor is held in an 'off' state that is relatively easily disrupted, as many different mutations can lead to the receptor being constitutively active. This leads to the conclusion that if the 'off' state of the receptor is relatively unstable, and given the extremely large number of G protein-linked receptors, there may be many more diseases, including neurological diseases, as yet unrevealed, that are due to these types of constitutively activating mutations.

6.13 Dopamine receptors have been implicated in schizophrenia

The dopamine hypothesis of schizophrenia, in its simplest form, proposes that the psychotic symptoms in schizophrenia are related to an excessive action of dopamine in the brain. This suggestion is based on the effects of a number of ligands acting on the dopaminergic system in both psychotic and normal subjects.

The symptoms of schizophrenia have been classified as either positive or negative. Positive symptoms include delusions, thought disorders, hallucinations, mood disorders and increased motor function, whilst negative symptoms include poverty of speech, lack of emotional responsiveness, reduced motor function, apathy and social withdrawal.

The observations, in the 1950s, that chlorpromazine (CPZ) could reduce the positive symptoms of schizophrenic patients encouraged the study and development of the phenothiazine class of antipsychotics. Although CPZ blocks the action of many neurotransmitters, including histamine, catecholamines, ACh and serotonin, it is thought to be its action on dopamine receptors which determines its antipsychotic action. Other more specific dopamine antagonists, including haloperidol, raclopride and spiperone, are now used clinically as antipsychotics to reduce the positive symptoms of schizophrenia, and all have been shown to be effective at concentrations which correlate with their binding to dopamine D2 receptors (see below). The indirectly acting sympathomimetic, amphetamine, which causes the release of noradrenaline but also serotonin and dopamine from nerve terminals in the brain, when given to normal subjects, may produce the positive symptoms of schizophrenia and exacerbates the symptoms of schizophrenics. Another link with the dopamine system was shown by some of the side effects of antipsychotics. In Parkinson's disease there is a reduction in dopamine release from cells originating in the substantia nigra, with symptoms which include tremor and rigidity (see Chapter 9). Schizophrenic patients treated with anti-psychotics commonly develop these same symptoms and with a severity which is related to the dose used.

In order to substantiate the dopamine hypothesis, attempts have been made to measure the levels of dopamine in the brains of schizophrenics. These studies are fraught with difficulty, mainly due to the problems associated with accessing the brain. Firstly, the levels of dopamine or its metabolites and synthetic enzymes in the post-mortem brain may be affected by not only the long-term effects of the illness and the changes brought about by drug treatment but also by death itself. Levels of these compounds in urine, blood or cerebrospinal fluid may not reflect levels in the brain, and in any case specific levels in different areas of the brain may be critical.

Positron emission tomography (PET) using labeled dopamine ligands allows the direct visualization of receptor density in living subjects, but again there may be the problem of the effects of medication on receptor numbers. Also, in the light of the discovery of a number of different receptor subtypes, the specificity of some ligands is of importance in interpreting the results.

Dopamine and dopamine receptors are found in their highest concentrations in the putamen, the caudate nucleus and the nucleus accumbens, with lower levels in the amygdala, the median eminence and the cerebral cortex. Two major dopaminergic pathways travel from the substantia nigra in the brainstem to the corpus striatum and from the ventral tegmental area of the brainstem to the limbic system. The first of these pathways is involved in the control of movement and is the pathway that degenerates in Parkinson's disease. It has been suggested that the second pathway, with its links to the areas of the brain involved in the regulation of emotional behaviors, may have a role in schizophrenia.

After the early discovery of D1 and D2 receptors, which respectively activate and inhibit adenylate cyclase, the method of homology cloning using probes derived from the sequence of the known dopamine receptors has now increased the number of known dopamine receptors to five, D1–D5, with some of them, particularly D4, having a large number of variants. These receptors are classified into two groups, D1-like, which consists of the D1 and D5 receptors, and D2-like, which consists of D2, D3 and D4 receptors. These five receptors have been shown to have diverse effects on multiple signaling pathways. D1-like receptors activate adenylate cyclase and D1 receptors also increase levels of IP_3 and Ca^{2+}. D2-like receptors all inhibit adenylate cyclase and reduce Ca^{2+}. D2 receptors increase levels of IP_3, whilst all three D2-like receptors have differing effects on AA metabolism. Using *in situ* hybridization to locate the cells expressing the specific mRNAs, it is of particular interest in the study of schizophrenia that D3 and D4 receptors are located specifically in the cortical and limbic areas of the brain, and antipsychotic drugs block not only D2 but also D3 and, in some cases, D4 receptors.

There is some evidence to suggest that there may be an increase in D4 receptors in the post-mortem brains of schizophrenics which is not related to medication with antipsychotics, as it is not observed in patients who are not schizophrenic but are receiving antipsychotic medication for other conditions. However, as yet, it is not possible to visualize D4 receptors *in vivo* due to the lack of selective ligands, although one compound, clozapine, seems to be more selective for D4 receptors than for D2 or D3. The development of antagonists which are highly selective for each receptor subtype will be an important tool both in the study of dopamine receptors and in the possible avoidance of D2-related side effects. Interestingly, antagonists with some selectivity for the D3 receptor cause behavioral stimulation in rats, suggesting that they may be active against the negative symptoms of schizophrenia.

The different receptor subtypes show a wide range of variants. Whilst the D1 receptor seems to have no variants, there are five types of D2 receptor (A, S, C, short and long). The A, S and C genes are rare, but the short and long versions are splice variants of a single gene which coexist in different proportions in different cell populations. D3 has two variants, one of which seems to be nonfunctional, whilst D5 variants only seem to consist of two pseudogenes, again producing apparently nonfunctional proteins.

In humans, it is the D4 receptor which shows the largest diversity. This is due to the presence of a repeat sequence of 16 amino acids in the third cytoplasmic loop. This repeat sequence occurs between two and 10 times, hence these receptors are called D4.2 to D4.10. Another level of complexity occurs because not all the repeats are identical. At least 25 different coding sequences have been identified for the 16 amino acids. Although, in humans, the first, second and last repeats are always the same, there is still scope for the production of an enormous number of different receptors by varying the repeat sequences at the other positions. The genetic basis of schizophrenia has been hotly disputed, with varying weight being given to genetic vs. environmental arguments. If the inheritance of one or more of these different receptors underlies a susceptibility to schizophrenia it is no wonder that, as yet, a tight genetic linkage has not been found.

Box 6.1. Antisense knockouts

One way of removing a protein is by inhibiting the expression of the gene by blocking the transcription and translation of the mRNA coding for that protein.

Antisense knockouts are produced by introducing oligonucleotides into the cell which are complementary to the mRNA coding for the protein. Recall that DNA has sense and antisense strands and that mRNA is transcribed from the antisense strand, giving it the same sequence as the sense strand, except that thymidine (T) replaces uracil (U) (*Figure 1a*). Thus, oligonucleotides which are complementary to this mRNA will have the same sequence as the antisense DNA, hence their name **antisense oligonucleotides**. They are thought to work in a number of possible ways, either by binding to the sense strand of the DNA forming a triple helix which blocks transcription, by blocking the translation of the mRNA into protein by binding to the mRNA, or by inducing the digestion of the double-stranded RNA by RNase H.

Antisense oligonucleotides are usually about 15–18 bases in length and can be introduced into the cell in two main ways. In the direct approach, they can either be injected into the cell or liposomes containing the oligonucleotides are added to the cells. An advantage of cell injection is that in experiments where the cell will be studied using single-cell approaches, such as patch clamping and single-cell fluorescence, it is certain that the cell actually has been treated. However, the use of liposomes enables larger number of cells to be treated simultaneously. An indirect approach is to transfect the cell with the target gene in the reverse orientation which, when expressed, produces mRNA which is antisense to the normal mRNA (*Figure 1b*). Because this method uses the gene it is also called **antigene suppression**.

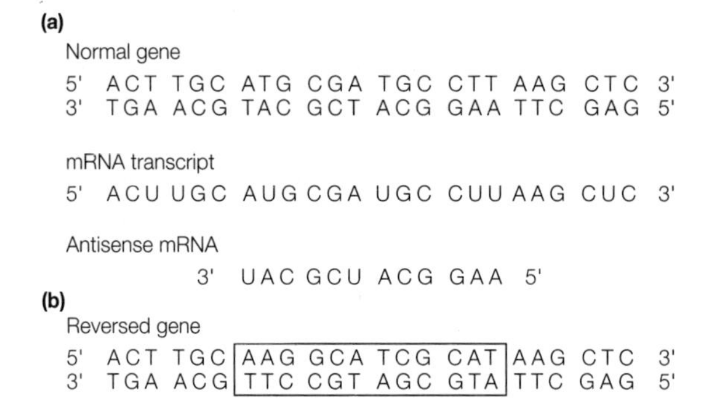

Figure 1. (a) Antisense mRNA has the same sequence as the antisense DNA strand. (b) Transcription of the reversed gene (boxed) gives mRNA which is complementary to the sense strand of the normal gene.

Chapter 7

Neurotransmitter release

7.1 Introduction

The process of neurotransmitter release involves the regulated **exocytosis** of vesicles containing the neurotransmitters. Vesicles are found in a variety of shapes and sizes which correlate with the type of neurotransmitter they contain. Small vesicles (<60 nm) with clear electron-lucid centers (SSVs) are found to contain the 'fast' neurotransmitters in either spherical (excitatory neurotransmitters) or flattened (inhibitory neurotransmitters) vesicles. Other small (40–60 nm) dense-core vesicles contain catecholamines (CAs), and large (120–200 nm) dense-core vesicles (LDCVs) contain either CAs or neuropeptides. Whatever the type of vesicle involved, it is thought that all transmitters are released via a process which involves fusion of the vesicle with the plasma membrane.

This event can be monitored directly using patch clamping. When a vesicle fuses with the plasma membrane of a cell, the insertion of the vesicle membrane increases the surface area of the cell and hence its capacitance, as total cell capacitance depends on the surface area of the membrane. Using whole-cell patch clamping, step increases in cell capacitance have been observed, corresponding to the exocytosis of vesicles. The extra membrane is later retrieved by a process called **endocytosis**, where the extra membrane is pinched off to form an intracellular vesicle.

However, neurotransmitter release is not identical for all neurotransmitters and all synapses. The **neuromuscular junction** was the first synapse to be studied, and it is still the most well known. The arrival of an action potential produces a large release of neurotransmitter from the nerve terminal, corresponding to the simultaneous release of several hundred vesicles. This release is rapid (200 μsec) and invariably depolarizes the underlying muscle to cause contraction. When compared with synapses in the CNS, this behavior is unusual as central synapses may require the arrival of more than one action potential before releasing a single vesicle. Furthermore, the release of some neurotransmitters, particularly neuropeptides, occurs relatively slowly (10–100 msec) and requires repetitive stimulation for release to occur. An even slower type of secretion is observed in the chromaffin cells of the adrenal medulla, which release CAs from LDCVs after hundreds of milliseconds, a process more reminiscent of the stimulated secretion from exocrine and endocrine cells.

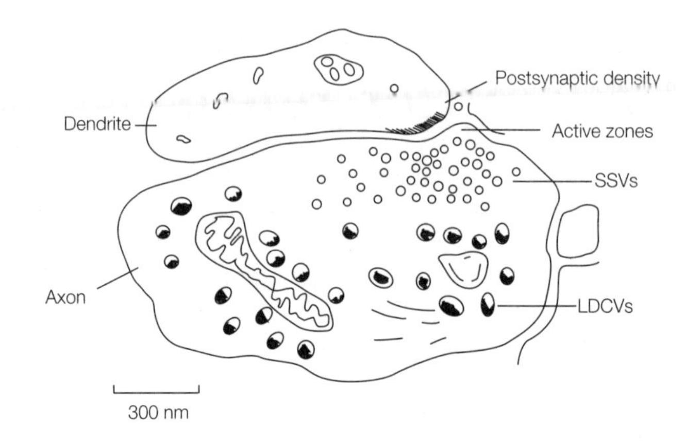

Figure 7.1. Diagram of a synapse showing SSVs, LDCVs and active zones. Note the scale bar.

One of the reasons for the difference in the rate of release of SSVs and LDCVs lies in the fact that SSVs are found close to the synaptic membrane at specialized areas called active zones, whereas LDCVs occur further away in the body of the terminal (*Figure 7.1*).

Because of variations in the number of vesicles released, the rate of secretion and the location of the vesicle to be released, whilst the mechanism of neurotransmitter release is likely to be essentially similar at all synapses, there are certain to be substantial differences depending of the type of synapse; this must be borne in mind when comparing results from diverse sources.

Apart from the neuromuscular junction, other preparations which have been used as model systems in the study of neurotransmitter release include the squid giant synapse, chromaffin cells of the adrenal medulla, synaptosomes, brain slices and yeast.

7.2 Neurotransmitter release at the neuromuscular junction is quantal

Experiments carried out in the 1950s showed convincingly that the release of neurotransmitter could be explained by a model in which neurotransmitter is packaged in discrete packets, named **quanta**, and that these are released independently of each other. Evidence from microscopy identified the potential identity of the quanta as membrane-bound intracellular vesicles which contained neurotransmitter and which could fuse with the presynaptic membrane to release their contents by exocytosis. Under normal conditions, large numbers of vesicles are released simultaneously, leading to a large depolarization of the postsynaptic muscle membrane and the generation of an action potential. Each vesicle was assumed to contain approximately the same amount of neurotransmitter, since each quantum released produces approximately the same postsynaptic depolarization. These can be observed as small depolarizations of about 0.5 mV; they are called **miniature end-plate potentials** (mepps) and occur spontaneously at the neuromuscular junction at the rate of about 1 per second under resting conditions.

7.3 Neurotransmitter release at central synapses may consist of a single quantum

Although central synapses also contain identifiable vesicles, on stimulation of the axon the amount of neurotransmitter released is much smaller and the postsynaptic response is much more variable. This has led to the idea that only one quantum is released by the arrival of a single action potential, but with a probability of less than 1. Values quoted for the probability of release vary from about 0.4 to 0.8, in other words, at central synapses an action potential often fails to trigger transmitter release. There are also suggestions that at central synapses a single quantum is sufficient to saturate the postsynaptic receptors, and thus the variation in size of the postsynaptic depolarization may be due to a variation in the number of postsynaptic receptors. Both pre- and postsynaptic responses may be modified under certain conditions, a subject which will be followed up in Chapter 8. The firing of the postsynaptic cell in the CNS therefore requires the spatial and temporal summation of a large number of inputs, confirming the neuron's role as an integrator.

7.4 Neurotransmitter release is calcium dependent

Experiments from the squid giant synapse have shown that external calcium is required for neurotransmitter release and that calcium enters the nerve terminal through voltage-gated calcium channels. Any procedure which lowers intracellular calcium, including removal of extracellular calcium, injection of a calcium chelator such as ethylene-glycol-bis-(β-aminoethyl)-*N*,*N*′-tetraacetic acid (EGTA) into the nerve terminal or nonspecific blockade of calcium channels by magnesium, can reduce neurotransmitter release.

Electron micrographs of the synapse show SSVs close to the presynaptic membrane, and freeze-fracture studies of the presynaptic membranes show parallel rows of particles at the **active zones** (*Figure 7.2a*). After stimulation, under conditions which enhance neurotransmitter release, in addition to the rows of particles, pits can be seen between the rows (*Figure 7.2b*).

The rows of particles have since been identified as rows of calcium channels and the pits as the open mouths of vesicles undergoing exocytosis. In transverse section, these pits can be seen as open vesicles with the characteristic 'omega' profile (*Figure 7.2c*). When frog muscle was incubated with toxins conjugated to fluorescent dyes, the pattern of binding of an α-BuTX–rhodamine conjugate, which binds to the nAChR on the post-synaptic membrane, exactly corresponds to the binding of ω-CTX–Texas red conjugate, which binds to calcium channels. Both showed linear bands similar to the rows of particles seen in freeze-fracture, showing that the calcium channels found at active zones are immediately opposite the postsynaptic receptor sites.

After the opening of the voltage-gated calcium channels and subsequent calcium influx, release of neurotransmitter from SSVs is very rapid. This implies that the calcium acts at a site very close to the active zones. If it is assumed, from the distribution of calcium channels, that the calcium entry is confined to the area of the active zones, the local calcium concentration can be estimated using mathematical models based on the rate

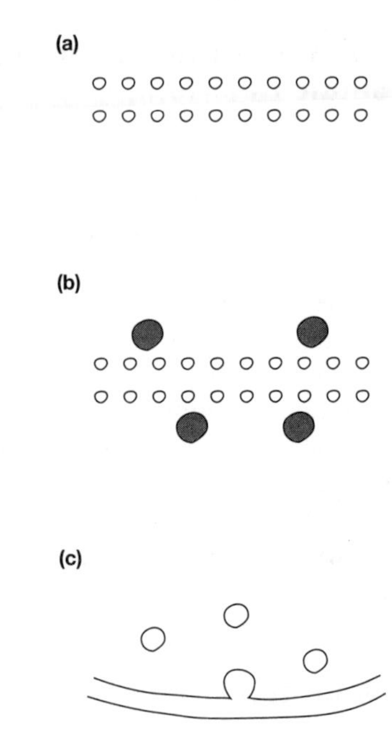

Figure 7.2. Active zones and pits from freeze-fracture studies (a) before and (b) after stimulation. (c) A cross-section through a synapse showing omega profiles.

of calcium entry, the diffusion of calcium within the terminal and the spatial distribution of the channels. Using this type of model (*Figure 7.3*), it has been estimated that the calcium concentration can increase, in the area immediately adjacent to the calcium channels, from its resting level of about 0.2 μM to steady-state of about 400 μM, within a few hundred microseconds, returning to submicromolar levels with similar speed.

Direct visualization of calcium influx during stimulation has been shown using a number of preparations where the synapses are large, such as the squid giant synapse, using a technique called **fluorescence imaging** (***Box 7.1***, p. 147). This method has been used to monitor many intracellular events, but has been most widely developed to measure changes in intracellular calcium. It is based on the use of fluorescent compounds which have different excitation and emission properties depending on whether the compound has calcium bound to it or not.

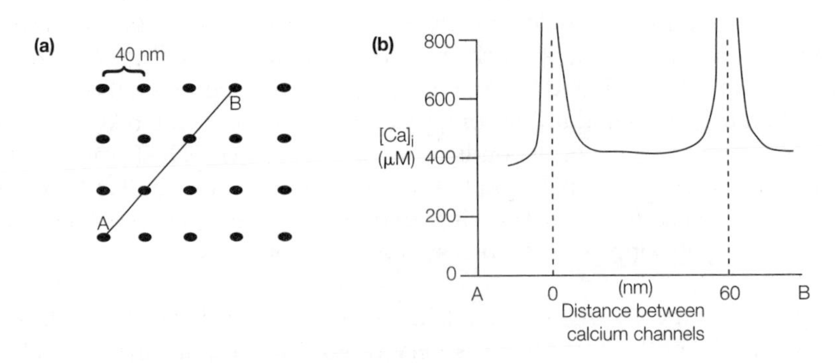

Figure 7.3. (a) Theoretical square array of calcium channels with a spacing of 40 nm. (b) Calculated intracellular calcium concentration along the line AB 1 msec after channel opening.

In squid giant synapses which have been injected with **aequorin**, a protein from the jellyfish, *Aequora forkalea*, which is chemiluminescent on binding calcium, there were specifically localized increases in intracellular calcium during neurotransmitter release. These areas, seen by the presence of brief flashes of light, were called microdomains, as they were very small (0.5 μm diameter) and well localized. On depolarization, the calcium concentration in these areas reaches 200–300 μM, with a peak at about 0.3 msec after stimulation and an overall duration of about 0.8 msec.

Direct measurement of the calcium dependence of neurotransmitter release in nerve terminals has been carried out on goldfish retinal bipolar neurons. Intracellular calcium levels were monitored using the calcium indicator, furaptra (Magfura-2). This particular calcium dye has a lower affinity for calcium than the more commonly used **fura-2** and can thus be used to follow large calcium transients without becoming saturated. In these experiments, the exocytosis of neurotransmitter was measured using patch clamping to measure membrane capacitance. The levels of intracellular calcium were raised in a controlled manner using a light-sensitive calcium chelator, DM-nitrophen, an example of a 'caged' molecule.

Caged molecules are based on chelators which bind the desired molecule, in this case calcium. These chelators change their conformation in response to UV light, supplied by an arc lamp, and on doing so release the bound molecule. As the amount of calcium released is proportional to the amount of UV light, by changing the flash intensity of the arc light the amount of calcium released can be varied. By measuring the changes in capacitance with varying flash intensities and with simultaneous measurement of intracellular calcium, it is possible to measure the direct dependence of exocytosis on calcium concentration, enabling an estimate to be made of the number of calcium ions required and the affinity of the calcium target.

The results were consistent with a model in which there is cooperative binding of four calcium molecules to an intracellular target, which is followed by secretion. From this model, the concentration of calcium at half-maximal was 194 μM, which is a relatively low affinity, and the maximal rate of secretion was high. These two properties allow exocytosis to be turned on and off rapidly if the calcium levels rise and fall rapidly as has been shown experimentally.

As the rise in intracellular calcium is very localized, the fall in calcium can occur largely by diffusion and by binding to calcium-binding proteins in the terminal, thus allowing calcium to dissociate from the target. Thus exocytosis can be terminated without the requirement for large amounts of calcium to be pumped out of the terminal. The speed of neurotransmitter release and the location of the calcium signal mean that the intracellular target for calcium must be closely associated with the vesicles at the presynaptic membrane. The identity of this target will be discussed further in Section 7.7.

7.5 Which calcium channels are responsible for neurotransmitter release?

A number of techniques have been used to identify the type of calcium channel involved in the regulated release of neurotransmitter. These have included the measurement of pre- and postsynaptic potentials and their potential reduction in the presence of different calcium channel blockers. There is a particular problem with this approach, in that the identification of the different classes of calcium channel has involved the examination of calcium current blockade in cell bodies since synapses are generally too small to record from them directly. There is, therefore, the assumption that calcium channels at synapses have the same properties as those on the cell body, which from evidence of calcium channel distributions may not be the case. This is compounded in the CNS by difficulties in accessing the synapse and differentiating between actual neurotransmitter release and the clearance of neurotransmitter by diffusion and transport. Electrical measurements from the postsynaptic cell body may also be affected by the problem of assessing the number of activated inputs and the geometry of the dendritic tree.

In another technique, tissue slices or synaptosomes, pre-loaded with radiolabeled neurotransmitters, are stimulated both with and without calcium channel blockers. Other potential problems, with both this and the previous approach, are the degree of blockade by a particular antagonist, as blockers may vary in their affinity for different channels and thus act at different concentrations on different channels, and the specificity of the blocker for the particular channel types, which may vary between tissues and species.

Immunofluorescence methods have also been used to identify calcium channels and their distributions. Molecular cloning and sequencing of the different calcium channel subtypes (see Section 4.5) and subsequent identification of regions which are specific for each subtype have allowed the development of antibodies which are specific for each subtype and, in some cases, even specific for a particular subunit.

7.5.1 P-type calcium channels are responsible for neurotransmitter release at the mammalian neuromuscular junction

At the frog neuromuscular junction, transmission can be blocked by ω-CTX GVIA, an N-type channel blocker (see *Table 4.1*) but this is not effective at blocking neuromuscular transmission in mammals, neither are dihydropyridines (DHPs), which block L-type channels. But, FTX, a blocker of P-type channels, abolished nerve-stimulated muscle contraction. The P-channel blocker, FTX, has been analyzed and has been identified as a polyamine. This molecule, which occurs naturally in the venom of the funnel web spider, *Agelenopsis aperta*, is difficult to synthesize, so a similar but more easily made molecule was developed, sFTX, which has a similar blocking activity. This synthetic toxin was then used in affinity gels to isolate and purify a protein which, when reconstituted in lipid bilayers, displays P-channel-like activity.

7.5.2 Central neurotransmitter release depends on a number of different calcium channel types

A number of studies have shown that whilst L-type calcium channels are involved in the regulated secretion of hormones from a variety of neuroendocrine cells, they are not involved in neurotransmitter release from SSVs in the CNS as this is DHP insensitive.

Secretion of a variety of neurotransmitters including glutamate, ACh, dopamine (DA) and NA from a range of brain regions is reduced in the presence of the N-type channel blocker, ω-CTX GVIA. However, this blockade is only partial, with a maximum inhibition of 30% of the total calcium-dependent release. The effect of the toxin is confirmed by a reduction in the synaptic transmission as measured by postsynaptic responses. The fact that the same toxin can produce a 100% blockade of the N-type calcium current shows that other calcium channels must also be involved in neurotransmitter release.

Synapses which release different neurotransmitters may contain different populations of channels. The calcium-dependent release of glutamate from rat striatal synaptosomes could be reduced by the P-channel blocker, ω-Aga IVA, whereas DA release was reduced by both ω-Aga IVA and the N-channel blocker ω-CTX GVIA. The combination of both ω-Aga IVA and ω-CTX GVIA reduces calcium-dependent DA release still further in a synergistic manner, but DA release was not blocked entirely, suggesting that channels other than N and P may also be involved in DA release.

The component of neurotransmitter release not assigned to N- and P-type channels can be blocked by the nonspecific toxins, ω-CTX MVIIC and ω-GTX SIA (from the spider *Grammastola spatulata*), which may indicate that Q channels may also be involved or another unidentified channel type, possibly R.

Immunofluorescence studies using subtype-specific antibodies have so far shown that L-type channels are found mainly on cell bodies and on proximal dendrites. The identification of L channels at the base of dendritic spines of hippocampal pyramidal cells has suggested a role in the development of long-term potentiation (see Chapter 8). In contrast, N-type channels are located in dendritic shafts and in punctate synaptic structures which were thought to be nerve terminals. The distribution of P-type channels has been investigated using a polyclonal antibody raised against the FTX-binding protein. The antibody labeling showed that P-type channels are found in discrete regions of the CNS, with the strongest labeling being found in the Purkinje cell dendrites and axon terminals.

The accumulated evidence thus suggests that, in the CNS, neurotransmitter release is controlled by more than one type of calcium channel and that at least P- and N-type and, possibly, Q-type channels all contribute to the calcium influx required for regulated neurotransmitter release.

7.6 A large number of proteins have been identified in nerve terminals

The isolation and characterization of proteins found in the nerve terminal, especially those associated with SSVs, has led to a bewildering array of proteins thought to be involved in the different stages of neurotransmitter release. *Table 7.1* shows a small selection of proteins implicated in neurotransmitter release. Many of them have alternative names, sometimes referring to a homologous protein from a different species, which are given in parentheses.

Table 7.1. Some of the proteins implicated in neurotransmitter release

Protein	M_r (kDa)
Vesicular or vesicle-associated proteins	
Synapsin	70–74
Synaptobrevin (VAMP)	13
Synaptophysin (p38)	34
Synaptotagmin (p65)	48
Rabphilin	78
Dynamin	94–99
Plasma membrane proteins	
Syntaxin (HPC1)	35
Munc-18 (n-sec1)	67
SNAP-25	25
Physophilin	36
Neurexins	160–200
Cytoplasmic proteins	
NSF	76
SNAPs (α/β/γ)	35–39
Rab3A	25

7.7 Neurotransmitter release involves a number of calcium-dependent steps

Initially, synaptic vesicles are normally tethered to cytoskeletal proteins some distance from the active zones. Vesicle **recruitment** is a calcium-dependent step which frees the vesicles, which can then move to the active zones on the presynaptic membrane. Various lines of evidence suggest that after mobilization of vesicles there are three distinct phases before neurotransmitter can be released from a given vesicle. Once the vesicle is released from the cytoskeleton, it must be closely bound to the presynaptic membrane, a process called **docking**.

This is followed by an event called **priming**. Experiments which have measured the time course of the different stages of neurotransmitter release indicate that after docking there is a relatively slow, ATP-dependent priming event. After which, given the appropriate calcium stimulus, there is a rapid **fusion** of the primed vesicle and **exocytosis** of the neurotransmitter. The different steps have different calcium dependencies, and it is only the final fusion event which requires very high local calcium concentrations (*Figure 7.4*).

7.8 Vesicle recruitment may involve synapsin

In the nerve terminal, the SSVs are found close to the presynaptic cleft and some are aligned at the active zones. These two pools of vesicles correspond to the reserve and releasable pools. Those vesicles at the active zones participate in cycles of exocytosis

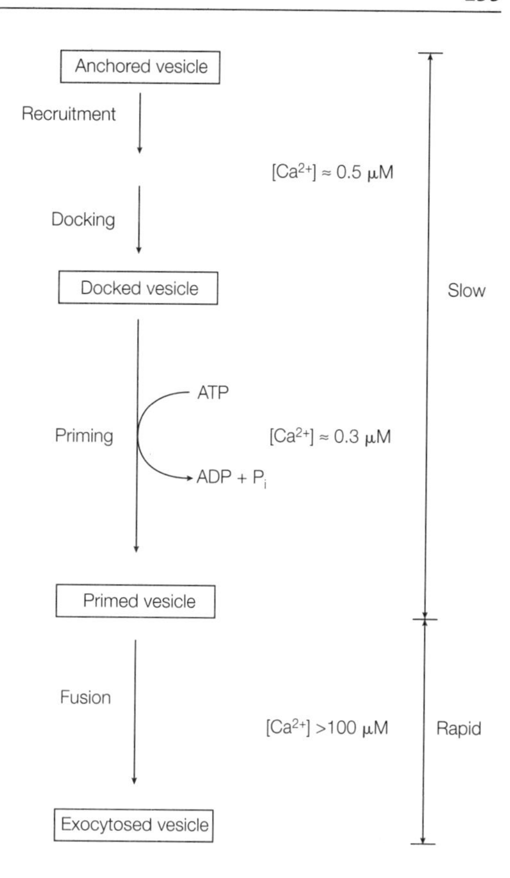

Figure 7.4. A model for the key steps required for exocytosis in neurons.

and endocytosis at low frequencies of nerve stimulation, whereas those vesicles found further away can be mobilized during repetitive stimulation. It has been shown that the vesicles are attached to the cytoskeleton, which consists of actin and fodrin (a spectrin-like molecule found particularly in nerve terminals).

Vesicles are attached to fodrin and actin via proteins called synapsins. These proteins account for about 9% of the total vesicle protein. Synapsins are a family of four homologous proteins named Ia, Ib, IIa and IIb which are derived from two genes by alternative splicing. They exist as dimers Ia–b and IIa–b which have identical N-terminal ends, with potential phosphorylation sites in the head region for both calmodulin-dependent protein kinase I (CaMKI) and cAMP-dependent protein kinase. Synapsin I but not synapsin II also has other phosphorylation sites in the tail region at the C terminus for calmodulin-dependent protein kinase II (CaMKII). Synapsin I is bound to vesicles by interactions with the phospholipids and vesicle-associated CaMKII (*Figure 7.5*). *In vitro*, synapsin I also binds to cytoskeletal proteins, such as actin and fodrin, via two binding sites in the head region, and thus forms a link between the cytoskeleton and the vesicle. When synapsin I is phosphorylated by CaMKII, its binding to both actin and CaMKII is reduced, which could allow the vesicles to move to the active zone. Thus, the activation of CaMKII following a rise in intracellular calcium could be one of the links between calcium and neurotransmitter release. However, the level of calcium required to produce neurotransmitter release is much higher than that required to activate CaMKII, which suggests that whilst intracellular calcium can

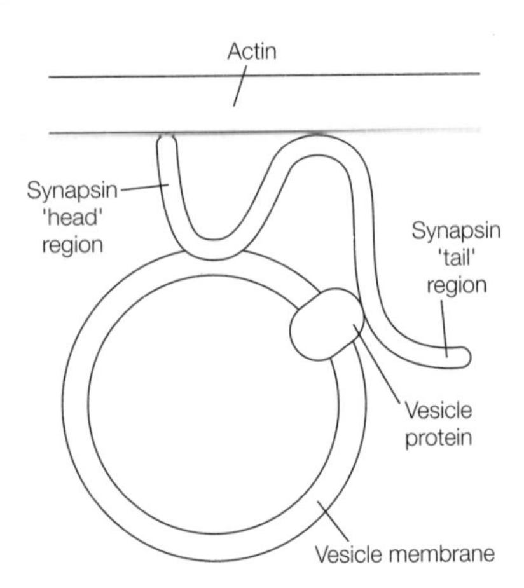

Figure 7.5. Attachment of SSVs to actin filaments via synapsin.

influence the movement of vesicles between the cytoskeleton and the release sites, other calcium-dependent processes are needed for exocytosis.

Some experiments have argued against the involvement of synapsins in SSV release but suggest that synapsin I has an essential role in presynaptic terminal development and longer term synaptic vesicle recruitment which could be regulated by both calcium and cAMP. In neuroblastoma cells in culture, clones which overexpressed rat synapsin IIb showed an increase in the number of varicosities, and these varicosities contained a greatly increased number of synaptic vesicles compared with normal cells.

7.9 SNAPs and SNAREs are involved in docking and then priming vesicles for release

Vesicles which are recruited from the reserve pool must be docked at the appropriate sites on the plasma membrane at the active zones, and the targets for the docking of vesicles involves a group of proteins called **SNAREs**. After the docking of the vesicle at the presynaptic membrane, there is a second calcium-dependent step called priming. During priming, a number of cytoplasmic proteins called NSF (*N*-ethylmaleimide-sensitive factor) and **SNAPs** (soluble NSF attachment proteins) bind to the SNAREs to form a 20S complex which seems to be central to the priming of vesicles prior to fusion and release of neurotransmitter. NSF is a trimeric ATPase of 76 kDa which has two ATP binding and hydrolysis sites. It binds to SNAREs via SNAPs and is essential for membrane fusion *in vitro*. SNAPs are cytoplasmic proteins which bind to both NSF and to SNAREs. Three different SNAPs have been isolated from brain called α (35 kDa), β (36 kDa) and γ (39 kDa). NSF attachment to SNAPs is dependent on the pre-attachment of SNAPs to distinct sites in membranes, and for a time their targets were unknown.

The identification of SNAREs as the target used a technique which employed NSF and SNAPs as ligands for the purification of SNAREs (*Figure 7.6*). This purification technique also used a property of NSF which enabled the formation of stable 20S complexes. When NSF is bound to SNAPs and SNAREs, the 20S complex is stable only

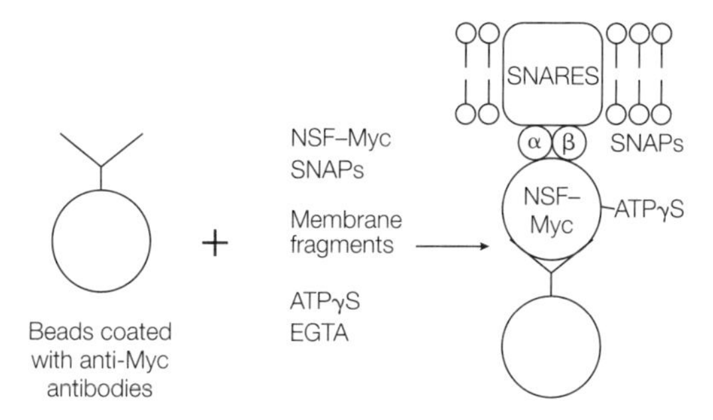

Figure 7.6. Purification of SNAREs using NSF and SNAPs.

in the absence of Mg-ATP. In the presence of Mg-ATP, the complex rapidly dissociates with ATP hydrolysis, but if the complex is formed with either magnesium-free ATP or a nonhydrolyzable analog of ATP, Mg-ATPγS, then the 20S complex is stable.

The method used a modified, but active, form of NSF which had a peptide, known as Myc, attached to the C terminus. A column was packed with beads coated with anti-Myc IgG antibodies and added to the top of the column were NSF-Myc, SNAPs and detergent-extracted membranes containing the SNAREs. ATPγS was also present to ensure ATP-dependent, but nonhydrolyzable binding, and the presence of EGTA ensured that any contaminating magnesium was chelated. After washing with Mg-ATPγS to remove any nonspecific binding, the column was eluted with Mg-ATP, which would allow the dissociation of the 20S complex and the recovery of the NSF, SNAPs and the specifically captured SNAREs. The proteins in the eluate were then separated by gel electrophoresis. Specific bands were excised, digested with trypsin and microsequenced. The sequences were then compared with the sequences of previously identified synaptic proteins.

The SNAREs were thus identified as the vesicle protein, synaptobrevin, and two plasma membrane proteins, syntaxin and, the confusingly named, SNAP-25. At rest, synaptobrevin is bound mainly to another vesicle protein, synaptophysin, whilst syntaxin is bound to munc-18 on the plasma membrane (*Figure 7.7a*). The binding of munc-18 to syntaxin prevents the interaction between syntaxin and SNAP-25 and, whilst synaptobrevin is bound to synaptophysin, it cannot bind to the syntaxin–SNAP-25 complex. In this way, the formation of a 20S complex is inhibited until the appropriate stimulus occurs. After vesicle docking (*Figure 7.7b*), syntaxin and synaptobrevin dissociate from munc-18 and synaptophysin, although the trigger for this is unknown. Syntaxin and SNAP-25 can then associate with synaptobrevin and formation of the 20S complex can proceed. Evidence suggests that the role of munc-18 is to ensure that the 20S complex is only assembled at the active zones.

The stable combination of SNAREs and SNAPs in the 20S core complex and the subsequent NSF-stimulated hydrolysis of ATP disrupts the core complex. This disruption is thought to leave it in a different conformation, which leads to the fusion of one of the two leaves of the bilayer, called **hemifusion**. This hemifusion has been shown to be

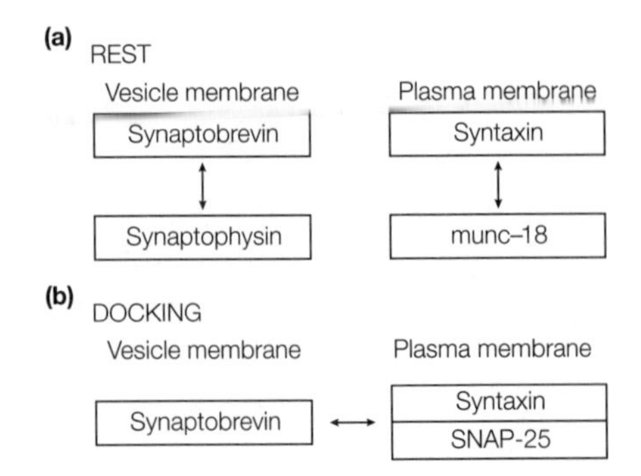

Figure 7.7. Binding of vesicle and plasma membrane proteins (a) at rest and (b) during docking.

stable and would allow subsequent fusion to occur very rapidly as half of the fusion reaction is already complete. These events have all been studied in yeast, where they underlie constitutive calcium-independent secretion. In order to produce the regulated secretion found in neurons and neuroendocrine cells, other processes must inhibit the final fusion event until it is triggered by calcium influx, and it has been suggested that total fusion and subsequent exocytosis is prevented by some kind of calcium-sensitive brake.

The importance of the 20S complex is shown by the effects of a number of neurotoxins which disrupt constitutive exocytosis in all cells and regulated neurotransmitter release in neurons. All the clostridial neurotoxins, **tetanus** and **botulinum** A–F, target the SNAREs of the 20S complex (*Table 7.2*). These toxins have a common action, and all act as peptidases, cleaving either cellubrevin, a protein homologous to synaptobrevin which is found in all cells and is involved in constitutive exocytosis, or the neuronal SNAREs, synaptobrevin, SNAP-25 and syntaxin.

Table 7.2. Some neurotoxins and their targets

Neurotoxin	Target
Tetanus	Synaptobrevin Cellubrevin
Botulinum A	SNAP-25
Botulinum B	Synaptobrevin
Botulinum C	Syntaxin
Botulinum D	Synaptobrevin Cellubrevin
Botulinum E	SNAP-25
Botulinum F	Synaptobrevin Cellubrevin
Botulinum G	Unknown

7.10 Synaptophysin and physophilin may form the fusion pore

The existence of a transmembrane fusion pore which later expands to allow the release of the vesicle contents is suggested by electrophysiological experiments where an initial event has been observed which corresponds to the formation of a type of ion channel with a conductance of the same order of magnitude as a single gap junction. In the gap junction, the ion channel is formed by the conjunction of transmembrane proteins called connexons from both of the opposing membranes. A candidate molecule for the fusion pore which is present in the vesicle membrane is synaptophysin. This 34-kDa molecule has four TM segments and forms hexamers in the synaptic vesicle membrane which resemble the appearance of gap junctions. When reconstituted in planar lipid bilayers, single-channel recordings showed voltage-sensitive ion channel activity with clearly defined channel openings.

Binding studies which have shown interactions between synaptophysin and a plasma membrane protein, physophilin, led to suggestions that these two molecules could form the fusion pore, with oligomers of the two molecules forming the two parts of the pore. However, purification of the physophilin has shown that it may be a form of H^+-ATPase, and although this does not explain its binding to synaptophysin it may not after all be involved in neurotransmitter release.

7.11 Synaptotagmin may be the calcium sensor involved in the final exocytotic event

One of the proteins which has elicited a great amount of interest is synaptotagmin. A comparison of the properties of the hypothetical calcium target and synaptotagmin shows many similarities. The protein is inserted into the synaptic vesicle membrane by a single TM segment with a short intravesicular N-terminal domain. The large cytoplasmic domain has two repeats of a sequence which is similar to the calcium-binding domain (C2) of protein kinase C (PKC) (Chapter 6). Like PKC, synaptotagmin binds calcium in a phospholipid-dependent manner. The calcium affinity is approximately 10–100 μM and there are multiple calcium-binding sites, both criteria for the calcium sensor.

An experiment showing the importance of synaptotagmin in neurotransmitter release involved the injection of peptides based on the C2 domain into the squid giant synapse, which inhibit the binding of synaptotagmin to its target. The neurotransmitter release was completely inhibited, and observation of the synapses after treatment showed an accumulation of vesicles at the active zones, demonstrating that inhibition had occurred downstream of the mobilization and docking of vesicles.

A technique which involved more direct manipulation of synaptotagmin in a whole animal was the production of mice known as synaptotagmin **knockouts** with a mutant gene for synaptotagmin. These **transgenic** animals were produced by **homologous recombination** in embryonic stem (ES) cells using the positive/negative selection process (described in ***Box 7.2***, p. 148) to select for clones containing the neomycin resistance (*neo*^{r}) gene in place of part of the synaptotagmin I gene. Synaptotagmin I expression in wild-type (+/+) and mutant mice (+/– and –/–) was tested by Western blotting, showing that the mutant gene disrupted the production of synaptotagmin.

Although (–/–) mutants were indistinguishable from wild-type (+/+) and heterozygous (+/–) litter mates at birth in that they were able to breathe and respond to tactile stimulation, they became noticeably weaker and usually died within 48 h postnatally. When examined, they had no gross abnormalities, and this was confirmed at the level of brain and retinal morphology. Neurons from the hippocampus of mouse embryos were cultured and appeared indistinguishable by light microscopy. However, electrophysiological studies of evoked synaptic responses between pairs of neurons using whole-cell patch clamping showed large differences between the mutant (–/–) and the other (+/+ and +/–) cultures.

Evoked responses were greatly reduced in mutant cultures, and a range of tests of postsynaptic responsiveness indicated that this was due to a reduction in presynaptic neurotransmitter release. The calcium binding of the mutant (–/–) was unchanged. Both calcium-independent release and the frequency of mepps was unchanged. Neurotransmitter release at central synapses has been shown to have two kinetically distinct components, a fast component which accounts for more than 80% of release and which is synchronous with the stimulus, and a slower component which occurs slightly later. The fast component is thought to be controlled by a low-affinity calcium sensor which could be activated at the peak of the calcium influx, whilst the slower component has a higher calcium affinity and could be active at lower calcium concentrations during the post-stimulus decay phase. In the synaptotagmin mutants, the slow component was unchanged, but the fast component was largely absent. The high concentration of calcium found at active zones after the opening of calcium channels could activate synaptotagmin, which has a low calcium affinity, and allow fast synchronous release.

Synaptotagmin has been shown to bind the plasma membrane proteins, syntaxin, neurexin and N-type calcium channels, thus locating it clearly in the neurotransmitter release zone. This localization of synaptotagmin at the active zones, combined with its calcium sensitivity and its involvement in neurotransmitter release, has led to a theory which suggests that synaptotagmin acts as a calcium-sensitive brake, inhibiting total fusion until it binds calcium.

7.12 Rab3A is involved in recycling of neurotransmitter vesicles

Another molecule implicated in the trafficking of synaptic vesicles with the appropriate site at the active zones is the small G protein, Rab3A, and rabphilin, its associated binding protein, found on vesicles. Rab3A is one of a number of small G proteins found in the cytoplasm, and a suggested model involves only vesicles with GTP bound to Rab3A associating with the fusion site (*Figure 7.8*). After exocytosis and subsequent endocytosis of the vesicle, hydrolysis of GTP to GDP would allow dissociation of Rab3A and recycling of the vesicle. The level of GTP-associated Rab3A could depend on another protein, as Rab3A in the cytosol is held mainly in the GDP-bound state by a protein called GDI (GDP dissociation inhibitor). The discovery that Rab3A knockout mice have normal neurotransmitter release, except in cases of repetitive stimulation when there is impaired enhancement of neurotransmitter release, showed that Rab3A is required for maintaining a normal reserve of synaptic vesicles. The role of the Rab3A-binding protein, rabphilin, is as yet undetermined, but it binds Rab3A in a GTP-dependent manner, can be phosphorylated and has two C2 domains which can bind phospholipids and calcium.

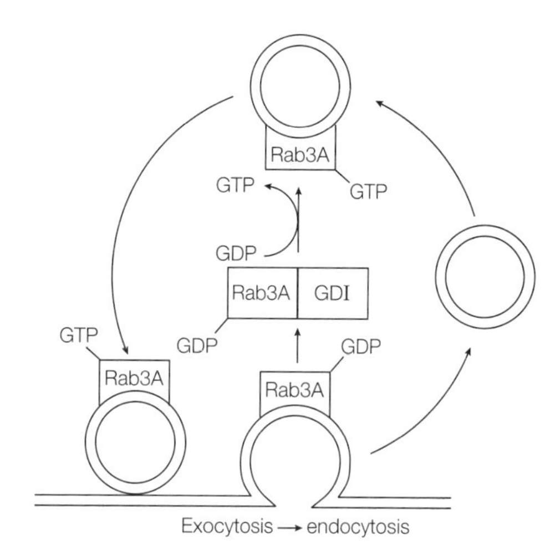

Figure 7.8. Interactions between vesicles, GTP/GDP and Rab3A may control vesicle recycling. The exact timing of the GTP hydrolysis is undetermined, but it occurs before vesicle endocytosis.

7.13 Neurexins are a target for neurotoxins

A family of brain-specific proteins called neurexins has also been implicated in neurotransmitter release because the first member of the family identified, neurexin 1α, binds α-latrotoxin with high affinity in a calcium-dependent manner. This toxin from the black widow spider, *Latrodectus mactans*, causes massive exocytosis, which leads to depletion of neurotransmitter and subsequent irreversible block of nervous transmission. Evidence for the involvement of neurexin in neurotransmitter release is the binding of the α-latrotoxin receptor to synaptotagmin and its modulation of synaptotagmin phosphorylation. However, toxin-induced release is calcium independent, which poses the question as to whether neurexin is the only target of the toxin or if another as yet unidentified molecule is responsible.

Neurexins may have a role as cell recognition molecules. Examination of the genes coding for neurexins has shown that there are three separate genes (I, II and III) each with two promoters, α which gives a long form of neurexin and β which gives a much shorter form, giving six different mRNAs in total. These genes are expressed differentially in different cells. For example, in the hippocampus, pyramidal cells of CA3 coexpress all six isoforms, whereas pyramidal cells in CA1 lack neurexins Iβ and IIIα. However, the largest source of diversity in neurexins is provided by the alternative splicing of these six mRNA transcripts. By probing a cDNA library with a probe derived from the whole of the coding region of neurexin Iα, out of 102 positive clones, 87 of them encoded 47 different splice variants of neurexin Iα, as well as 15 clones of neurexin Iβ and other nonidentifiable clones. Analysis of the splice variants showed the presence of five different splice sites in α-neurexin (with two present in β-neurexin), and that while some inserts were preferred, the different splice sites are used independently and occur in all combinations. Furthermore, the alternative splicing seems to be regulated, in that neurexins with different splicing patterns are found in different brain regions. Estimates of the number of potential neurexins varies from about 600 to almost 3000, depending on the degree of variation allowed in the insert sequence. This means that the only other protein families with such a large diversity are the antigen recognition molecules of the immune system, which are generated by somatic gene rearrangements, and the odorant receptors, which are produced from large gene families. If the assumption is made that this structural diversity is reflected in functional diversity, then there is clearly much more to learn about neurexins.

7.14 Recycling of synaptic vesicles may involve dynamin

At the neuromuscular junction, the docked vesicles represent about 1% of the reserve pool, and each stimulus releases about 10% of these. Thus, at each stimulus, the released vesicles represent 0.1% of the total number of vesicles. If the stimulation rate were 10 per second, which is not unreasonable, then nerve terminals should become depleted in about 100 sec. However, this is not seen under normal conditions as vesicles are regenerated by endocytosis.

Various hypotheses account for exactly how the vesicles are recycled. SSV endocytosis is very tightly coupled to exocytosis, and vesicle proteins such as synaptotagmin are sorted into SSV and away from endosomes, which implies that membrane retrieval is specific for the vesicle membrane. As nerve terminals do not have the necessary mechanisms for protein synthesis, the vesicle proteins must be retrieved from the plasma membrane in order for vesicles to be regenerated quickly without requiring protein transport from the cell body. Experiments with antibodies directed against the intravesicular N-terminal portion of synaptotagmin, which is exposed to the extracellular fluid after vesicle exocytosis, show that a substantial pool of SSVs fuses completely with the plasma membrane.

Synaptic vesicles are coated with neuron-specific isoforms of clathrin to form coated vesicles. Clathrin forms a complex with other cytoplasmic proteins called the assembly protein complex, AP2. Synaptotagmin has also been implicated in the endocytosis, as synaptotagmin-deficient mutants showed a large reduction in synaptic vesicles consistent with a defect in vesicle retrieval. Clathrin-coated vesicles subsequently are endocytosed and the clathrin removed.

Vesicle endocytosis is also calcium dependent, although the point of action is not as yet identified unequivocally. A neuron-specific protein which has been implicated in this process is dynamin. Dynamin is a GTP-binding protein with GTPase activity which can be phosphorylated by PKC with a very high affinity ($K_D = 0.35\ \mu M$). During exocytosis, dynamin is dephosphorylated in a calcium-dependent manner by the phosphatase, calcineurin. The intrinsic GTPase activity is reduced by dephosphorylation and increased by phosphorylation, which means that during calcium influx dynamin would be dephosphorylated and thus would tend to be in its GTP-bound form (*Figure 7.9a*). Dynamin can also be phosphorylated at a different site by casein kinase II, and although this has no effect on GTPase activity it prevents phosphorylation by PKC, giving a potential interaction between distinct signaling pathways (see Chapter 6). Although the exact role of dynamin in endocytosis has not been identified, receptor-mediated endocytosis is blocked in dynamin mutants that are deficient in GTPase activity, and it has been suggested that dynamin with GTP bound can activate endocytosis by an interaction with a target protein. Experiments with the nonhydrolyzable GTP analog, GTPγS, suggested a model in which dynamin-GTP interacts with the neck of clathrin-coated invaginations forming a collar around the vesicle. Hydrolysis of GTP

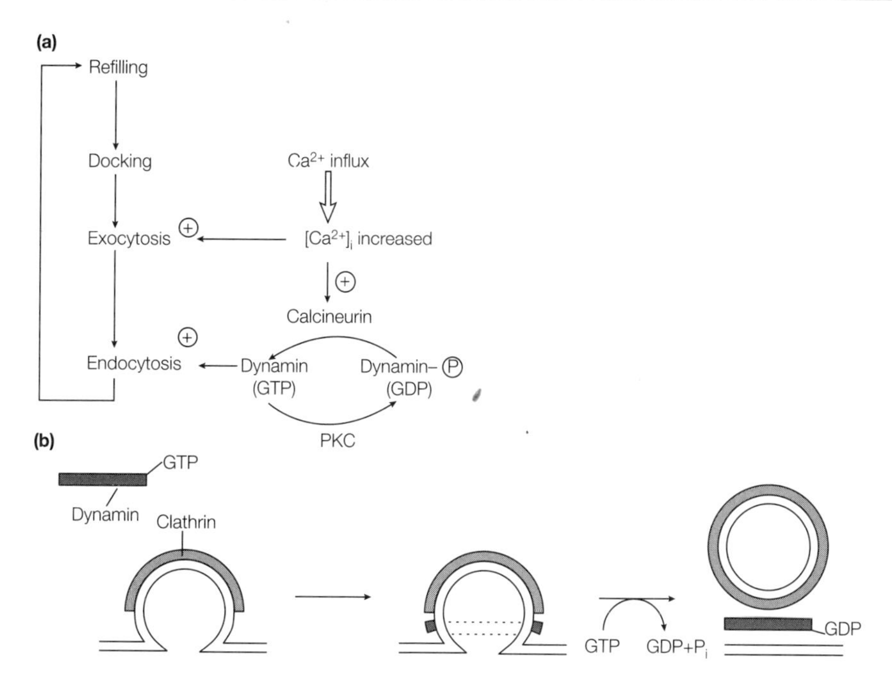

Figure 7.9. (a) Regulation of dynamin phosphorylation by calcineurin and PKC controls levels of 'active' GTP-bound dynamin. ⊕, stimulates. Note that phosphorylation of dynamin increases its intrinsic GTPase activity. In the dephosphorylated state, therefore, dynamin tends to be in its GTP-bound form. (b) Dynamin hydrolysis leads to fission of the endocytosed vesicle.

would produce a conformational change in the dynamin, leading to vesicle fission (*Figure 7.9b*)

7.15 Refilling of vesicles with neurotransmitter occurs via an antiport mechanism

SSVs are refilled with neurotransmitter in the terminals, unlike peptide-containing vesicles which are loaded whilst in the cell body and then transported to the terminals. The insides of neurotransmitter vesicles have a high concentration of protons produced by the activity of a H^+-ATPase. The transport of neurotransmitter into vesicles is then driven by the **antiport** of H^+ out of the vesicle (*Figure 7.10a*).

The vesicular transporter for ACh was discovered by a roundabout route involving a variety of species. A number of mutants of the nematode *Caenorhabditis elegans* show uncoordinated movements. The involvement of cholinergic transmission was suggested by the fact that these mutants were resistant to inhibitors of acetylcholinesterase despite synthesizing ACh. Investigations of the mutated genes showed that one of these, *unc-17*, coded for a protein with a high homology (40%) for the previously identified monoamine transporters, suggesting that the *unc-17* product was a transporter, possibly for ACh. Immunostaining showed that *unc-17*-encoded protein was found in some but not all neurons, and was colocalized with synaptotagmin, showing that the *unc-17* product is a vesicle protein. As *Torpedo* electroplax was known to contain large numbers

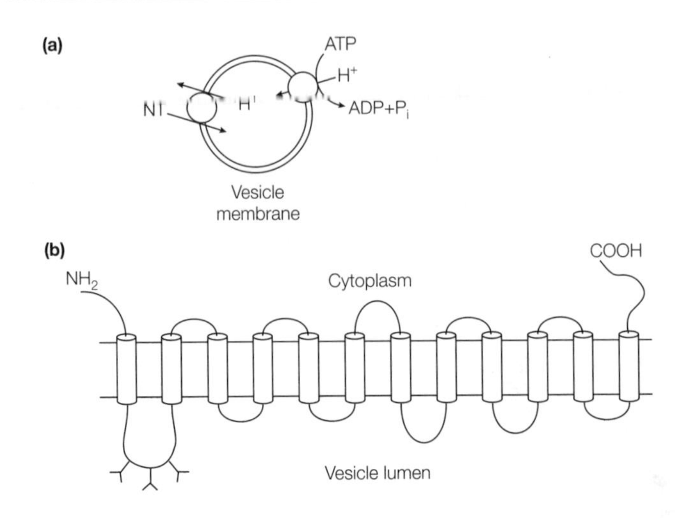

Figure 7.10. (a) Diagram of H^+/neurotransmitter (NT) antiport and H^+-ATPase in synaptic vesicles. (b) Proposed structure of vesicular neurotransmitter transporter.

of cholinergic nerve terminals, a cDNA library was constructed from the cell bodies that innervate the electric organ. Probes derived from the sequence of *unc-17* were used to isolate and clone the equivalent *Torpedo* gene. This gene, in turn, was used to probe rat and human cDNA libraries, leading to the isolation of the mammalian versions. The final evidence that this was the vesicular acetylcholine transporter (VAChT) was shown by expressing the cloned gene in a cell line which then transported ACh and which could be inhibited by known inhibitors of ACh transport such as vesamicol.

From analysis of the homologous sequences of the known VAChTs, the suggested structure gives 12 TM segments with both C and N termini in the cytoplasm (*Figure 7.10b*). This structure is also proposed for the vesicular monoamine transporters (VMATs) with about 40% homology between the rat VAChT and VMAT1 and 2. Comparisons with other 12TM transport proteins shows no such homologies. This leads to the proposal that the vesicle transporters form a distinct family.

In the step before ACh is transported into the vesicles, the enzyme choline acetyltransferase (ChAT) synthesizes ACh from acetyl-CoA and choline. Interestingly, further investigation of the *unc-17* gene showed it to be very closely associated with the gene for ChAT, *cha-1*. Since then, it has been shown that not only in *C. elegans* but also in rats and humans, both VAChT and ChAT are produced by the same gene. The two separate proteins are produced by alternative splicing, possibly with multiple transcription initiation sites, at least in humans.

7.16 Termination of neurotransmitter action is due to diffusion and binding

After its release into the synaptic cleft, the action of the neurotransmitter is terminated by two major mechanisms, diffusion and uptake (or, in the case of ACh, hydrolysis by acetylcholinesterase). Firstly, simple diffusion away from the release site will rapidly reduce the concentration of neurotransmitter. This diffusion will be affected by the geometry of the cleft and adjacent spaces. The observed variation in synaptic morphology could produce large differences in the importance of diffusion. Secondly, released neurotransmitter will bind not only to postsynaptic receptors but, depending on their affinity, to transporters and, in the case of ACh, to cholinesterases. The concentration of neurotransmitter is then reduced by either transport out of the synaptic cleft or, in the case of ACh, by cleavage into inactive compounds.

Recently developed computer models which incorporate both cleft geometry and neurotransmitter binding have been used to estimate the clearance rates of neurotransmitter and the contribution of the different components. Diffusion is thought to be the rate-limiting step immediately after neurotransmitter release, but if the density of transporters is high the clearance will be accelerated. Uptake, especially into glial cells, will restrict diffusion of neurotransmitter to adjacent terminals and prevent non-specific interactions.

At high levels of neurotransmitter release, uptake may be saturated and there may be 'spill-over' and activation of nearby receptors, especially high-affinity presynaptic receptors. This could well occur in cases of multivesicular release. It has been calculated that at central synapses the number of postsynaptic receptors is low and that a single vesicle is sufficient to saturate them. In this case, the observed effects of multivesicular release will be entirely due to spill-over of neurotransmitter to adjacent synapses.

The uptake of neurotransmitter from the synaptic cleft is carried out by high-affinity sodium-dependent transporters. These form part of two gene families, one a family of Na^+- and K^+-dependent glutamate transporters and the other a family of Na^+- and Cl^--dependent neurotransmitter transporters which have been shown to transport a wide range of neurotransmitters, including the biogenic amines as well as GABA, glycine and the breakdown product of ACh, choline.

The structure of the glutamate transporters is still speculative but has been proposed as an 8TM segment structure with intracellular C and N termini and a large extracellular loop between segments 3 and 4. The other neurotransmitter transporters are all 12TM segments, again with intracellular C and N termini and a large extracellular loop between segments 3 and 4. As mentioned previously, despite superficial similarities, there seems to be no significant homology with the vesicle 12TM neurotransmitter transporters. The extracellular loop of these transporters is thought to be implicated in the binding of the appropriate neurotransmitter and also in interactions with a number of important drugs and toxins. The reuptake of DA is blocked by cocaine ($K_i = 1\ \mu M$) and the DA transporter is thought to be the main brain 'cocaine receptor' and, because of this, is of interest in the problem of cocaine addiction. The tricyclic antidepressants bind to 5-HT and NA transporters, and of recent interest is the discovery of fluoxetine (Prozac) which is a selective blocker of 5-HT uptake.

Susceptibility of neurons to the neurotoxin **MPTP** (1-methyl-4-phenyl-1,2,3,4-tetrahydropyridine), which has proved to be a useful model of Parkinson's disease, is affected by both the uptake and vesicular storage of MPP^+ (1-methyl-4-phenylpyridinium), the toxic form of MPTP (see Section 9.4.2). After oxidization from MPTP to $MPDP^+$ and subsequent rearrangement to MPP^+ in astrocytes, the MPP^+ is taken into neurons by the DA uptake system where it can damage mitochondria. The differential susceptibility of particular neurons to the effects of MPP^+ may be due to two factors. Firstly, the ability of some cells to sequester MPP^+ in vesicles via the vesicular transporters where it cannot damage mitochondria, and secondly, the MPP^+ may be concentrated in some neurons by binding to proteins such as neuromelanin, which is in particularly high concentrations in the neurons of the substantia nigra which are selectively lost in Parkinson's disease.

7.17 Lambert–Eaton myasthenic syndrome is a failure of neurotransmitter release

Lambert–Eaton myasthenic syndrome (LEMS) is a disorder of neuromuscular transmission. It is an autoimmune disease characterized by impaired evoked release of ACh at the neuromuscular junction. The number of quanta released per nerve impulse is reduced by more than 80% so that many muscle end-plates may not release sufficient ACh to produce an action potential and hence muscle contraction.

When the neuromuscular junction in LEMS was observed using freeze-fracture electron microscopy, it was found that there were fewer active zones than normal and that the existing active zones were disorganized, which seems to correspond to the disruption of calcium channels in the active zones. In LEMS, the response of the postsynaptic membrane is unaffected. The size of mepps is the same as at the normal neuromuscular junction, and the number of ACh receptors, as measured by α-BuTX binding, is unchanged. The target for the LEMS antibodies has been located at the active zones, and it has been suggested that calcium channels may be the target molecule. However, there is some evidence which suggests that synaptotagmin may also be a target for LEMS antibodies. The sera from LEMS patients have been shown to contain antibodies to a number of proteins, including L-, N-, P- and Q-calcium channel subtypes and synaptotagmin. However, not all patients had anti-synaptotagmin antibodies, showing that disruption of synaptotagmin function is not a necessary component of LEMS. A large proportion of LEMS patients (60%) also have small-cell lung cancer (SCLC), and it has been suggested that antibodies initially raised against the tumor cells cross-react with voltage-gated calcium channels at the synapse, disrupting the active zones and leading to the failure of exocytosis.

Box 7.1. Imaging calcium in cells

Direct evidence for the involvement of calcium in neurotransmitter release has been provided by the technique of calcium imaging, which makes visible how calcium signals spread in time and space through cells.

The technique uses fluorescence microscopy and relies on dyes which, on binding of a specific ligand, in this case calcium, alter their fluorescence properties. One of the most commonly used of these dyes is fura-2, a synthetic chemical structurally related to the calcium chelator, EGTA. It binds calcium with a high affinity (K_D = 225 nM) and, unlike previously used compounds, like aequorin, is unaffected by magnesium ions, at least at physiological concentrations. Although fura-2 itself cannot cross cell membranes, cells can be loaded with fura-2 by using the acetoxymethyl (AM) ester form which is lipid soluble. Once inside the cell, the ester groups are hydrolyzed by esterase enzymes, which are present in the cytoplasm of most cells, to give the active calcium-sensitive form. This effectively traps the dye in the cytoplasm.

The technique is independent of dye concentration because fura-2 is a ratiometric dye, that is the emission is measured at two excitation wavelengths, and the ratio of these two measurements gives the changes in calcium concentration. In the absence of calcium, fura-2 absorbs UV light maximally at a wavelength of 385 nm and emits at 540 nm (*Figure 1*). On binding calcium, the absorption spectrum shifts so that the peak absorbance becomes 345 nm whilst it still emits at 540 nm. By measuring the intensity of emission at 540 nm in response to excitation by UV at alternately 345 and 385 nm, it is possible to estimate the changes in calcium concentration within the cells.

The actual calcium concentration can be calibrated by measuring the emission after the addition of a calcium ionophore, such as A23187, which allows the extracellular calcium, in which the calcium concentration is known, to fill the cell. The background emission, at zero calcium, can then be measured by adding a large amount of EGTA to chelate

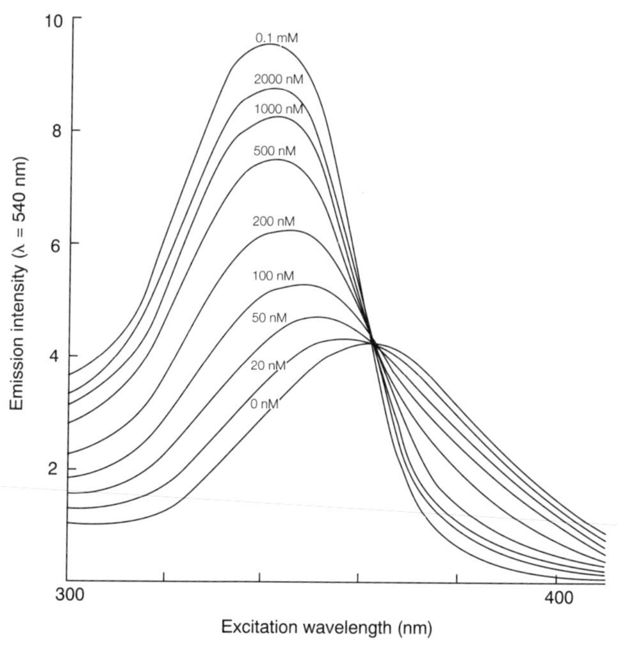

Figure 1. Excitation spectrum of fura-2 as a function of calcium concentration. Reprinted with permission from Tsien, R. Y. and Poenie, M. (1986) Fluorescence ratio imaging: a new window into intracellular ionic signaling, *TIBS* **11**: 450.

Continued

all the calcium both inside and outside the cell.

The method can be used to measure changes in calcium concentration over the whole cell either by measuring the light emitted in a window covering the whole cell, a technique called single-cell fluorescence, or by using a video or digital camera to measure the light emitted from different areas of the window, thus creating an image of the calcium changes in different areas of the cell. This technique called calcium imaging has been used to show the influx of calcium through voltage-dependent calcium channels in squid giant synapses and how this calcium entry is highly localized.

Box 7.2. Production of transgenic animals

Transgenic animals are those which have had a synthetic gene, a transgene, inserted into the genome, allowing the function of that gene to be examined in the context of the whole animal. The insertion of DNA into mammalian cells is called transfection, because the process is similar to viral infection. Transgenes can be inserted in two ways, either by random insertion, via heterologous recombination, of one or more copies of the gene into the genome where they may be inappropriately regulated or, more usefully, by targeted replacement by homologous recombination of the original gene at the specific gene locus. The new gene will stop production of the original protein and thus these are called gene knockouts.

One method for producing transgenic animals by heterologous recombination uses fertilized eggs which are harvested from a female mouse which has been induced to hyperovulate and then mated normally. The transgene is microinjected into the male pronucleus before it fuses with the nucleus of the egg. When the eggs reach the two-cell stage, they are then implanted in foster mothers. The offspring with the recombinant gene will be heterozygous (+/–) for the inserted gene. Breeding these mice will produce litters containing a mixture of homozygotes (+/+ and –/–) and heterozygotes (+/–) unless the insertion of the transgene proves fatal, in which case only the –/– and +/– will survive. In extreme cases, only the –/– will survive, in which case the experiment is a complete failure. In this method, the transgene integrates into the mouse genome at random and many copies of the gene may be integrated.

A more useful method allows the insertion of the transgene at the appropriate site, by homologous recombination, and relies on the observation that transfected DNA will occasionally line up with its homologous sequence in the genomic DNA and replace the host gene with the injected DNA by recombination. The first step is the construction of the targeting vector (*Figure 1*). The vector contains the gene of interest with the gene for neomycin resistance (*neo*^{r}) inserted in one of the exons. This insertion has two functions, firstly, it inactivates the gene by interrupting the normal coding sequence and, secondly, it acts as a marker for recombination as it confers resistance to the neomycin analog, G418. At either side of the gene there are flanking sequences which are homologous to the target locus where the

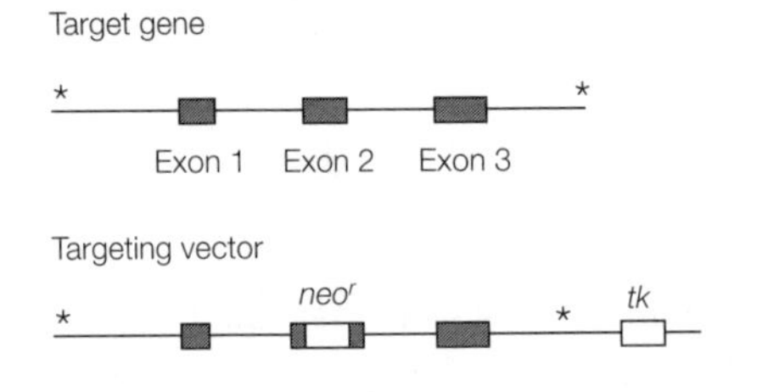

Figure 1. Structure of the targeting vector. The flanking sequences (*) are homologous to the genomic DNA.

Continued

vector can line up with its homologous counterpart in the genomic DNA. Finally, at one end of the vector there is a second marker, the herpes thymidine kinase (*tk*) gene. Any cells which carry this gene are killed by the drug, ganciclovir. There must be a large degree of homology between the vector carrying the transgene and the target locus, especially at the flanking sequences. Any part of the vector lying outside the homologous flank-ing sequences, such as the *tk* gene, will not be inserted during homologous recombination.

Early mouse blastocysts contain an inner cell mass of embryonic stem (ES) cells which can be removed and placed in culture, whilst retaining the ability to eventually develop into a mouse; that is they are pluripotent. The vector DNA is introduced into the ES cells, either by electroporation or microinjection, and, after transfection, it is necessary to select those cells which have been transfected correctly. The cells are grown in the presence of both G418 and ganciclovir. Those cells which contain no recombinant DNA will not contain *neo*^{r} and thus will be killed in the presence of G418. Those cells which have been transfected via heterologous recombination will contain the whole of the vector. In this case, the cells will contain both the *neo*^{r} and *tk* genes. They will not be killed by G418 but are sensitive to ganciclovir. The third outcome is in those cells which have undergone homologous recombination. They will have integrated the part of the vector including the flanking sequences, but not the *tk* gene. These cells will grow in the presences of G418 as they contain the *neo*^{r} gene but, as they do not contain the *tk* gene, they are not affected by ganciclovir. Thus, only those cells which have undergone homologous recombination will survive the dual positive/negative selection process. Homologous recombination is an extremely rare event in mammalian cells, so the selection of correctly inserted transgenes is a critical step in the success of the method.

The correctly transfected cells are then placed in a second, fresh, blastocyst. This is implanted in a foster mother where it will form part of the developing embryo. The resulting mouse will be a chimera that is part host and part transfected ES cell derived. In order to select those mice which contain the transgene in their germ cells and can therefore be used to breed a population of transgenic mice, the host and transfected ES cells come from mice with differing coat colors. The transfected cells are from mice which are homozygous (+/+) for a gene called *agouti*, which produces a brown coat color even in heterozygotes. The host blastocyst is from a mouse with a black coat (–/–). The chimeric mice will have a coat which is patchy, with both black and brown areas, because some of their cells are derived from (+/+) transfected ES cells and others from the (–/–) host blastocyst ES cells. Some of these chimeras will have germ cells carrying the transfect and will thus also carry the *agouti* gene (+). Mating of normal and chimeric mice will result in a mixture of heterozygous (+/–) offspring or normal homozygotes (–/–) (*Figure 2a*). The heterozygotes will contain both the transgene and the *agouti* gene, and will be brown. The homozygotes will be black. Further mating between heterozygotes (+/–) will result in a mixture of homozygous (+/+, brown), (–/–, black) and heterozygous (+/–, brown) offspring (*Figure 2b*).

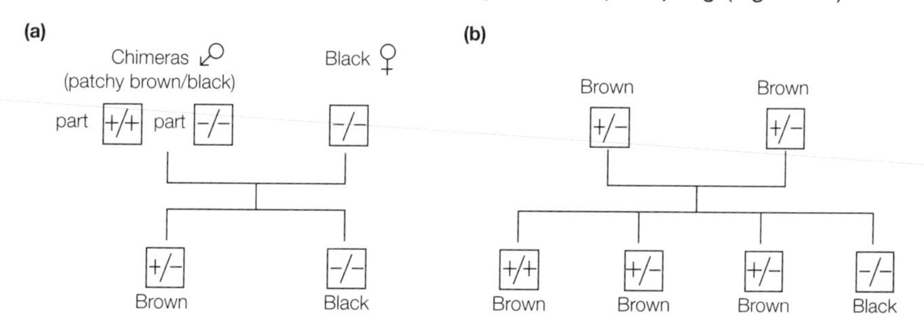

Figure 2. Selection of germline transfected animals by coat color. + = *agouti* (brown coat) gene and transgene.

Chapter 8

Mechanisms of plasticity

8.1 The hippocampus is required for episodic memory

That the hippocampus is crucial for certain types of learning and memory has been accepted for almost 50 years. Patients who have had surgical ablation of the hippocampus or who have suffered transient hypoxia, destroying just a single population of hippocampal cells, experience profound **retrograde amnesia**. Whilst their short-term and remote memories are unimpaired, they lack the ability to consolidate new long-term memories. Lesions of the hippocampus or its inputs in nonhuman primates impair learning in a variety of tasks that require animals to remember specific episodes, especially those in which events are associated with each other or with a particular place. For this reason, the hippocampus is said to process **episodic memory**. Rats are good at **spatial navigation learning** (SNL) which enables them to locate their position in their immediate environment – such as a four-arm maze or a pool of water – by remembering the locations of distal cues. It is argued that SNL is a special case of episodic memory since a rat must learn specific conjunctions of cues and place. Hippocampal lesions severely impair SNL. Chemical lesioning of the hippocampal granule cells, at varying times after training rats on an SNL task, reveals that by 12 weeks post-training, the hippocampus is not required for successful recall of the task. The memory for the task has clearly shifted to another locus, probably the neocortex. The prevailing view is that the hippocampus is involved in initial encoding and storage of a given episode, but over a period of a few months this is transferred to the neocortex and memory of the episode is lost from the hippocampus.

Recording from several hippocampal pyramidal cells simultaneously in conscious behaving rats while they are engaged in SNL has revealed the existence of **place cells**; cells which fire at relatively high frequency when the animal is in a particular location and are quiescent elsewhere. These cells have **place fields** which are analogous to the **receptive fields** of afferent neurons.

Place cells, investigated for over 20 years by John O'Keefe and colleagues at UCL, London, appear to use a combination of vision and movement-evoked proprioceptor input that enables a rat to localize itself in relation to salient features of its environment, e.g. the walls of its enclosure. The rat hippocampus thus seems to be the seat of a cognitive map; a representation of spatial relations which the animal learns during its explorations.

A central tenet of contemporary neuroscience is that plasticity arises by nonrandom alterations in synaptic connectivity. The time scale and extent of the changes are thought to be very variable, from more or less permanent gain or loss of synapses, to transient fluctuations in the strength or **weighting** of individual synapses which alter the efficacy with which they can excite or inhibit the postsynaptic cell. One idea that has acted as a great stimulus to research first appeared in Donald Hebb's *Organization of Behaviour* published in 1949 and is now usually referred to as **Hebb's rule**. Simply put, it states that if two neurons are excited at the same time, then any active synapses between them will be strengthened. These Hebbian synapses are acting as coincidence detectors. A corollary of Hebb's rule is that in the absence of neuron coactivation, synaptic weighting is diminished. Hence Hebbian synapses are equipped to monitor the degree of correlation between firing of pre- and postsynaptic cells. For this reason, a Hebbian mechanism is postulated to underlie associative learning such as that needed for episodic memory.

A striking feature of the hippocampus is that it is essentially laminar and each layer contains the same relatively simple circuitry (*Figure 8.1*). A thin (e.g. 300 μm) hippocampal slice thus contains all the components appropriately connected and so is thought to be a valid substitute for the whole structure, and the regularity of the circuitry facilitates attempts at functional modeling.

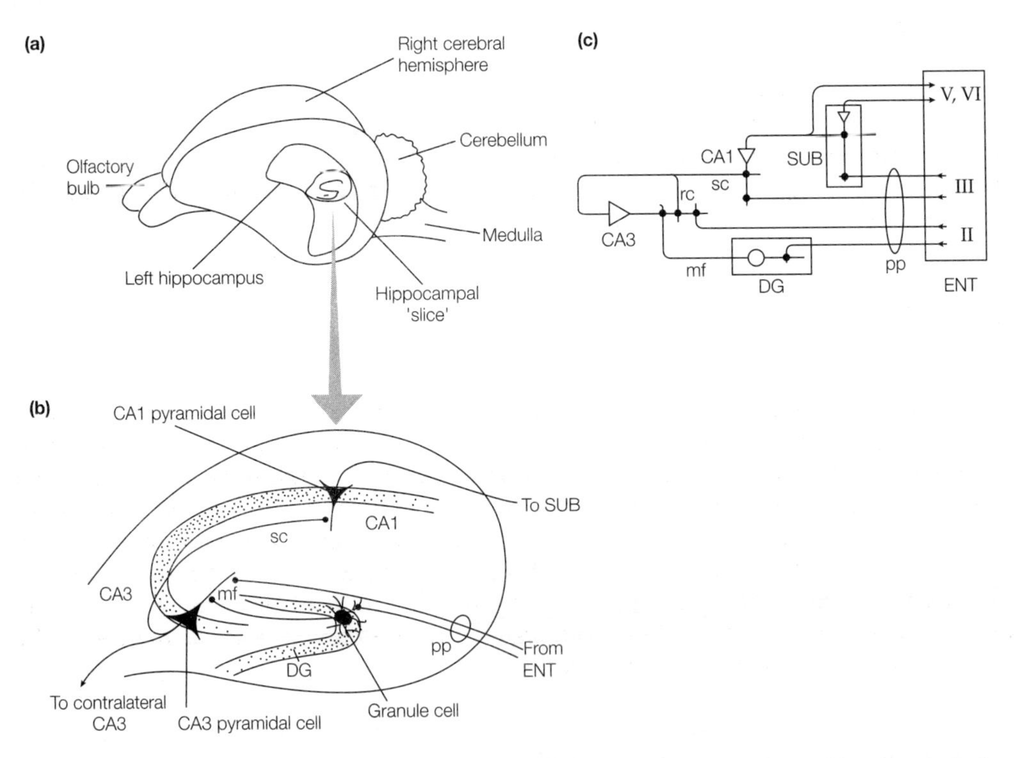

Figure 8.1. Hippocampus. (a) Location of the left hippocampus in rat brain; a hippocampal slice is at right angles to the long axis of the hippocampus. (b) Structure of a hippocampal slice showing the principal excitatory neurons. (c) Hippocampal circuitry. DG, dentate gyrus; ENT, entorhinal cortex; mf, mossy fiber; pp, perforant pathway; rc, recurrent collateral; sc, Schaffer collateral; SUB, subiculum.

The hippocampus receives input from association neocortex via the perforant pathway which synapses with granule cells of the dentate gyrus. Axons of these cells (mossy fibers) synapse with dendrites of the large pyramidal cells in the CA3 field of the hippocampus proper. CA3 cells also receive direct inputs from the perforant pathway. Branches of CA3 pyramidal cell axons termed Schaffer collaterals synapse in turn with smaller pyramidal cells in CA1 whose axons return to the neocortex. Each of these cells is glutamatergic; glutamate receptor density is high throughout the hippocampus with autoradiography of NMDA receptor binding highest in zones containing granule cell dendrites and CA1 pyramidal cell dendrites. The hippocampus also contains numerous GABAergic interneurons which receive recurrent axon collaterals from the glutamatergic neurons and mediate inhibition, via $GABA_A$ receptors which produce fast ipsps and $GABA_B$ receptors which result in slow ipsps.

Precisely how the circuitry of the hippocampus encodes and enables the recall of memory is not known, but computer modeling which retains fidelity to hippocampal architecture suggests that the CA3 region acts as a network where arbitrary associations between conjunctive stimuli are stored as a pattern of modified synaptic strengths. It is this pattern of synaptic weights that is the neural representation of the episode. Excitation of a small part of the network involved in encoding a given episode is sufficient to activate the whole. This retrieval of the representation subserves both recall and its gradual transfer to the neocortex. The characteristics of the hippocampal wiring that allow the CA3 region to act as an associative network are firstly that CA3 pyramidal cells have extensive recurrent collaterals, axon branches that form synapses with other pyramidal cells including the cell of origin, and that the CA3 region is the site of convergence of input from widely disparate cortical regions funneled through the dentate gyrus, which permits associations between different sensory modalities. Crucially, in this modeling, the changes in synaptic strengths that encode the associations are generated by Hebb's rule.

8.2 Long-term potentiation is an activity-dependent increase in synaptic strength

First described in detail in 1973 by Tim Bliss and Terji Lømo of the Medical Research Council in London, long-term potentiation (LTP) refers to a persistent increase in efficacy of synaptic transmission in response to brief volleys of high-frequency stimulation of excitatory pathways. It can be elicited in all pathways in the hippocampus, either in hippocampal slices or via implanted electrodes in anesthetized or conscious behaving animals. LTP in other brain regions is also well documented. A large research effort to understand how LTP works is motivated, in part, by the belief that it may be the cellular substrate for some types of learning; at certain synapses (such as the CA3–CA1 connections), LTP is Hebbian in nature.

8.2.1 LTP can be induced by physiological patterns of stimulation

Commonly, LTP is induced by applying a tetanic stimulus (a train of 100 stimuli at 100 Hz for example) to a pathway and recording from either individual neurons or populations of cells. LTP typically is monitored as a rise in the slope of the field epsp which lasts for hours in anesthetized animals or *in vitro*, but days in freely moving animals (*Figure 8.2*).

Interestingly, the firing pattern that occurs in the hippocampus of rats exploring novel environments turns out to be remarkably effective in triggering LTP. This theta rhythm, normally generated in the septal nucleus which has an extensive projection to the hippocampus, consists of bursts of four or so action potentials at 100 Hz occurring at intervals of about 200 msec.

8.2.2 NMDARs are implicated in associative LTP

LTP can be both **associative** (Hebbian) and **nonassociative**, though with rather different mechanisms. Glutamatergic synapses in the hippocampus exhibit one or the other type. LTP may be NMDAR dependent or independent. Usually, but not invariably, LTP dependent on NMDARs is associative, whereas NMDA-independent LTP is nonassociative. Associative LTP is seen at most synapses in the hippocampus, but not those between granule cell mossy fibers and CA3 pyramidal cells.

The unambiguous involvement of NMDARs in associative LTP is shown by the fact that induction of LTP is blocked by NMDAR antagonists, e.g. AP5. LTP induction is enhanced *in vitro* by having a low Mg^{2+} concentration in the bathing medium, and blocked by low external Ca^{2+} concentration. Normally, NMDAR activation requires both ligand and depolarization. It is not surprising then that LTP induction can be prevented by properly timed hyperpolarizing pulses and facilitated by blockade of postsynaptic inhibition with $GABA_A$ antagonists. Once induced, the subsequent expression of LTP cannot be blocked by AP5, so clearly the maintenance of LTP depends on receptors other than NMDARs.

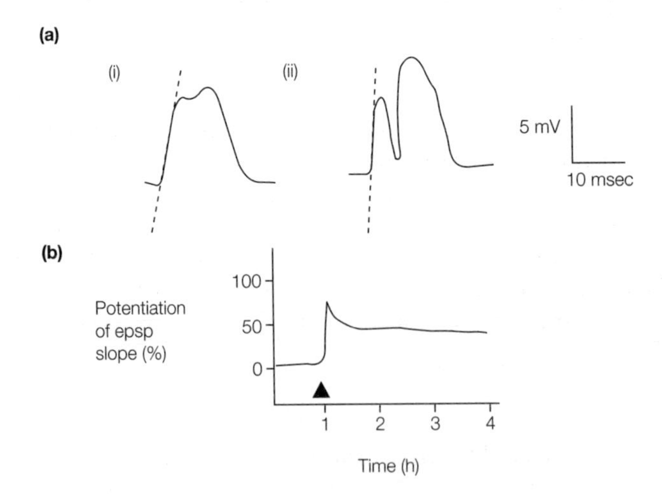

Figure 8.2. LTP *in vivo*. (a) Extracellular recordings from the CA3 pyramidal cell layer (i) before and (ii) after tetanic stimulation (- - - -, slope of epsp); (b) potentiation of the rising phase of the field epsp after tetanus (▲).

In order to appreciate the events that trigger LTP, it is necessary to review what occurs during low-frequency stimulation of, say, the Schaffer collateral (*Figure 8.3a*). A single stimulus evokes an epsp in the postsynaptic CA1 cell which can be blocked by AMPA/kainate receptor antagonists such as 6-cyano-7-nitroquinoxaline-2,3-dione (CNQX). This is curtailed rapidly by a biphasic ipsp, caused by excitation of GABAergic interneurons which results in activation, first of $GABA_A$ receptors, then of $GABA_B$ receptors. Under such circumstances, there is little chance for activation of NMDARs because the brief depolarization is insufficient to lift their voltage-dependent Mg^{2+} blockade.

What of a high-frequency tetanic stimulus? This results in a sustained depolarization for which there are several possible explanations, but the end result is that now the conditions for activating NMDARs are met; glutamate binding to the receptor and a fall in the membrane potential large enough to drive Mg^{2+} from the open channel,

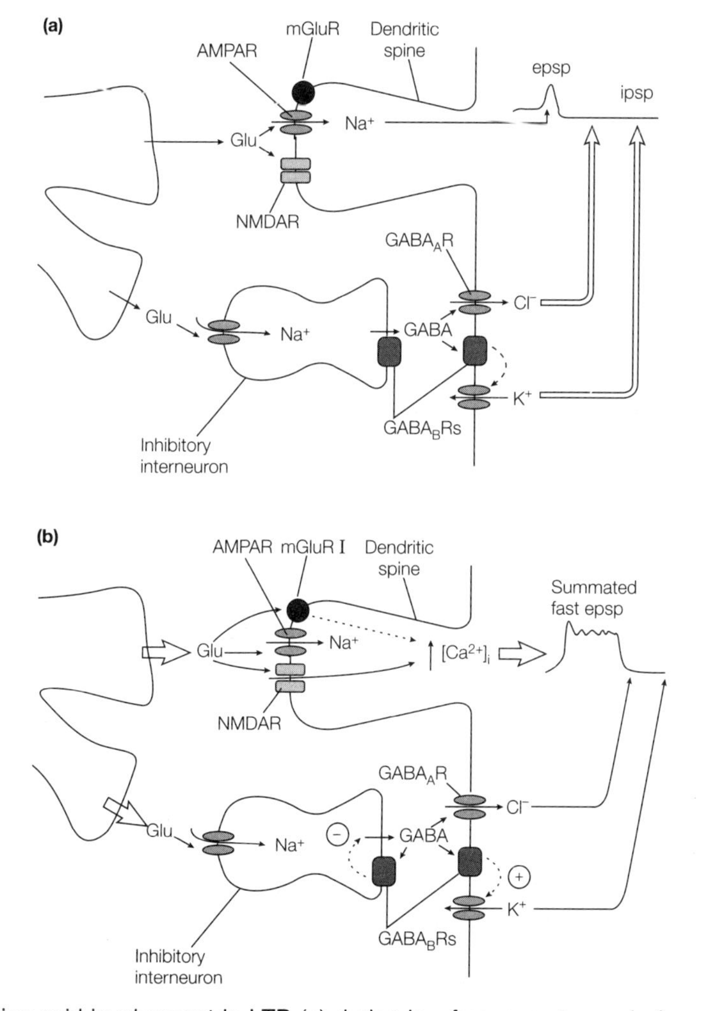

Figure 8.3. Amino acid involvement in LTP (a) during low-frequency transmission and (b) during high-frequency (tetanic) stimulation. Redrawn with modification from Anwyl, R. (1996) The role of amino acid receptors in synaptic plasticity. In Fazeli, M.S. and Collingridge, G.L. (eds) *Cortical Plasticity: LTP and LTD*, pp. 9–34. BIOS Scientific Publishers.

allowing Ca^{2+} influx. Note that these conditions are precisely those demanded by Hebb's rule, namely simultaneous coactivation of a presynaptic cell (glutamate release) and a postsynaptic cell (large depolarization). Thus it is the behavior of the NMDAR which confers the Hebbian nature on associative LTP.

There are several reasons which may account for the large depolarization generated by tetanic stimulation (*Figure 8.3b*). Firstly, high-frequency stimulation, by producing sustained glutamate release, causes summation of AMPAR-mediated epsps. Secondly, accompanying GABAergic inhibition may become less effective because raised intracellular Cl^- and extracellular K^+, due to prolonged activation of $GABA_A$ and $GABA_B$ receptors respectively, cause the equilibrium potentials for these ions to shift in a depolarizing direction. Thirdly, $GABA_B$ **autoreceptors** on inhibitory nerve terminals when activated by GABA result in reduced GABA release from the terminals, so reducing inhibition. This mechanism is not activated by low-frequency stimulation since it takes over 10 ms to develop.

8.2.3 *Calcium is the second messenger mediating LTP*

Evidence for the involvement of Ca^{2+} in LTP is very strong. Induction of LTP can be blocked by intracellular injection of EGTA, a Ca^{2+} chelator. Ca^{2+} imaging (***Box 7.1***, p. 147) coupled with whole-cell patch clamping (***Box 4.2***, p. 72) shows that tetanic stimuli which activate NMDAR currents produce increases in Ca^{2+} concentration in dendritic spines. However, Ca^{2+} influx via NMDARs is not the only way in which the intracellular Ca^{2+} concentration increases during LTP. Ca^{2+} release from internal stores also contributes, since thapsigargin, a drug which depletes Ca^{2+} stores, also blocks LTP. This is probably related to activation of G protein-linked receptors coupled to phosphatidylinositol metabolism such as mGluRs, which also have been implicated in LTP. Moreover, Ca^{2+} influx through L-type Ca^{2+} channels in pyramidal cell dendrites (see Section 7.5.2) may be important in LTP. These matters are dealt with in more detail later in this chapter.

8.3 Both pre- and postsynaptic events appear to be involved in the maintenance of NMDAR-dependent LTP

A great debate has flourished about whether LTP is maintained largely by presynaptic means – by increased glutamate release, on average, per action potential – or postsynaptic means, that is by an increase in receptor number or receptor sensitivity to glutamate. Differences have arisen in part because there are different forms of LTP (nonassociative and associative), and because LTP even at one type of synapse can have different characteristics depending on the precise protocol for triggering it. Single tetanic trains delivered to Schaffer collaterals result in LTP which lasts about 3 h, which cannot be blocked by protein synthesis inhibitors. However, three tetanic volleys can produce LTP which is stable for many hours that does depend on protein synthesis. In conscious behaving animals, LTP persisting for days seems to involve gene expression since it can be prevented by drugs which interfere with both transcription and translation. There is no *a priori* reason for supposing that the synaptic locus of all these types of components are the same. Indeed, a consensus is emerging that nonassociative

LTP at mossy fiber synapses is largely a presynaptic phenomenon, and the best guess favors a postsynaptic locus for NMDAR-mediated LTP at CA3–CA1 synapses.

8.4 Protein kinases are activated during LTP

A central tenet for a postsynaptic hypothesis for LTP is that the number, or sensitivity, of receptors is increased, probably by a change in the phosphorylation state of the receptors and/or cytoskeletal proteins, which would be brought about by activating second messenger systems. Both AMPA and NMDA receptors have consensus sequences for phosphorylation by a variety of kinases, and there is a great deal of evidence which suggests that many kinases are activated during LTP.

The first hints came in 1988 with the discovery that induction and maintenance of LTP in hippocampal slices could be blocked by nonspecific inhibitors of protein kinases with potencies implying that PKC was the vital target. More convincing evidence came from later findings using **inhibitor peptides** which block the enzyme-binding sites. The PKC inhibitor peptide, PKC_{19-31}, blocks the induction of LTP, as do antibodies directed at several PKC isoforms. Furthermore, injecting active PKC into postsynaptic neurons produces synaptic potentiation.

A key feature of PKC is that once activated by DAG, it may remain active long after the second messenger concentrations have fallen to baseline. This **autonomous activation** is exactly the property expected of a molecule which is needed to transduce a transient signal into a stable change. Several mechanisms have been identified which result in persistent activation of PKC independently of DAG and Ca^{2+}. That PKC inhibitors – including the inhibitor peptide PKC_{19-31} – can *reverse* LTP even several hours after induction is good evidence for sustained PKC activity. This is supported by reports that PKC substrates remain phosphorylated long after LTP induction.

The role of PKC substrate phosphorylation is not yet understood in LTP, but both AMPARs and NMDARs have PKC consensus sequences, and an obvious hypothesis to test is that potentiation of the NMDAR component in LTP is due to PKC phosphorylation. In this regard, it is interesting that 48 h after LTP induction, as shown by *in situ* hybridization with a radiolabeled probe for exon 21, there is a selective rise in mRNA encoding those NMDAR splice variants with the C1 cassette. At face value, this suggests an increase in the number of receptors that can be phosphorylated by PKC since most of the consensus sequences for this kinase are in C1; however, deletion of N1 rather than C1 seems to be the most important determinant of PKC potentiation (see Section 5.11), and increased message could just represent increased turnover of receptor.

Recent work, however, sheds light on what probably activates PKC in LTP. Group I mGluRs, coupled to PLC-β (see Section 6.4.1), occur both pre- and postsynaptically, and activation of mGluRs is now known to be an absolute requirement for induction of LTP, an issue taken up later in this chapter.

Calcium/calmodulin-dependent protein kinase II (CaMKII) is an obvious candidate for involvement in LTP because it is activated by Ca^{2+} and because, like PKC, it can remain persistently active after the Ca^{2+} signal has decayed. Inhibitors of CaMKII, including the inhibitor peptide, $CaMKII_{273-302}$, and inhibitors of calmodulin prevent the induction of LTP. CaMKII activity is already second messenger independent as early as 5 min after tetanic stimulation. Moreover, both *in vitro* and *in vivo* there is an increase in the mRNA for the α-, but not the β-isoform, of the enzyme, initially in the cell body but subsequently in proximal and then distal dendrites with a time course consistent with dendritic transport of the message. The α-isoform of CaMKII is a major component of the **postsynaptic density** (PSD) so the enzyme is well placed for both activation by Ca^{2+} influx and phosphorylation of targets, including receptors, in the postsynaptic membrane. Hippocampal slices from transgenic mice lacking αCaMKII show severe deficits in LTP.

The mechanism of autonomous activation of CaMKII is well understood. It involves perpetual autophosphorylation which can sustain the active state regardless of the turnover of the individual subunits of which the holoenzyme is composed. CaMKII could clearly maintain stable alterations on a time scale of years.

Early attempts to show changes in postsynaptic receptors in LTP have met with mixed rewards. Radioligand binding studies have suggested an increase in AMPAR numbers. Electrophysiological experiments suggest that LTP enhances the responsiveness of the postsynaptic cell to iontophoretically applied AMPA, a result consistent with either an up-regulation of receptors or altered properties of pre-existing ones.

Molecular biological studies of AMPARs have been particularly instructive. GluR1 AMPARs are heavily phosphorylated both by purified CaMKII and by PSD preparations. The current through AMPARs in whole-cell patched hippocampal neurons is enhanced by having activated CaMKII in the pipette. GluR1 expressed in *Xenopus* oocytes exhibit currents enhanced by CaMKII, and the enhancement was not seen in mutant receptors which lack a serine phosphorylated by CaMKII. All of this is *prima facie* evidence for CaMKII enhancement of AMPAR responses in LTP.

8.4.1 PKA is implicated in nuclear signaling in plasticity

The cAMP second messenger system is activated in a wide variety of learning paradigms including long-term facilitation in the sea hare, *Aplysia californica*, olfactory learning in *Drosophila* and imprinting in chicks, and LTP is no exception. The concentration of cAMP rises briefly after tetanic stimulation, and this has been shown to activate PKA. The increase in cAMP requires NMDAR activation and is brought about by stimulation of the calcium/calmodulin-sensitive isoform of adenylate cyclase (see Section 6.7.1). Studies using PKA inhibitors for an hour after tetanic stimulation show that the enzyme is not needed for induction or early events in LTP, but is needed for late LTP which requires protein synthesis and may involve gene expression.

The link between PKA activation and nuclear events is well established. Inactive PKA is a heterotetramer consisting of a dimeric regulatory subunit (R) and two catalytic subunits (C). By binding to the regulatory subunit, cAMP causes its dissociation from the catalytic subunits which now have kinase activity. These, on entering the nucleus (*Figure 8.4*), phosphorylate a transcription factor (TF) called CREB, a **cAMP response**

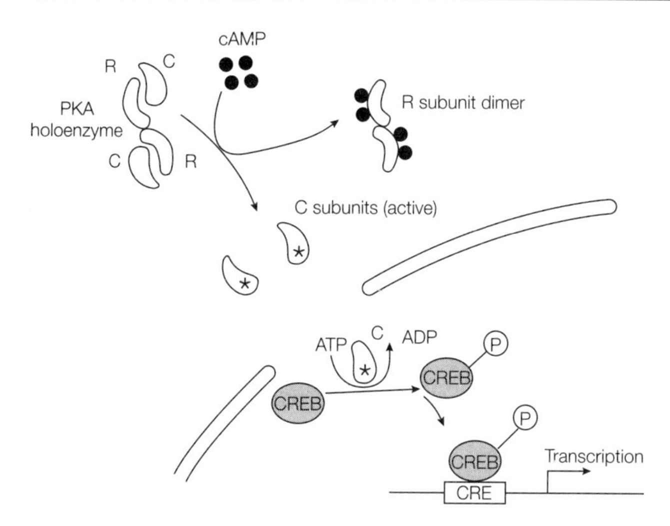

Figure 8.4. Transcriptional activation mediated by cAMP.

element-binding protein, which engages with cAMP response elements (CREs), short sequences of DNA found upstream of genes activated in response to raised intracellular cAMP.

8.4.2 *Transcription factors are important in plasticity*

TFs fall into several families; two such families have attracted particular attention in the context of neuronal plasticity, but it is highly likely that members of other families will also turn out to be involved.

The family to which CREB belongs is known as the basic leucine zipper family (bZIP) because it contains one region with a cluster of basic amino acids and a second with regularly spaced leucine residues (*Figure 8.5*). The basic region mediates interaction with the promotor of the target genes, while the leucine-rich region allows dimerization to occur, hence the name leucine zipper. The DNA sequence, CRE, to which CREB binds is TGAACGTCA, but a second very similar sequence (TGA[C/G]TCA), the AP-1 site, is the target for the other bZIP family members, the Fos and Jun proteins. Interest in these proteins first arose because they are encoded by cellular oncogenes, c-*fos* and c-*jun*; These and related TFs are activated by growth-stimulatory molecules (mitogens and numerous agents associated with cellular stress (cytokines, toxins, radiation). Fos and Jun associate to form heterodimers called activator protein 1 (AP-1) and, as there are several types of Fos and Jun proteins, there are a number of possible heterodimers.

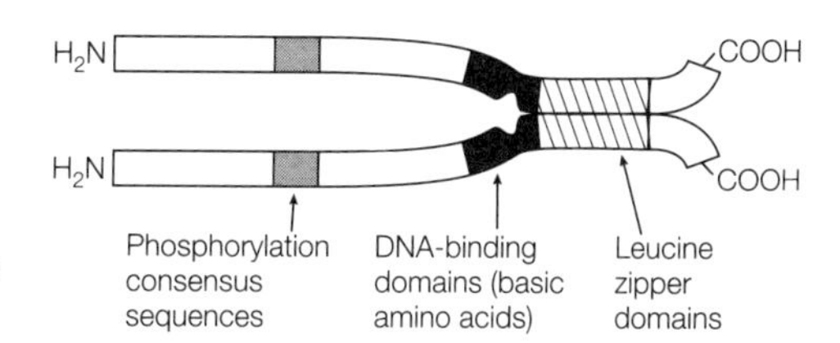

Figure 8.5. bZIP TFs dimerize via their leucine zipper regions.

The dimers are thought to straddle the DNA so that its N-terminal ends can interact with the transcription machinery. The affinity with which distinct heterodimers bind to AP-1 sites *in vitro* depends in the exact Fos–Jun combination, and there is reason to suspect this has consequences for efficiency of transcription *in vivo*.

A second TF family contains a series of loops, each stabilized by coordinate bonds between four cysteine residues and a zinc atom, and which interact with 5′-GCGGGGGCG-3′ DNA sequences. These are called zinc finger TFs, at least one of which, NGF1-A (zif/268) seems to play a part in LTP.

8.5 Invertebrates currently provide the most complete evidence for nuclear signaling via cAMP in plasticity

Mutant *Drosophila* with defects in classical conditioning have proved helpful. Mutations affecting cAMP-dependent phosphodiesterase (*dunce*), a calcium/calmodulin-dependent adenylate cyclase (*rutabaga*) and the catalytic subunit of PKA (*DCO*) all disrupt learning and memory. Transgenic flies in which CREB function could be switched off at will, by inducing expression of a gene encoding an inhibitory isoform of CREB with a brief heat shock, suffered permanent disruption of long-term memory. Moreover, heat shock induction of an activator CREB isoform allowed subthreshold training protocols to produce long-term olfactory conditioning.

The role of the CREB family of proteins has been explored in *Aplysia* by Eric Kandel and colleagues. A light touch to the siphon elicits a defensive withdrawal reflex of the gills and siphon. If a noxious stimulus (an electric shock) is delivered to the tail, subsequently the withdrawal reflex is facilitated. This is an example of a type of nonassociative learning called sensitization. The underlying physiology and biochemistry have been explored extensively.

The synapse between the tactile sensory neuron and the motor neuron is plastic, and subject to presynaptic heterosynaptic facilitation by release of 5-HT (serotonin) from an interneuron which is excited by noxious stimuli applied to the tail (*Figure 8.6*).

Release of 5-HT from the facilitatory neuron causes an increase in neurotransmitter release from the sensory nerve terminal, thus enhancing motoneuron excitation to gill muscles. Like other types of learning, this facilitation has at least two phases: a short-term memory which decays after a few moments and is unaffected by inhibition of protein synthesis, and long-term facilitation (LTF) that can persist for over 24 h and can be blocked by protein synthesis inhibitors. LTF can be induced *in vivo* by giving

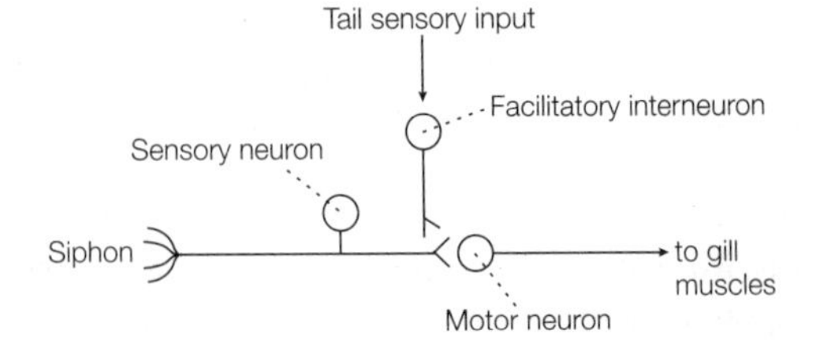

Figure 8.6. The minimum circuitry required to demonstrate facilitation of the gill withdrawal reflex in *Aplysia*.

repeated noxious stimuli; in isolated parts of the *Aplysia* nervous system or neurons in culture, prolonged (2 h) exposure to 5-HT is effective.

Some of the mechanisms involved in facilitation have been deduced. They are summarized in *Figure 8.7*. The effect of electrically stimulating the interneuron, or exposing the sensory neuron to 5-HT, is mimicked by injecting cAMP or catalytic subunits of PKA into the cell body of the sensory cell. Patch clamping of the sensory neuron shows that 5-HT acting via the cAMP second messenger system causes closure of potassium (K_s) channels. The reduced K^+ efflux allows the presynaptic action potential in the sensory neuron to broaden, and this, by allowing more Ca^{2+} influx, promotes greater transmitter release. LTF *in vitro* is associated with a prolonged closure of K_s channels, and is likely to be sustained by altered gene transcription.

After a 2-h exposure to 5-HT, there is a protracted reduction in the number of regulatory subunits of PKA. Presumably this results in increased phosphorylation of K_s channels by the relative excess of catalytic PKA subunits.

What causes the fall in R subunits? CREB proteins seem to be involved. CRE sequences, when injected into the sensory cell nucleus, prevent 5-HT from triggering the LTF, by sequestering phosphorylated CREB and preventing it from binding to CRE sequences in the genome. One isoform of CREB in *Aplysia* (*Ap*CREB1) is constitutively expressed in sensory neurons and, when activated by PKA, one of it actions is thought to be the stimulation of expression of proteins which degrade the regulatory subunits of PKA. The net effect is persistent activation of PKA, triggered by its own action.

Other CREB family proteins are also involved in LTF in *Aplysia*. In the sensory neurons, a 2-h 5-HT exposure induces expression of a TF called *Ap*C/EBP. LTF is reduced by antibodies to *Ap*C/EBP, and by antisense oligonucleotides which, by hybridizing to the DNA sequence recognized by *Ap*C/EBP, block its effects. *Ap*C/EBP is required for the stable structural changes, such as synaptogenesis, that accompany LTF. Activation of the *Ap*C/EBP gene is assumed to be mediated by *Ap*CREB1.

Additional complexity is revealed by the discovery of a second CREB protein in *Aplysia*, *Ap*CREB2, which is constitutively expressed in sensory neurons. To investigate the role of *Ap*CREB2, mouse cells were transiently cotransfected with a CRE-lacZ plasmid plus combinations of plasmids containing the cDNA for either *Ap*CREB1 or *Ap*CREB2. Any combination that caused activation of the CRE domain would increase β-galactosidase activity which was measured. These experiments showed that when given together, *Ap*CREB2 repressed the CRE-mediated transcription by *Ap*CREB1. Use of an **electrophoretic mobility shift assay** (EMSA) (***Box 8.1***, p. 189) to demonstrate the DNA-binding specificity of *Ap*CREB2 revealed that it binds to some CRE domains (for example those associated with somatostatin or proenkephalin genes) but not others (such as that for the *Ap*C/EBP gene). By eluting transcription factors through an *Ap*CREB2 affinity column, it was shown that *Ap*CREB2 forms only weak homodimers but stronger heterodimers with *Ap*CREB1, *Ap*C/EBP and c-Fos.

*Ap*CREB2 appears to be phosphorylated by several serine/threonine kinases such as PKA *in vitro*, and in response to 5-HT *in vivo*. Antibodies against *Ap*CREB2, when injected into sensory neurons, permit a single brief application of 5-HT to produce

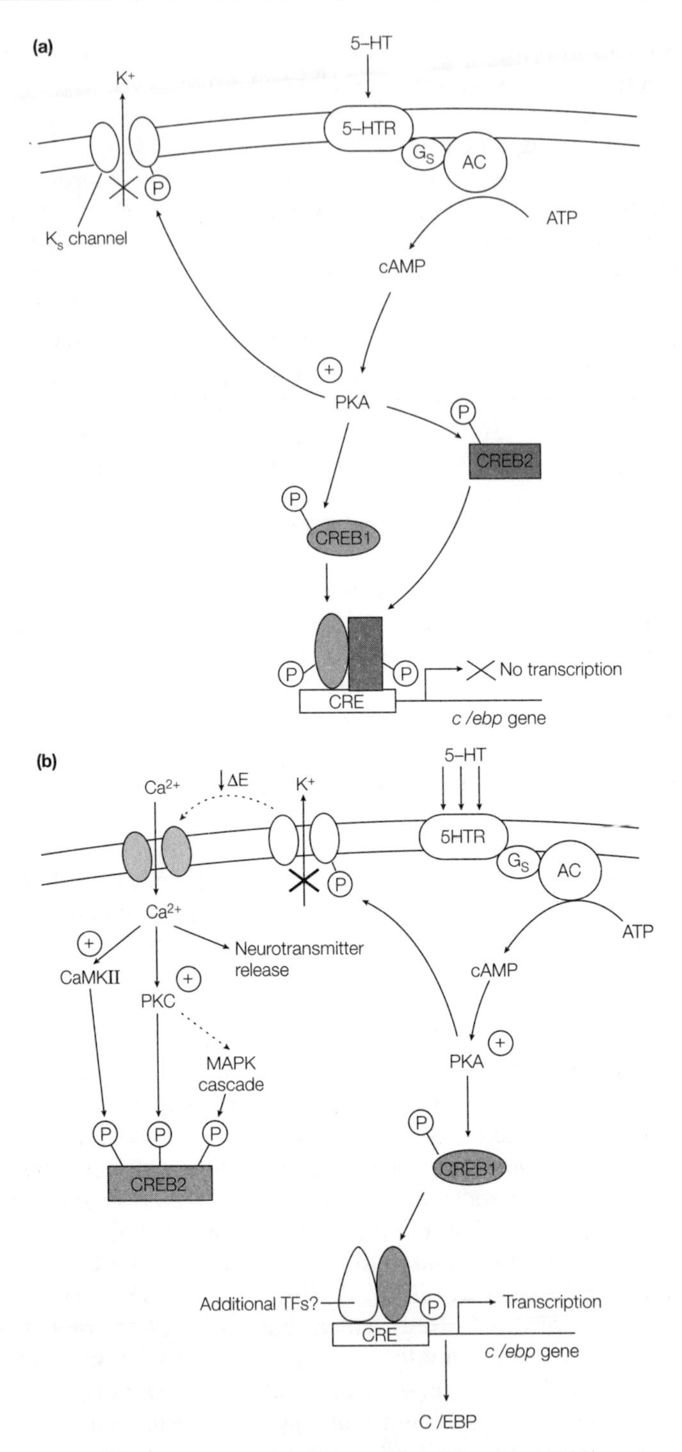

Figure 8.7. Long-term facilitation in *Aplysia* involves activation of cAMP and gene expression. (a) A single pulse of 5-HT has no effect on gene expression. (b) Persistent exposure to 5-HT causes transcription probably by removing CREB2 inhibition on CREB1 activity. How this occurs is not yet known, but phosphorylation of CREB2 by second messenger molecules activated by the increased Ca^{2+} entry is a possibility.

full-blown LTF, rather than just the short-term facilitation that might be expected. Thus, *Ap*CREB2 functions as a constitutive repressor of LTF. The most parsimonious explanation for its repressor action is that *Ap*CREB2 forms heterodimers with *Ap*CREB1, thus inhibiting transactivation of the *Ap*C/EBP gene (*Figure 8.7a*). LTF then would arise by relief of the *Ap*CREB2 repression. Because *Ap*CREB2 represses transcription by *Ap*CREB1 in the presence of PKA, another signaling pathway must be involved in derepression. Since *Ap*CREB2 is phosphorylated by PKC, CaMKII and MAPK, as well as PKA, there are many possibilities. The increased Ca^{2+} entry which occurs during 5-HT exposure and inhibition of K_s could activate both PKC and CaMKII and subsequently the MAPK cascade (*Figure 8.7b*).

That CREB and related proteins also play a part in hippocampal plasticity is suggested by four lines of evidence. Firstly, stimuli that provoke LTP phosphorylate CREB. Secondly, the mammalian version of *Ap*C/EBP, C/EBPβ, is expressed 2 h after LTP induction in granule cells in freely moving rats, as shown by an increase in their mRNA. Thirdly, CREB knockout mice cannot sustain LTP beyond 90 min. Finally, similarities between *Aplysia* and mammalian plasticity are suggested by the fact that human CREB2 represses transcription activation by human CREB1.

There is some evidence that tyrosine kinases play a part in LTP. At 24 h after LTP induction, there is increased expression of mRNA for the brain-specific RafB protein, a component of the serine/threonine kinase signaling cascade which couples membrane events to modifications in transcription. This cascade normally is activated by ligands, such as growth factors or hormones, binding to receptors that have intrinsic tyrosine kinase activity (see Section 6.10.2); the RTK superfamily. Binding of ligand leads to dimerization of the receptor (in most cases), then autophosphorylation which permits the receptor to couple to a small guanine nucleotide-binding protein with intrinsic GTPase activity, Ras, with properties rather like the α-subunit of G proteins. In its GTP-bound form, Ras somehow activates (the mechanism is unknown) Raf, and hence MAP kinases (*Figure 8.8*). The end result is phosphorylation of TFs, such as Fos proteins or C/EBPβ, and hence altered transcription.

It has been proposed that this pathway is switched on by growth factors during LTP. However, it now seems likely that the Ras–Raf–MAPK cascade may be activated by a nonreceptor tyrosine kinase, PYK2 – localized in hippocampal pyramidal cell dendrites – in a Ca^{2+}-dependent manner. The notion is that the Ca^{2+}-sensitive γ-isoform of PKC will be activated by Ca^{2+} entry via NMDARs, by DAG and raised intracellular Ca^{2+} provoked by mGluRs, and that PKC-γ activates the Ras–Raf–MAPK cascade via PYK2 (*Figure 8.9*).

The significance of this lies in the fact that it shows clear cross-talk between signal pathways coupled to ion channels and G protein-linked receptors which modulate intracellular Ca^{2+}, and signaling pathways activated by RTKs which, via transcription events, bring about long-term changes in growth and structure. In other words, neurotransmitters can have direct access to nuclear events previously thought to be regulated only by growth factors.

Does activation of the MAPK cascade occur in plasticity? Studies of unilateral LTP induced at perforant pathway–granule cell synapses in awake rats have shown

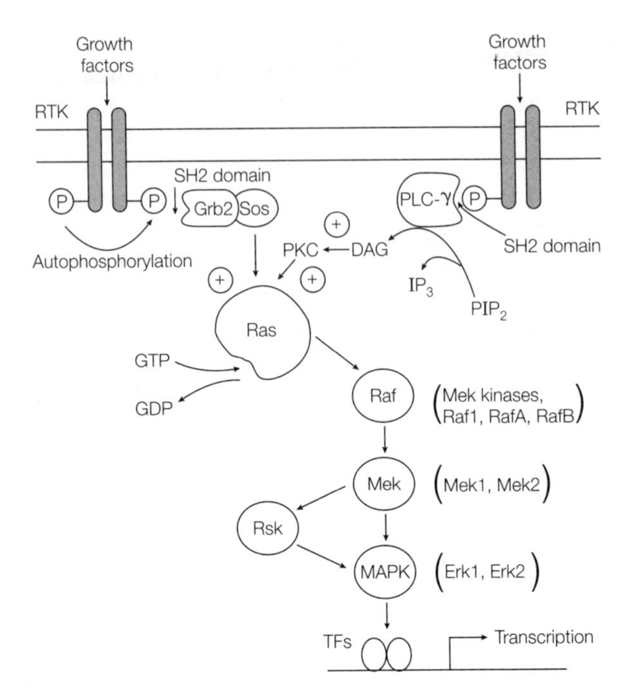

Figure 8.8. The MAPK cascade of serine/threonine protein kinases couples membrane and nuclear events. Ras protein is activated via proteins which bind to phosphotyrosyl residues by **SH2** (Src-homology 2) domains. One such adaptor protein, Grb2, forms a complex with Sos, a guanine nucleotide exchange protein which promotes exchange of GDP or GTP. In an alternative pathway PLC-γ is activated through its SH2 domain, resulting in Ras activation via PKC. Several isoforms of the kinases exist. These are listed in parentheses. Sos, son of sevenless; MAPK, mitogen-activated protein kinase; Mek, MAPK kinase; Erk, extracellular signal-regulated kinase; Rsk, ribosomal S6 protein kinase.

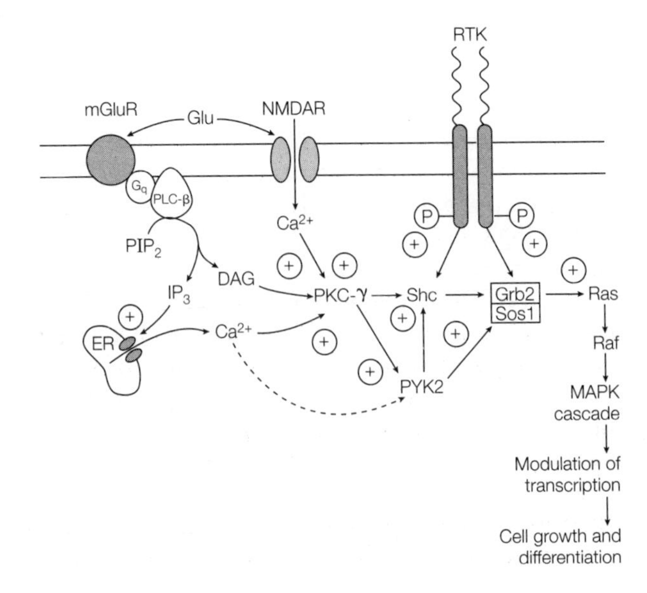

Figure 8.9. Cross-talk between signaling pathways may allow neurotransmitters to modulate cell growth in plasticity.

increases in mRNA for PKC-γ, RafB and Erk2. Also C/EBPβ is phosphorylated by MAPK *in vitro*, which increases its binding affinity for the target DNA sequences thus increasing its transcription capabilities. If, by analogy with *Aplysia* LTF, LTP arises by release of CREB2 repression on CREB1 transcription, is this derepression mediated by phosphorylation of CREB2 by MAP kinases?

Given the cross-talk between signaling pathways alluded to above (*Figure 8.9*), the MAPK cascade could be triggered in hippocampal LTP by Ca^{2+} entry through NMDARs or VDCCs, or Ca^{2+} mobilization via mGluRs, just as well as by activation of RTKs.

Interestingly, LTP in the amygdala *in vitro* which depends on muscarinic cholinergic receptors, not NMDARs, is abolished by knockout of the gene for a neuron-specific isoform of a MAPK cascade component, Ras-GFR. In this case, the link between receptor and Ras-GFR is the G protein βγ-subunit. Surprisingly perhaps, hippocampal LTP in these knockouts was unimpaired despite the fact that Ras-GFR is highly expressed in CA1 pyramidal cells.

Tyrosine kinase-mediated phosphorylation of membrane proteins may also play a part in plasticity. The NR2B subunit of NMDARs is up-regulated 48 h after LTP induction, and this subunit is phosphorylated by tyrosine kinases. This may allow NMDARs to interact with other PSD proteins, perhaps anchoring them to cytoskeletal proteins. NR2B knockouts have not proved helpful in these investigations since they die soon after birth.

8.6 A metabotropic receptor switch is required for LTP

As mGluR agonists enable weak subthreshold tetani to induce LTP, and the specific mGluR antagonist (+) α-methyl-4-carboxyphenylglycine (MCPG) prevents induction of LTP at Schaffer collateral–CA1 synapses in hippocampal slices, it is reasonable to argue a role for mGluRs in LTP. The activation of PKC and the apparent need to mobilize internal Ca^{2+} stores in LTP suggests the necessary mGluRs might be the group I that are coupled to the phosphoinositide second messenger cascade. Unfortunately, knockout mice lacking mGluR1 have not been helpful since while one study showed defective LTP, in a second LTP was normal.

A classical approach has revealed the part played by mGluR. LTP is not induced by mGluR alone. Prior induction of LTP with a single tetanus (which does not produce maximal LTP), which activates both NMDAR and mGluR, prevents MCPG from blocking the induction of further LTP (*Figure 8.10*). The simplest explanation for this is that a single activation of mGluR is required to trigger LTP since it produces a long-term change, a conditioning that means that their subsequent reactivation is not needed. A single tetanus is sufficient to produce full conditioning, but maximal LTP requires several bursts of NMDAR activation by tetanic stimuli. It seems that mGluRs throw a metabolic switch necessary for LTP. Interestingly, this switch can be turned off by prolonged low-frequency stimulation which causes the synapses to revert to a deconditioned state in which MCPG would block induction of LTP. A baroque caveat to these experiments was the observation that the deconditioning itself was blocked by MCPG and hence was mGluR dependent. Hence, the metabolic switch required for

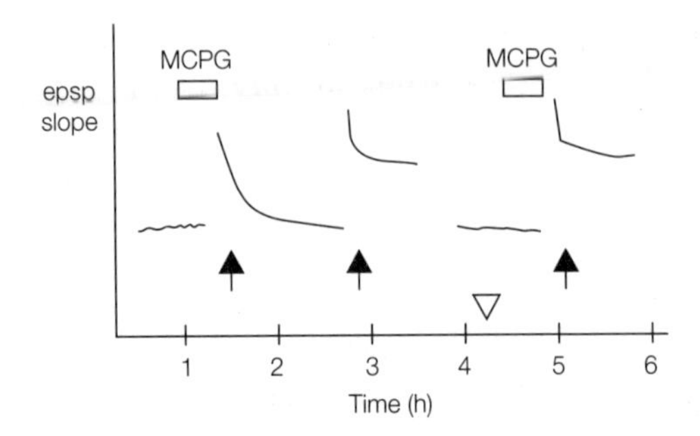

Figure 8.10. Evidence for conditioning by mGluR. The first tetanic stimulus (↑) in the presence of MCPG produces only a short-term potentiation but no LTP. The second tetanic stimulus after the drug has been washed out produces LTP. At ▽, the strength of the single stimulus used to test the preparation is reduced so as to restore the baseline. A third tetanic stimulation evokes further LTP even in the presence of MCPG. Redrawn with permission from Bortolotto *et al.* (1994) *Nature* **368:** 740–743.

LTP is turned on and off by mGluRs, though presumably by different subtypes. That conditioning can be prevented by a kinase inhibitor is consistent with its being caused by mGluR-mediated activation of PKC and subsequent phosphorylation.

Although subsequent events in the metabolic switch are not known, an idea worth testing involves a 17-kDa PKC substrate localized postsynaptically, called neurogranin. This protein binds CaM, but releases it when phosphorylated, as has been shown in LTP. It is postulated that normally free CaM levels are low. Activation of PKC via mGluRs causes a sudden release from neurogranin of CaM, which is then available to activate CaMKII. In this mode, the mGluR switch throws the synapse from a 'read only' mode to a 'read–write' mode, primed to adapt plastically.

8.7 Presynaptic mechanisms for induction of LTP

Even the advocates for a presynaptic locus for maintenance of LTP accept that postsynaptic NMDAR activation is needed for induction. What a presynaptic hypothesis must explain is how an initial postsynaptic event causes a subsequent increase in presynaptic neurotransmitter release. The current idea is that a diffusible second messenger is generated postsynaptically, which crosses the synapse the 'wrong' way to arrive at the presynaptic cell where it alters release. Several candidates have been proposed for such a **retrograde messenger**. A presynaptic mechanism implies that both NMDA and AMPA receptor components of the response be enhanced, since both types of receptor would be exposed to any increased glutamate released. This has been confirmed, although the AMPAR component is potentiated to a greater extent than the NMDAR component. This is consistent with postsynaptic modifications to AMPA receptors.

Various approaches have been attempted to demonstrate increased transmitter release. These include depolarization-evoked release of radiolabeled glutamate from synaptosomes or hippocampal slices following LTP induction *in vivo* or *in vitro*. Alternatively, measuring extracellular glutamate concentrations during LTP has been tried with glutamate-sensitive electrodes or using an outside-out patch on a patch electrode, so that the receptors in the patch act as detectors of glutamate release, a technique known as **patch sniffing**.

The problem with the neurochemical approach is that it is hard to distinguish a genuine increase in release from reduced reuptake. Dialysis has failed to provide evidence for increased release in CA1, though it has in the dentate gyrus, a result apparently confirmed by sniffer patches, although doubts have been raised about whether receptor populations in excised patches are sufficiently stable to act as reliable markers of neurotransmitter release.

One strategy that has been used extensively is to attempt to deduce whether the locus for LTP is pre- or postsynaptic by analyzing the average amplitude of responses to release of a single quantum, a miniature excitatory postsynaptic potential (mepsp), and the frequency with which different numbers of quanta are released simultaneously. This **quantal analysis** is predicated on the basis that neurotransmitter release is binomial; for a given release site, the arrival of an action potential may, or may not, trigger the release of a single quantum with a probability, p_r. With n, the number of release sites, the **quantal content**, $m = np$, is taken as a measure of the number of quanta released per action potential, and is classically thought of as a presynaptic marker. Another measure thought to reflect presynaptic function is the proportion of failures, that is when an action potential does not trigger release, since this reflects release probability, p_r. By contrast, the amplitude of the mepsp, called the **quantal size** (q), is traditionally thought of as being determined by the number of *postsynaptic* receptors.

Now some studies have shown that LTP at CA3–CA1 synapses results in a reduced number of failures, implying an increase in p, without a change in quantal size; all of which, at face value, suggests a presynaptic locus. Other experiments have found increases in both quantal content and quantal size within 30 min of induction, but these changes were sometimes seen in different cells, indicating independent pre- and postsynaptic mechanisms. Unfortunately, there are serious problems with the standard interpretation of this sort of data.

Firstly, it is possible that quantal amplitude might reflect the degree of vesicle filling, a presynaptic phenomenon, rather than anything to do with receptors. Secondly, changes in quantal content might be explained by alterations in the number of postsynaptic receptors; failures may reflect the existence of **silent synapses**, active zones with no functional receptors beneath. These synapses have NMDARs but not AMPARs and so effectively are nonfunctional at normal resting potentials. These silent synapses gain AMPARs during LTP and would thus transmit, reducing the number of failures. Such synapses were suspected by discovering that the number of failures is greater in trials when CA1 cells are whole-cell voltage clamped at negative potentials (–55 to –65 mV) compared with positive potentials (+40 to +60 mV). This difference in failure rate could be accounted for by having some synapses equipped only with NMDAR. At negative potentials, such synapses would fail when stimulated, but at positive potentials they would transmit because of the relief of the voltage-dependent Mg^{2+} blockade on the NMDAR. This idea was confirmed by finding that in the presence of D-AP5, failures were just as high at depolarizing potentials as at negative ones. Using a stimulating protocol that would elicit NMDAR responses only from the silent synapses, it was then shown that LTP activated these silent synapses. After LTP, the NMDAR responses of the silent synapses were unchanged, but the failure rate decreased at negative potentials, suggesting a recruitment of AMPARs at these silent synapses. Interestingly, the failure rate at depolarizing potentials after LTP was the same as before; another nail in the coffin for an increase in presynaptic release.

8.7.1 Morphological changes may explain some characteristics of LTP

Despite the great diversity in paradigms thought to induce plastic changes in the brain, the reported structural alterations are rather similar; increases in area of the PSD and in the number of **perforated synapses**. Dendritic spines come in three basic morphological types, one of which – the mushroom spine – can have a gap in the PSD and a corresponding gap in the presynaptic grid. The discontinuities correspond either to a narrowing of the synaptic cleft or a finger-like projection (spinule) of the postsynaptic cell into an invagination of the presynaptic membrane (*Figure 8.11*). Such synapses are said to be perforated. Spinule formation, studied in cone triad synapses in the retina, can occur remarkably rapidly, within 30 min, implying that functional changes even early on in LTP could arise by altered morphology.

A model has been proposed in which, following LTP induction, simple synapses become perforated, which accounts for some of the perplexing quantal analysis results. The notion is that initially up-regulation of AMPARs gives rise to the increase in quantal size reported in some studies.

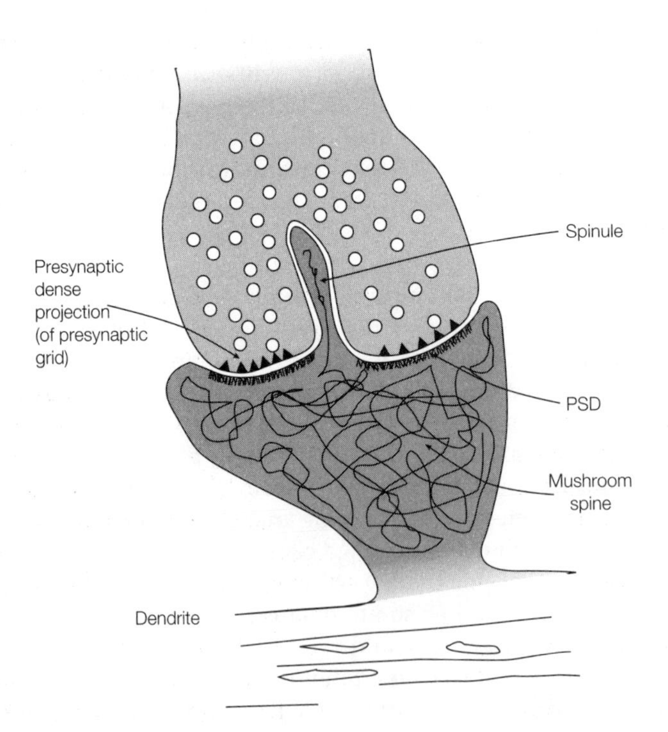

Figure 8.11. Perforated synapses can be visualized by electron microscopy in thin sections stained with phosphotungstic acid.

By the completion of perforation, two or more release sites are generated (where before there was one); this increase in *n* could account for the increase in quantal content and the reduction in failures reported in LTP and sometimes held to have a purely presynaptic origin. At intermediate stages of the process, before barriers to diffusion have been completely raised, overflow of transmitter between nascent sites would cause increases in quantal size. Discrepancies between laboratories in quantal analysis might be accounted for by exactly how advanced perforation is when measurements are made, and this is likely to be sensitive to the exact stimulating and testing protocol adapted. Precisely what biochemistry causes the morphological changes leading to perforation is not known. Rearrangements of cytoskeletal elements via altered phosphorylation states controlled by Ca^{2+}-sensitive second messenger cascades seem likely.

8.7.2 *There are a number of possible retrograde messengers*

Several candidates for retrograde messengers have been proposed. The first was arachidonic acid (AA), partly because of the success of nordihydroguaiaretic acid (NDGA), an inhibitor of phospholipase A_2 (PLA_2), in blocking LTP. It transpired subsequently that AA fails an important test for any retrograde messenger, namely that it should induce LTP when applied during NMDAR blockade.

However, a molecule produced downstream of AA by the PLA_2-evoked second messenger cascade, platelet-activating factor (PAF), passes the same test, in that PAF coupled with either 1-Hz (weak, subthreshold) or theta stimulation would support generation of LTP in the presence of AP5, and, moreover, it does so by a presynaptic mechanism. Presumably, the inhibitory effect of NDGA on LTP is caused by the block of PAF production. PAF is released from neurons in culture, PAF receptors are present in the hippocampus and PAF antagonists inhibit LTP after a brief delay; all of which provides a case for PAF as a retrograde, transsynaptic messenger.

A molecule that has attracted a wealth of attention in neuroscience and elsewhere in recent years is nitric oxide (NO). NO is synthesized from arginine in a reaction catalyzed by nitric oxide synthetase (NOS), is freely diffusible and has a half-life of just a few seconds. Glial cells express a Ca^{2+}-independent form of NOS (iNOS), the transcription of which can be induced by inflammatory mediators. By contrast, neurons express a constitutive version (ncNOS) which is Ca^{2+} dependent. The retrograde messenger argument for NO is summarized in *Figure 8.12*. The ncNOS is activated, via calmodulin, by Ca^{2+} influx through NMDARs during tetanic stimulation. The resulting NO diffuses out of the postsynaptic cell, across the synapse and activates (amongst other targets) guanylate cyclase in the presynaptic terminal which subsequently activates PKG. The evidence for this scheme seems quite good. Neurons in culture challenged with NMDA release NO. Some, but not all, studies find that NOS inhibitors block LTP. Paradoxically, transgenic mice lacking neuronal NOS have normal LTP which is still blockable by NOS inhibitors. This has been explained to be because they continue to express endothelial constitutive NOS. Hemoglobin, a scavenger of NO, also blocks LTP, and is presumably in the extracellular space, showing that NO must be released into the extracellular compartment in order to act. Weak tetanic stimulation coupled with NO has been shown to induce LTP.

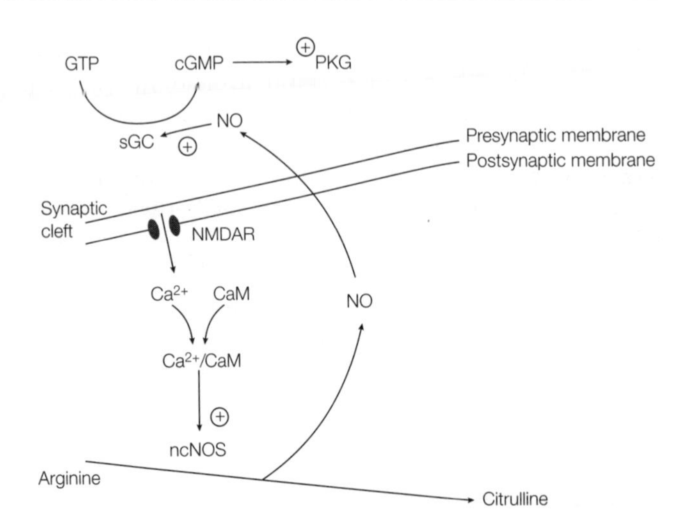

Figure 8.12. Cartoon depicting the proposed role of NO as a retrograde messenger in LTP. CaM, calmodulin; sGC, soluble guanylate cyclase; PKG, cGMP-dependent protein kinase.

Interestingly, NO released from a single pyramidal cell coupled with weak tetanus can produce LTP in neighboring cells, suggesting that the specificity originally claimed for LTP may not be as precise as first thought. Presumably the sphere of influence of NO will be determined by barriers to diffusion and the exact half-life of the molecule. Finally, cGMP is elevated in LTP, and PKG appears to be required for LTP. This said, there are several reports which demonstrate that NOS activity is not always required for LTP (and worse still, a report showing NOS inhibitors actually inducing LTP) and these serve as powerful counter-arguments to any proposal that NO is a *necessary* component of LTP.

All in all, given the ambiguous evidence relating to altered transmitter release, and that incontrovertible evidence for a retrograde messenger is lacking, a presynaptic locus for the maintenance of LTP at CA3–CA1 synapses has not been established convincingly.

8.7.3 Mossy fiber LTP probably is presynaptic

By contrast with CA3–CA1 synapses, evidence for a presynaptic locus for LTP between mossy fibers of the dentate gyrus granule cells and CA3 pyramidal cells is rather good. The first clue that mossy fiber LTP might be different was the discovery that NMDAR binding is extremely low where mossy fibers terminate. Considerable negative evidence seems to rule out a postsynaptic locus; LTP is not blocked by AP5, nor by injection of Ca^{2+} chelators postsynaptically, neither does postsynaptic depolarization induce LTP at mossy fiber–CA3 synapses. Use of Ca^{2+}-free media to prevent presynaptic Ca^{2+} influx blocks LTP at this site. Perhaps the clearest evidence comes from elegant experiments to measure release probability using the NMDAR open channel blocker, dizocilpine. It turns out that there is an NMDAR component to postsynaptic responses which can be exploited. Repeated stimulation of mossy fibers causes a progressive decline in the size of the NMDAR response, because each stimulus triggers the opening of some

NMDARs which are immediately blocked by dizocilpine and hence will not allow current to flow on subsequent stimulations. The crucial point is that the rate of decline of the response is a measure of p_r, since when p_r is larger more glutamate is released per stimulus and this opens more NMDARs. This study showed a faster decay of NMDAR responses in tetanized compared with unstimulated mossy fiber–CA3 synapses; in other words, evidence that LTP increases p_r. It is worth pointing out that similar studies of CA3–CA1 synapses fail to show changes in p_r, reinforcing the claim that LTP at these synapses is not presynaptic.

By what mechanisms is LTP generated at mossy fiber terminals? Firstly, it is clear that mGluRs are involved (*Figure 8.13*). Ibotenate induces, and mGluR antagonists 2-amino-4-phospono-butanoate (AP4) and MCPG block mossy fiber LTP. Moreover, one study of homologous recombination mGluR1 knockout mice show severe deficits in mossy fiber LTP, though LTP at other hippocampal sites is unimpaired. It appears that mGluRs other than mGluR1 are involved, or can substitute at synapses other than at mossy fibers. Now, mGluR1 receptors are coupled to phosphoinositide metabolism and so mobilize Ca^{2+} from internal stores. The type I Ca^{2+}-sensitive isoform of adenylate cyclase is expressed at very high levels in granule cells, and cAMP analogs potentiate mossy fiber–CA3 responses, whilst inhibitors of PKA block mossy fiber LTP. The final link in the chain of events leading to mossy fiber LTP may be enhanced Ca^{2+} entry through presynaptic VDCCs, as use of selective toxins to both N and P channels prevents LTP. It is highly likely that PKA-catalyzed phosphorylation of N and P-channels causes increased Ca^{2+} influx and so enhanced transmitter release. N channel α_1-subunits are indeed phosphorylated by PKA and, in *Xenopus* oocytes injected with cerebellar mRNA, expressed P-type currents are enhanced by cAMP.

It is worth pointing out that this model allows long-lasting LTP of mossy fiber synapses to be produced by persistent activation of PKA via CREB/CRE. Whether this is the case remains to be tested.

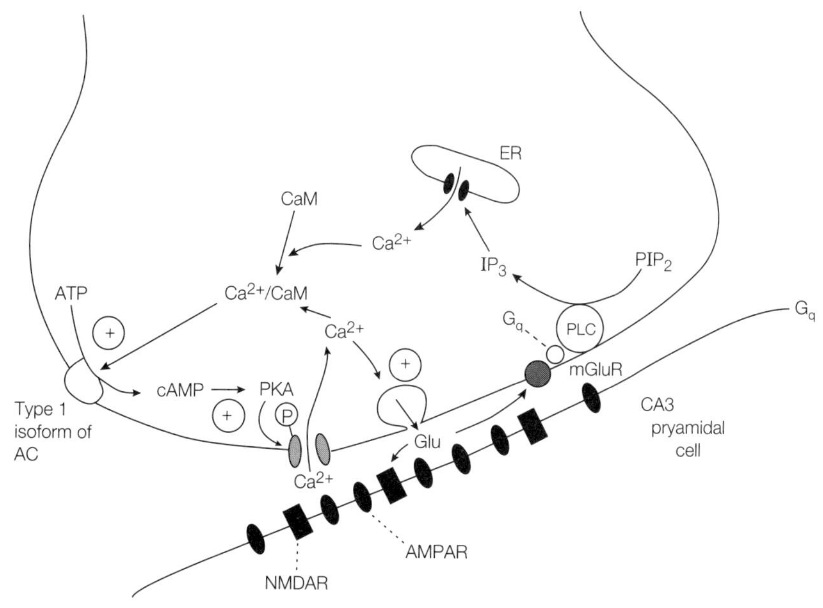

Figure 8.13. Mossy fiber–CA3 synapse LTP is presynaptic.

LIVERPOOL JOHN MOORES UNIVERSITY
LEARNING SERVICES

8.7.4 *The specificity of long-lasting LTP requires synaptic tags*

LTP which is stable for periods longer than 3 h requires protein synthesis. Protein synthesis occurs in the cell body, but as LTP shows input specificity (only synapses at or near the tetanized region of the cell are potentiated), some mechanism must ensure that only tetanized synapses are targets for the newly synthesized proteins.

Early on in LTP, tetanized synapses must acquire a marker or tag which singles them out. How this **synaptic tagging** works is not yet known, but its existence is not in doubt. The method used to demonstrate it was to stimulate one population of synapses with a train of three tetanic stimuli (S3 synapses), sufficient to trigger long-lasting LTP, and to stimulate a second population with a single tetanic stimulus (S1 synapses). Given alone, the single tetanus produced LTP that decayed within 2 h and was not blocked by anisomycin, a protein synthesis inhibitor, as expected. However, with the combined protocol, it was found that the S1 LTP was consolidated into long-lasting LTP. This is accounted for if the single tetanus tags the S1 synapses so that they are then modified by the protein synthesis generated by the LTP of the S3 synapses (*Figure 8.14*). The proteins synthesized in late LTP seem to be available to all synapses. It is the generation of the tag in early LTP which allows specific synapses to use these proteins for more permanent changes.

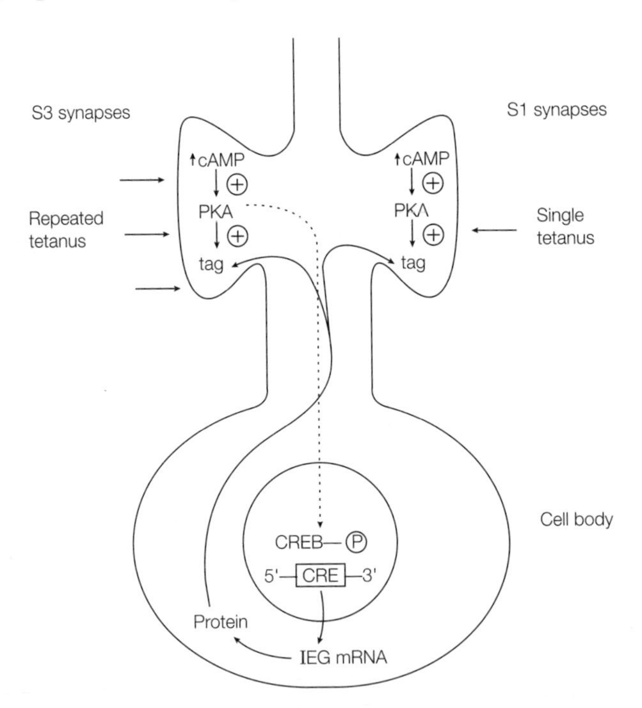

Figure 8.14. Persistent LTP requires a synaptic tag and *de novo* protein synthesis. cAMP may be the second messenger.

If a 2.5-h delay separated tetanic stimuli applied to the two populations of synapses, the protein synthesis produced by the later stimulus of one population was unable to rescue decaying LTP induced by the earlier stimulation of the other population of synapses in the presence of anisomycin. The synaptic tag had vanished by 2.5 h, so consolidation of early to late LTP requires the temporal coincidence of a neuron-wide protein synthesis with a localized, transient synaptic tag.

The nature of the tag, or the protein with which it interacts, is presently a mystery, but one guess is that the tag is a substrate for PKA and the protein the product of an immediate-early response gene (IEG) under the control of the CRE. Membrane-permeable analogs of cAMP produce a synaptic potentiation that resembles late LTP. PKA activity is critical for late LTP, and CREB knockout mice do not show long-lasting LTP.

It could be then that the rise in cAMP seen early in LTP both tags synapses and stimulates the gene expression that produces the protein which subsequently interacts with the tag to generate persistent LTP. If this view is correct, then PKA activation is required with single and repeated tetanic stimulation, and it is not obvious why this leads to nuclear events in the latter case only.

8.8 Long-term depression is a synaptic weakening

A type of plasticity called long-term depression (LTD), in which synaptic strength is reduced, has been discovered in several brain structures, most notably the hippocampus and the cerebellar cortex.

In the hippocampus CA1 region repetitive, low-frequency (1 Hz) stimulation of a synapse for several minutes causes a gradual reduction in the slope and amplitude of the epsp which is stable for longer than an hour (*Figure 8.15a*). This is an example of **homosynaptic LTD** since synaptic activation induces LTD in the same synapses. Similar homosynaptic LTD in neocortex, striatum and nucleus accumbens, however, requires higher frequency tetanic stimulation not unlike that needed to produce LTP (*Figure 8.15b*).

In some locations, homosynaptic LTD requires more than just appropriate stimulation of the affected synapses. In the CA3 region, theta stimulation of one input produces LTD of a second input excited by single shocks given within the succeeding 100 msec (*Figure 8.15c*). The input receiving theta stimulation is, not surprisingly, potentiated. This is an example of associative LTD. In CA1, using brief depolarizations of the pyramidal cell given out-of-phase with low-frequency stimulation of the Schaffer collaterals, it was found that the precise time delay between cell depolarization and synapse activation determined the size of the LTD. If these events were simultaneous, LTP occurred. If the delay was too long, no depression was seen. Thus, this sort of LTD is an anti-coincidence detector but does associate presynaptic activation with postsynaptic depolarization if it occurs within a critical time. Some experiments suggest that it is the hyperpolarization that follows the conditioning tetanic stimulation or depolarization that is crucial for associative LTD induction.

In the cerebellar cortex, LTD occurs at synapses between parallel fibers, which are the axons of cerebellar granule cells, and the large Purkinje cells, the inhibitory GABAergic neurons that provide the sole output of the cerebellar cortex. This LTD requires coactivation of parallel fibers *and* climbing fibers, which, because of their numerous

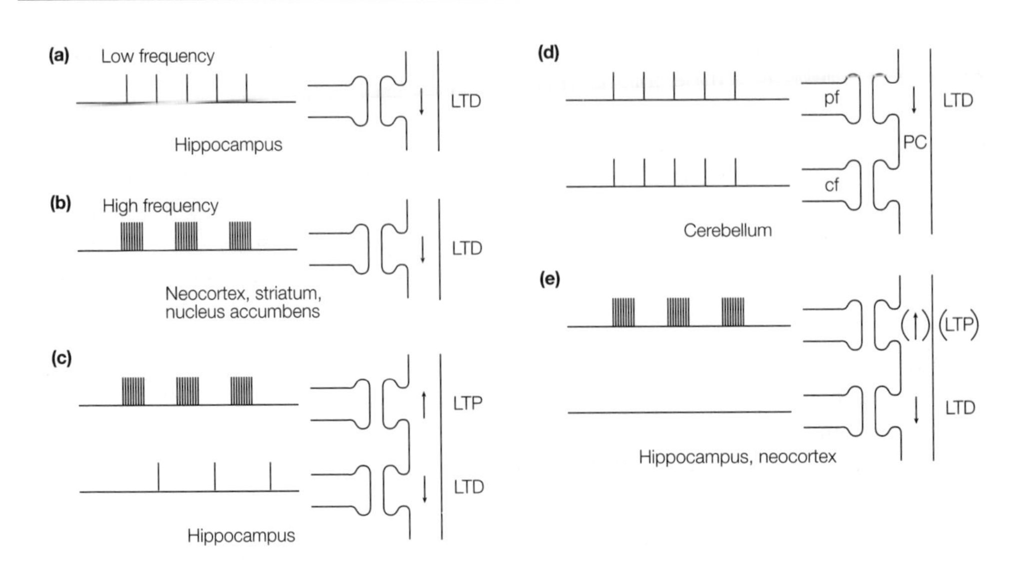

Figure 8.15. LTD can be produced in several ways. (a)–(d) are homosynaptic; (e) is heterosynaptic. (c) and (d) are associative. cf, climbing fiber; pf, parallel fiber; PC, Purkinje cell.

synaptic connections with their target Purkinje cells, cause it to depolarize extensively (*Figure 8.15d*).

Finally, **heterosynaptic LTD** is observed in the neocortex and hippocampus, in which powerful postsynaptic depolarization produced by tetanic stimulation of one input produces LTD in other quiescent synapses (*Figure 8.15e*). The first input is likely to experience potentiation, but this is not invariably so, and certainly not a necessary condition for LTD induction. Of course, heterosynaptic LTD is not input specific. It is intriguing that in both homo- and heterosynaptic cases, LTD at one population of synapses may be accompanied by potentiation of others, since the net result is to increase the *contrast* in synaptic weights.

8.8.1 LTD and LTP are closely related phenomena

LTP and LTD interact. Following induction of LTD, tetanic stimulation will impose LTP at the same synapse, and this can be reversed and LTD reimposed by prolonged 1-Hz stimulation. A wealth of evidence shows that both homo- and heterosynaptic LTD are mechanistically similar to LTP. Although there is no doubt that a rise in intracellular Ca^{2+} concentration is necessary for LTD – postsynaptic injection of Ca^{2+} chelators always blocks induction except perhaps in associative LTD – there are genuine differences and controversy about how this occurs. In the cerebellum, LTD is clearly not dependent on NMDARs. In homosynaptic LTD of hippocampal CA1 cells, some studies observed blockade by D-AP5 while others did not. The discrepancy may arise if LTD were independent of NMDAR *per se*, but requires a critical level of depolarization which is partly produced via NMDARs. Homosynaptic LTD of the visual cortex can be induced by tetanization after blockade of NMDA and nonNMDA receptors.

LTD at this location can be induced by quisqualate and prevented by heparin, which prevents IP_3-mediated Ca^{2+} release, suggesting that group I mGluRs are the vehicle for LTD here.

Associative LTD is problematic because different protocols seem to be needed at different synapses to induce it, and whether NMDA receptors are involved or not seems to be determined as much by the protocol as by the location.

The critical determinant of whether a given synapse exhibits LTD or LTP seems to depend on the exact extent of postsynaptic depolarization; induction of LTP needs a bigger depolarization than LTD. This is supported by intracellular recording from cells in slices of visual cortex in which a weak tetanic stimulation given alone has no effect on synaptic strength but leads to LTD in the presence of a low concentration of bicuculline (0.1–0.2 μM) and LTP at higher concentrations (0.3 μM). This suggests that critical threshold voltages must be reached to get LTD or LTP. So, a modest reduction of chloride current allows a small depolarization which crosses the threshold for LTD induction, whereas a larger disinhibition allows sufficient depolarization to reach the threshold for LTP induction. Similar conclusions can be drawn from studies in prefrontal cortex and striatum in which protocols that normally produced LTP gave rise to LTD when the Mg^{2+} concentration was increased, so reducing NMDAR-evoked depolarization.

Why is it that moderate depolarization induces LTD whereas large depolarization results in LTP? Currently, the view is that it is the amplitude of the Ca^{2+} signal – determined by the extent of the depolarization – that is crucial in whether a synapse is depressed or potentiated. Good evidence for this notion is that stimulus protocols that produce LTP result in LTD if the external Ca^{2+} concentration is lowered or when cells are injected with chelating agents to bring about only partial buffering of intracellular Ca^{2+}. The key question raised by all this is how a rise in intracellular Ca^{2+} is able to produce either LTP or LTD in the same synapse.

8.8.2 *Phosphatases are important in LTD*

A biochemical hypothesis explaining the dual role of Ca^{2+} in synaptic plasticity has been proposed by John Lisman at Brandeis University in the USA. His central idea is that the phosphorylation state of CaMKII determines synaptic strength; an abundance of the phosphorylated form results in LTP but when the dephosphorylated state predominates there is LTD. Now, CaMKII phosphorylation is controlled partly by a phosphatase, itself regulated by Ca^{2+}. At the intracellular Ca^{2+} concentrations seen at rest or in LTP, phosphatase I activity is low. At the moderate Ca^{2+} concentrations seen in LTD, however, phosphatase I activity is high, and this results in dephosphorylation of CaMKII.

The unusual Ca^{2+} dependence of phosphatase I arises because this enzyme is targeted by two second messenger systems with opposite effects on an inhibitor, inhibitor I (*Figure 8.16*). When phosphorylated, inhibitor I prevents phosphatase I activation, hence favoring the phospho-CaMKII form. Inhibitor phosphorylation is catalyzed by PKA, known to be activated in LTP. By contrast, calcineurin, a calcium/calmodulin-dependent phosphatase, by cleaving phosphates from inhibitor I, renders it incapable

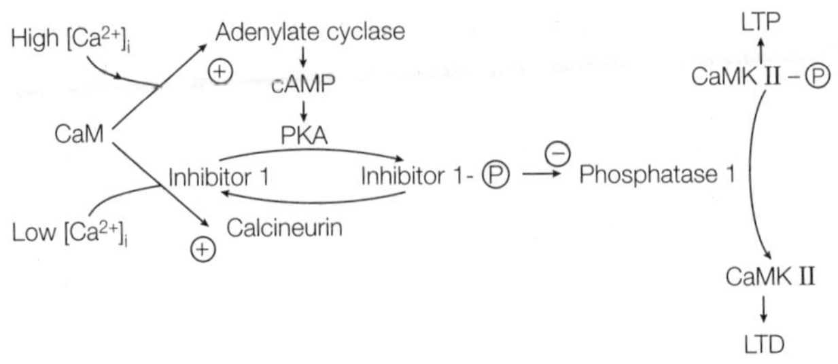

Figure 8.16. Phosphatase I is activated when inhibitor I is dephosphorylated.

of curtailing phosphatase I activity, so allowing net dephosphorylation of CaMKII and hence LTD.

Both of these second messenger systems are controlled by CaM, so how is it that bidirectional control can be effected? Probably, as the Ca^{2+} concentration rises, first calcineurin then adenylate cyclase are activated, either because calcineurin has the higher affinity for Ca^{2+}, or because it is closer to the point of Ca^{2+} entry. Certainly, large spatial gradients for Ca^{2+} exist within dendritic spines. As the Ca^{2+} concentration approaches 1 μM, not only will there be a net phosphorylation of previously inactive CaMKII molecules, but activation of adenylate cyclase causes deactivation of phosphatase I. The overall result will be high levels of stably phosphorylated CaMKII, and so synapse potentiation.

This model is supported by the finding that phosphatase inhibitors, like okadaic acid and calyculin A, block homosynaptic LTD induction or reverse pre-existing LTD in the hippocampus, and intracellular injection of calmodulin inhibitors also prevents LTD. That CaMKII is dephosphorylated in LTD has yet to be demonstrated and is clearly a crucial test of the hypothesis. However, even if the idea turns out to be flawed in detail, it seems reasonable to suppose that the mechanism which generates bidirectional plastic changes at a single synapse will not be too unlike that proposed.

8.8.3 *Cerebellar plasticity is atypical*

Axons of cerebellar granule cells bifurcate into parallel fibers, which traverse several millimeters in opposite directions, intersecting the planar dendritic fields of arrays of Purkinje cells (*Figure 8.17*). Each parallel fiber makes a single synapse with each Purkinje cell, though each cell receives inputs from some 100 000 parallel fibers. Purkinje cells have a second major input from climbing fibers which ascend from the inferior olive. Each Purkinje cell has just a single climbing fiber which branches to twine around the cell's dendrites, making numerous synapses, so that a climbing fiber action potential invariably triggers Purkinje cell firing. By contrast, parallel fiber inputs to Purkinje cells are weak. However, if a parallel fiber is active at the same time as climbing fiber input to the Purkinje cell, homosynaptic LTD of the parallel fiber–Purkinje cell synapse occurs.

This plasticity is thought to underlie motor learning in the cerebellum, and was first proposed on theoretical grounds by David Marr and James Albus. It is a Hebbian mechanism which associates neocortical motor commands, signaled by parallel fibers, with proprioceptive feedback relayed by climbing fibers, so that those synapses are depressed at which these two signals correspond. Presumably this conjunction of signals only occurs when motor output precisely matches motor cortex commands. Mismatch of motor intention and action will cause a failure in pairing climbing and parallel fiber activity in the same Purkinje cell. Re-excitation of depressed parallel fiber–Purkinje cell synapses will elicit optimal Purkinje cell output. Since this output is inhibitory, LTD results eventually in increased excitation of the corresponding motor output from cerebellar nuclei. The first experimental evidence for the Marr–Albus model of motor learning was provided *in vivo* in rabbit cerebellum by Ito and colleagues.

In contrast to cerebrocortical synapses, parallel fiber–Purkinje cell synapses show monotonic plasticity; they exhibit LTD but never LTP. The reasons for this are not clear. It may be because free Ca^{2+} concentrations in the Purkinje cell dendrites never reach the threshold required for potentiation, or because crucial components of the biochemical machinery required for LTP are missing from these cells.

The biochemistry of cerebellar LTD has been well researched. Climbing fibers are excitatory, and the large depolarization produced by their activation mediated by AMPARs produces Ca^{2+} influx via P and Q voltage-dependent Ca^{2+} channels. This has been visualized directly by Ca^{2+} imaging. The consequent rise in intracellular Ca^{2+} seems insufficient to produce LTD by itself, since Purkinje cell depolarization without parallel fiber stimulation does not cause synaptic depression. Parallel fiber synapses release glutamate which acts on AMPARs and mGluRs. Induction of LTD requires coincident Ca^{2+} influx and activation of both AMPARs and mGluRs. Pharmacological studies of cerebellar slices implicate the mGluR1 subtype. This is supported by: the abundance of mGluR1 transcript in Purkinje cells, anti-mGluR1 antibodies block LTD, and cerebellar slices from mGluR1 knockout mice show greatly attenuated LTD. Certainly, group I mGluRs are involved in cerebellar LTD since it can be induced by phorbol esters and abolished by PKC inhibitors.

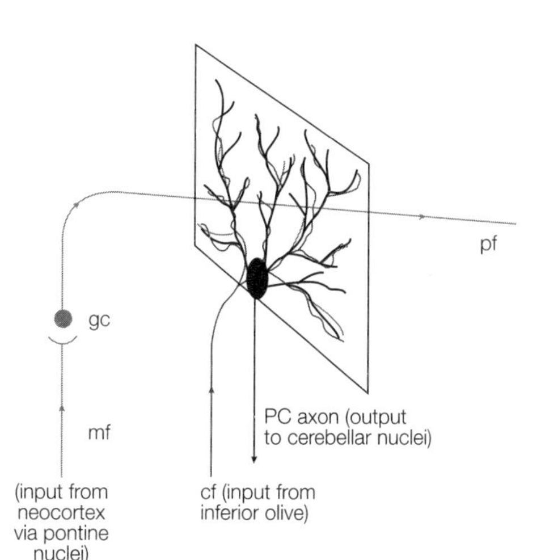

Figure 8.17. Purkinje cells receive two principal inputs. cf, climbing fiber; gc, granule cell; mf, mossy fibers; pf, parallel fiber.

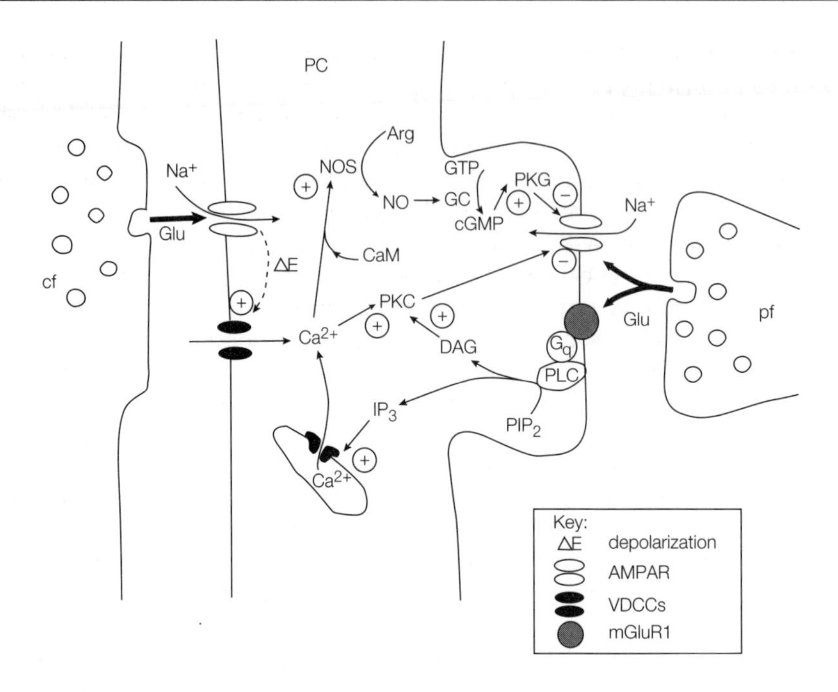

Figure 8.18. A model for the molecular events in cerebellar LTD. cf, climbing fiber; PC, Purkinje cell; pf, parallel fiber.

The final cause of the synaptic depression seems to be desensitization of the AMPARs. This can be produced in rat cerebellar slices by superfusion of 8-bromo-cGMP (but not cAMP analogs) and this is blocked by an inhibitor of PKG, suggesting that this kinase is responsible for AMPAR desensitization. As the NOS inhibitor, L-*N*-nitroarginine (L-NAME), blocks LTD, and PKG seems to decrease AMPAR responsiveness, it seems likely that NO is involved in LTD (*Figure 8.18*).

8.9 Voltage-dependent channels are implicated in plasticity

Studies on the populations of voltage-dependent ion channels in dendrites have revealed details of how the extensive depolarization needed as a coactivator of NMDAR for LTP may be generated and fine tuned. The classical view of dendrites as passive cables has been overturned. Calcium spikes have been identified in dendrites of both pyramidal and Purkinje cells. It appears that in many types of neuron, action potentials triggered at the axon hillock back-propagate across the soma to invade the dendritic tree. Unlike those in axons, action potentials in dendrites are mediated by both sodium and calcium channels. It seems then that spine synapses as well as being input devices also receive feedback on the cell's output; indeed they are optimally placed to detect any temporal coincidence of neuron input and output.

LTP at CA3–CA1 pyramidal cell synapses is prevented by blockade of sodium channels. This suggests that back-propagation of action potentials is a mechanism by which the depolarization needed to lift NMDAR Mg^{2+} blockade is produced. Moreover, the activation of dendritic calcium channels will augment Ca^{2+} entry via NMDARs and so contribute to the high Ca^{2+} concentrations needed to induce LTP. Recently, K_A-type potassium channels have been discovered in CA1 dendrites (see Section 4.6.2 for a discussion on K_A channels). Like all potassium channels, they serve to offset depolarizing influences over the voltage range at which they can be activated. K_A channels at dendritic spines are, however, likely to be *inactivated* by depolarization produced by glutamate-evoked epsps. This means that any back-propagated spike arriving at the spine will be larger – as the K_A current will not be able to dampen it – and in consequence there will be greater Ca^{2+} entry into the spine, enhancing the probability of synaptic potentiation.

Since inactivation of K_A channels is postulated to accompany LTP, what is the normal function of these channels? It seems likely that they impose limits on dendritic spiking. Dendrites have similar densities of VDSCs to axons, and may experience quite large depolarizations, due to summation of epsps, were it not for the hyperpolarizing effect of K_A currents.

In addition to back-propagation of action potentials causing Ca^{2+} signals, stimulation of synapses on distal regions of dendrites of CA1 cells in guinea pig hippocampal slices summates sufficiently to activate a large Ca^{2+} influx into proximal dendrites that cannot be blocked by AP5. Peak Ca^{2+} accumulations, revealed by fura-2 imaging, reaching 1 μM, occur at about 100 μm from the soma, 4 sec after stimulation, and fall off steeply at some 150 μm, suggesting that the density of the channels mediating the influx drops dramatically towards the distal region of the dendrites. This has been confirmed by mapping the distribution of L-type Ca^{2+} channels using a monoclonal antibody raised against the α_2-subunit. L-type channels are clustered at the zones where basal and apical dendrites emerge from the cell body and extend into the proximal regions of these dendrites, particularly the apical dendrite.

It seems that the Ca^{2+} signals in distal and proximal dendrites are produced in different ways. Whilst Ca^{2+} influx in distal regions augments NMDAR-mediated Ca^{2+} entry and is responsible for the localized events that result in LTP of specific synapses, the Ca^{2+} signal in proximal dendrites, which invades the soma where Ca^{2+} concentrations reach 0.4 μM, is clearly able to activate processes which affect the whole cell. What this means is that the same distal synaptic input that triggers local LTP by activating proximal dendrites may also initiate events in the cell body with cell-wide effects. A major challenge for neuroscience is surely to understand better the role of the global changes evoked by proximal dendrite Ca^{2+} signals and how these are orchestrated with localized distal events in plasticity.

8.10 LTP may be a molecular model for some types of learning

One of the factors motivating research in hippocampal, and other types of plasticity is that the mechanisms involved share much in common with those underlying learning and memory. So, considerable effort has been expended attempting to show that experimental procedures influence LTP and memory consolidation in similar ways. A number of correspondences which have compared hippocampal LTP with SNL tasks, known to require the hippocampus, are noteworthy.

Firstly, the ability of rats to remember the location of a hidden platform in a pool of milky water on the basis of distal clues, a commonly used test of spatial navigation skills (invented by Richard Morris), is disrupted by the NMDAR antagonists, AP5 and MK801. Control experiments show that the drugs do not affect performance by impairing visual discrimination or motor aspects of the task. Rats given extensive LTP *in vivo* subsequently perform poorly in the Morris water maze, which hints at similarities between LTP and SNL. Maybe hippocampal synapses already in some sense 'saturated' by prior LTP subsequently are unable to be modified plastically by learning.

Targeted knockout mutations are being used increasingly to allow the relationships between learning and cellular mechanisms to be studied. Hence, transgenic animals lacking the CREB gene show learning deficits as well as defects in plasticity. However, studies of mGluR1 knockouts are confusing. These animals have severe ataxia and impaired cerebellar LTD, which is exactly what would be predicted. They also have defects in hippocampal-dependent learning. However, one group finds normal NMDAR-dependent LTP at all locations whilst mossy fiber–CA3 LTP is lesioned, and another group finds impaired NMDAR-dependent LTP. Moreover, both groups find that an mGluR1 agonist apparently retains its normal pharmacology in the absence of receptor!

Some knockout studies have failed to support the idea that spatial learning and LTP are related. Thus, transgenic animals lacking the gene for the ubiquitous glycoprotein *Thy-1* fail to show LTP *in vivo* at the perforant pathway–granule cell synapse (though they do show normal LTP elsewhere), but have normal spatial learning in the Morris water maze. Lesion studies have shown that the perforant pathway is needed for spatial learning. Hence *Thy-1* knockout animals show a disconnection between LTP and learning. However, the perforant pathway has inputs to CA3 pyramidal cells as well as to granule cells, and it could be argued that it is the perforant pathway–CA3 connections that are crucial for learning.

More recent work has reinforced the link between plasticity and learning by showing that when plasticity is deranged by gene knockouts there are alterations in the properties of place cells in the hippocampus which parallel deficits in spatial learning. This is an important step since, while much research has attempted to find precisely what attributes of the environment or behavior of the rat determine the characteristics of place fields, little has been done to explore how synaptic plasticity shapes place cell properties at the cellular level. These latest studies begin to close the gap.

Transgenic mice which express a constitutively active form of CaMKII, because of a mutation which renders it Ca^{2+} independent, fail to show LTP in response to theta stimulation (though they do show LTP with higher frequency tetanus) and suffer severe deficits in spatial, but not nonspatial learning. Recording of place cell activity via chronically implanted electrodes was done whilst the mouse freely explored a 49-cm diameter cylinder with a white cue card covering 90°. The position of the mouse was monitored with a TV camera so that place cell firing could be examined in relation to where the animal's head is in the cylinder. By comparison with wild-type mice, the transgenic mice had fewer place cells, and these place cells fired at lower frequencies so that their place fields were less precise. Normally, place cells are quite reliable in that they will fire in the same way whenever the animal is returned to the cylinder. For transgenic mice, however, while place fields are stable during a 16-min recording session, they become unstable between recording sessions only 3 min apart. Apparently, any disruption to the animals exploration perturbs place cell behavior. This is reminiscent of the amnesia that afflicts humans with hippocampal damage. Whilst they can recall explicit events for short periods given the chance for rehearsal, any disruption results in forgetting.

That altering the function of a single protein produced deficits in plasticity and place cell firing is an exciting finding, and might be criticized – like most knockout experiments – on the grounds that during the development of the transgenic animals any number of secondary alterations may occur as a result of the primary mutation, and that these are responsible for the modified phenotype. However, suppression of the activity of the mutant CaMKII in adult mice rescues both theta-induced LTP and spatial learning, resulting in apparently normal mice, strongly suggesting that developmental changes have not confounded these particular experiments.

A thoroughly convincing way to overcome many of the valid objections made to conventional knockout technology is to engineer transgenic animals in which the function of a gene is deleted only when development is complete. Such an approach has been taken by Joe Tsien and colleagues at the Howard Hughes Medical Institute in the USA. They have used a homologous recombination system to delete the NMDAR1 gene in the third postnatal week, about 2 weeks after hippocampal development is completed. Moreover, the gene deletion is restricted to the CA1 pyramidal cells of the hippocampus. Recall that NMDAR2 subunits alone cannot form functional NMDARs and the deletion of the NMDAR1 gene effectively removes the slow epsp due to NMDARs from CA1. Clearly any alterations to phenotype can now be directly attributed to the loss of NMDAR at this single locus, whereas until now pharmacological approaches to block NMDAR function have inhibited all these receptors in a hippocampal slice, or with *in vivo* experiments in the whole brain. Attempts to work with conventionally produced NMDAR knockout mice have been thwarted by the fact that they die soon after birth. The method used to produce the CA1-specific NMDAR1 knockouts is called floxing (see ***Box 8.2***, p. 189).

Floxed NMDAR1 mice failed to show any NMDAR-dependent LTP in CA1, but had normal LTP in the dentate gyrus. Furthermore, they have deficits in SNL in the Morris water maze. Place cell properties were investigated, much as in the study with mutant CaMKII except that the animals explored linear and L-shaped tracks. CA1-NMDAR knockouts had larger and more diffuse place fields than controls. While individual cells of wild-type mice show direction selectivity, firing only when the animal is moving one way along a straight track, knockouts showed reduced direction selectivity. Finally, whilst the firing of place cells with overlapping place fields is highly correlated in normal mice, no correlation was seen in the knockout animals. Since up to half of all cells in the hippocampus may fire when an animal is in a particular environment, it is thought that spatial information is mapped by networks or ensembles of neurons. This type of coding requires correlated firing of neighboring place fields. That this is absent in the knockouts is sufficient to account for the impaired spatial navigation. What is perhaps surprising is that place fields (albeit degraded) exist for CA1 cells in the complete absence of NMDARs. They are presumably generated by activity in the normal CA3 region. Modeling shows that even with the high degree of convergence and divergence of CA3 connections onto CA1 cells, CA3 place-related input can produce exactly the sort of broad spatial tuning of CA1 cells observed experimentally.

Further understanding of the role of the hippocampus in learning and memory will doubtless follow if similar strategies are used to target precisely timed gene deletions in other regions, such as CA3 and entorhinal cortex.

8.11 Epilepsy is a chronic hyperexcitable state

For reasons that are not well understood, the CNS can switch from a normal to a hyperexcitable state capable of generating transient abnormal, synchronized rhythmic firing of large populations of neurons termed **seizures**. This hyperexcitable state is the hallmark of **epilepsy**, and individual seizures occur apparently spontaneously. The unknown processes which underlie the development of the hyperexcitable state are referred to as **epileptogenesis**, and have been the subject of considerable research effort. The motivation for this is readily understandable. About 1% of the population are affected and, although human epilepsy is not usually permanent, it is long lasting, with 40% of patients experiencing seizures for longer than 10 years. Moreover, despite the variety of drugs currently available to reduce seizure incidence, control is only acceptably good in 75% of individuals; 10% continue to fit at intervals of less than one a month, despite therapy.

Epilepsy research is beset by several problems. Firstly, epilepsy is a cluster of disorders – rather than a single disease – as shown by the variety of clinical manifestations, and the fact that some drugs are effective only in selective clinical types, all of which suggests that underlying mechanisms may be different. Secondly, although the modes of action of many drugs used to treat epilepsy are quite well understood, this has not provided much insight into the underlying pathophysiology. This is because most drugs currently used are anticonvulsants, which limit the spread of seizure activity, rather than antiepileptics, which prevent epileptogenesis. So, for example, two of the most commonly used drugs, phenytoin and carbamazepine, act as open channel blockers of voltage-dependent Na^+ channels, stabilizing them in the inactivated state and so curtailing high-frequency neuron firing. However, it has never been seriously suggested that primary defects in VDSCs are responsible for the epilepsies in which these drugs are efficacious. Thirdly, research aimed at identifying causes of epilepsy is undertaken using a variety of animal models of epileptogenesis both *in vitro* and *in vivo*. The relevance of these models to human epileptogenesis is often not obvious, would be hard to establish in any case, and, because of their diversity, the models throw up results which are inconsistent or even conflicting.

Despite these serious difficulties, a consensus is beginning to emerge that ionotropic glutamate receptors are implicated in epileptogenesis, and in the triggering and spread of individual seizures in the epileptic brain. Moreover, epileptogenesis may be thought of as a type of plasticity, it is seen and studied extensively in the hippocampus, and seems to share some features in common with LTP.

8.11.1 The hippocampus is an important focus for epilepsy research

One of the most extensively studied *in vivo* models, **kindling**, developed by Graham Goddard in the late 1960s, seems to closely resemble complex partial seizures during which the patient loses consciousness and seizure activity in the brain is localized. Electrodes are chronically implanted, usually into the amygdala, hippocampus or piriform cortex, and a brief high-frequency electrical stimulus (e.g. 1 sec, 60 Hz) is delivered once or twice each day for a couple of weeks. Although the first stimuli are subconvulsive, after 2 weeks the same stimulus triggers a behavioral fit (the rat shows forelimb clonus, it rears up and falls over) together with an electrographic seizure pattern. This pattern of kindling is long lasting, since a stimulus delivered after a long interval (e.g. 3 months) will often evoke a seizure. Kindling can also be produced chemically, for example by repetitive delivery via chronically implanted catheters of low doses of NMDA, or pentylenetetrazole, a convulsant $GABA_A$ antagonist.

Thin slices of hippocampus are popular for *in vitro* work. Epiletiform activity can be generated in slices by a variety of manipulations including hypoxia, raised K^+, low Ca^{2+} or Mg^{2+} concentrations in the bathing medium, kainate, NMDA, $GABA_A$ antagonists or trains of electrical stimuli delivered every 5–10 min; a sort of *in vitro* fast kindling. These acute preparations are only valid if epileptic activity continues long after the chemical manipulation has stopped; not always the case! Recently it has become possible to maintain hippocampal slices in culture for 7–15 days so that longer lasting kindling *in vitro* can be achieved.

8.11.2 The signature of the epileptic brain is abnormal burst firing

In both human epilepsy and its animal models, two types of electrographic activity can be distinguished. Of course, in the animal models, the behavior of individual neurons and populations of neurons that correlates with the electrographic signals can be revealed. This suggests that the electrographic record of a tonic–clonic seizure is caused by the synchronized activity of a large population of neurons. Each one suffers prolonged depolarization on which rides a high-frequency volley of action potentials (tonic phase) followed by rhythmic depolarizations with bursts of action potentials (clonic phase). More remarkable, however, is that the epileptic cortex shows abnormal electrographic activity even between seizures. Underlying these brief **interictal spikes** is the synchronized activity of a population of neurons in which each cell depolarizes for up to 80 msec, generating a burst of action potentials; events called **paroxysmal depolarizing shifts** (PDS).

Spontaneous burst firing is part of the normal repertoire of CA3 cells. CA1 cells do not usually show burst firing, although they will do so if directly electrically stimulated. When CA1 cells are being driven physiologically by CA3 cells via the Schaffer collaterals, they display brief epsps rapidly curtailed by a fast ipsp which results in a single or well-spaced action potentials. Hence, burst firing of CA1 cells normally is prevented by recurrent GABAergic inhibition.

During epileptogenesis in hippocampal slices, PDSs occur in both CA3 and CA1 cells, but CA1 activity is driven by CA3 since cutting the Schaffer collaterals silences CA1 cells. Thus the epileptic hippocampal slice differs from controls in that normal bursting is perverted into PDSs which resemble large epsps, have Ca^{2+} as a major ionic component, and reversal potentials similar to currents flowing though iGluRs. Moreover, this abnormal bursting is not inhibited in the CA1 region. All of this supports the commonly held contention that epileptogenesis is a matter of too much excitation and too little inhibition.

Direct excitatory connections exist between CA3 pyramidal cells, each of which activates about 5% of its neighbors. Computer modeling of a virtual hippocampal network shows that once inhibition is reduced an epsp in a CA3 pyramidal cell can spread via recurrent excitation of its neighbors and so throughout the CA3 region. Re-excitation will allow this to continue. No CA1 recurrent excitation exists, so once CA1 cells are disinhibited they must be synchronized by the aggregate activity of CA3 cells.

8.11.3 *Loss of inhibition in epilepsy may be due to faulty GABA release*

Although loss of glutamic acid decarboxylase (GAD)-positive terminals on pyramidal cell bodies has been reported in epileptic monkeys, GABA-containing neurons and GABA receptors are preserved in temporal lobe epilepsy. Indeed, induction of temporal lobe seizures in rats by injection of kainate into the CA3/CA4 region resulted in increases in GAD mRNA, as shown by *in situ* hybridization using an antisense RNA probe. The increase was found not only in the hippocampus but also in the thalamus and cortex and, except in the dentate gyrus, was bilateral despite the injection being given into only one hemisphere. The increase was not due to a change in the number of GAD-labeled neurons. Clearly this, and other studies which show raised GAD activity in kainate-evoked seizures, does not lend support to the idea of a loss of GABAergic function in epilepsy. However, microdialysis probes in patients show a smaller rise in extracellular GABA concentrations in the epileptic hippocampus during seizures than the nonepileptic hippocampus, suggesting that a deficit in GABA release may underpin the loss of inhibitory function. Apart from the classical vesicular Ca^{2+}-dependent route, GABA can be released in a nonvesicular Ca^{2+}-independent manner that depends on reversal of an electrogenic GABA transporter. It is during a seizure that conditions for reversing GABA transport may pertain. However, microdialysis in humans shows that nonvesicular release (in response to raised extracellular K^+ or glutamate challenge) is reduced in the epileptic compared with the nonepileptic hippocampus. Further, the number of GABA transporters is halved in electrically kindled rats. Thus, reduced reverse GABA transport might account for disinhibition in epilepsy.

8.11.4 *Ionotropic glutamate receptors are implicated in epileptogenesis*

Most of the evidence that iGluRs are important in epilepsy comes from investigation of the effects of antagonists. It is important to distinguish between the antiepileptic and anticonvulsant effects of drugs. By an antiepileptic effect is meant the prevention of epileptogenesis, that is it reduces or blocks the progressive development of seizure capability. A pure antiepileptic drug would have no influence on a concurrent seizure. By contrast, a pure anticonvulsant would be expected to inhibit seizure incidence or severity but have no effect on the capacity of the brain to generate seizures in its absence. Most real drugs exhibit a mixture of these effects.

The majority of studies show that NMDAR antagonists are good antiepileptics but rather weak anticonvulsants. They will completely block hippocampal kindling and are partly successful in blocking amygdala kindling. The inference is that NMDARs are implicated in epileptogenesis.

So, NMDAR blockade by AP5 or dizocilpine prevents the development of epileptiform activity in brain slices during kindling, although it does not stop such activity in slices already kindled; in these it will only attenuate individual bursts.

Similar findings are reported in rat neocortical slices made epileptogenic by incubation in an Mg^{2+}-free medium. Interictal bursting and seizure activity develop, which persist even if the slices are returned to Mg^{2+}-containing medium. Whilst an NMDAR antagonist prevented development of epileptiform activity, the AMPAR antagonist, CNQX, abolished the persistent seizure discharge. Hence, there seem to be two distinct phases to kindling; induction or epileptogenesis mediated by NMDARs and maintenance requiring AMPA receptors.

A characteristic of epileptiform activity in kindling is the occurrence of **afterdischarges** which follow more or less spontaneously after the kindling stimulus train. They are assumed to be equivalent to the events that cause a full-blown seizure in epilepsy. Using a combination of intracellular recording from hippocampal slices and a computer simulation of the CA3 region constructed from 1000 pyramidal cells and 100 inhibitory neurons, Roger Traub at Columbia University and his colleagues have developed detailed models for the precise electrophysiological behavior of this network of neurons, both normally and when made epileptic. This work shows that afterdischarges, which are prolonged synchronized bursting of CA3 cells, come about by activation of AMPA/kainate receptors which cause initial synchronization, and this is followed by NMDAR activation. This generates a slow depolarization of dendrites via Ca^{2+} influx which eventually triggers repetitive Ca^{2+} spikes, via L-type calcium channels in proximal dendrites, and burst firing of the soma.

Spontaneous bursting of CA3 cells in response to kainate or elevated extracellular K^+ concentration is eliminated by CNQX. NMDAR antagonists act to shorten the bursts but do not abolish them. Hence, unlike the situation with epileptogenesis, individual seizures in tissue with established hyperexcitability seem to be triggered by AMPAR and prolonged by NMDAR activation.

8.11.5 *Epilepsy may involve altered receptor sensitivity*

Kindled slices show a greater uptake of Ca^{2+} when challenged with NMDA. CA3 neurons in kindled hippocampal slices have a dramatically increased response to iontophoretically applied NMDA. Moreover, single-channel patch clamping shows NMDA to be more potent in opening NMDARs in neurons in kindled than in control slices. This last result means that increased sensitivity occurs at least partly at the level of single receptors. Although one study of human epileptic brain tissue removed during surgery indicated an increase in NMDA receptor binding, this is hard to reconcile with several studies that have failed to show long-lasting alterations in NMDAR subunit mRNAs, nor incidentally is there evidence for protracted changes in transcription of AMPA/kainate or $GABA_A$ receptors.

If altered sensitivity of individual NMDAR molecules is crucial to epileptogenesis, it is important to elucidate the events that bring this about. Unfortunately, progress in uncovering the second messenger involvement in animal models of epilepsy has so far been less revealing than corresponding studies in LTP or LTD. The role of an obvious candidate, CaMKII, in kindling is ambiguous. Both increases and decreases in activity of the enzyme have been reported. Furthermore, the idea that CaMKII may act similarly in LTP and epileptogenesis is contradicted by αCaMKII knockout mice. These animals have a deficit in LTP, and might be expected therefore to have a high threshold for seizure induction. In complete contrast, however, these knockouts can develop **status epilepticus** lasting up to 2.5 h as a result of a single tetanic stimulus.

Apart from NMDARs, altered sensitivity of $GABA_A$ receptors has also been found to accompany kindling. For example, in guinea pig hippocampal slices, delivery of five trains of tetanic stimuli every 30 min resulted in a decrease in spontaneous and stimulus-evoked GABAergic ipsps, associated with a fall in responsiveness of CA1 cells to iontophoretically applied GABA. This was matched by enhanced synaptic excitability and, eventually, after ipsps dropped below 10% of control values, by epileptiform activity including PDS. Significantly, with NMDA receptors blocked by D-AP5, the tetanic stimulus trains had no effect on ipsps or epsps, and there was no alteration in response to applied GABA. The inference is that kindling causes an NMDAR-mediated reduction in inhibition by $GABA_A$ receptors. This is an exciting finding because it couples increased excitation to a decreased inhibition; the two processes can be seen as mutually dependent.

The link between NMDAR activation and reduced $GABA_A$ inhibition involves Ca^{2+}. The Ca^{2+} flux through VDCCs has been shown to reduce the Cl^- current through $GABA_A$ receptors in dorsal root ganglion cells. Whole-cell patched CA1 cells show a massive reduction in $GABA_A$ currents in response to NMDA which is blocked by inhibitors of calcineurin. The significance of this is reinforced by the fact that calcineurin inhibitors greatly retard epileptogenesis in amygdala kindling (*Figure 8.19*).

Interestingly, epileptogenesis in kindling can now be understood in the following way: whilst low-frequency stimulation activates only AMPARs, and resultant epsps are rapidly curtailed by recurrent $GABA_A$-mediated inhibition, high-frequency stimulation with its larger AMPAR-mediated depolarization causes NMDAR activation which by preventing GABAergic inhibition allows a hyperexcitable state to emerge.

This pattern of iGluR involvement is reminiscent of LTP at CA3–CA1 synapses where induction is NMDA-mediated and maintenance requires AMPARs. These parallels hint that epileptogenesis is a type of plasticity, albeit maladaptive. There are, however, substantial arguments *against* the notion that kindling occurs via a conventional LTP mechanism. Some pathways, e.g. entorhinal–granule cell synapses, involved in *in vivo* kindling do not show epsp potentiation that inevitably accompanies LTP. Secondly, LTP *in vivo* usually decays within days or weeks, whereas kindling lasts for at least 3 months and may be permanent.

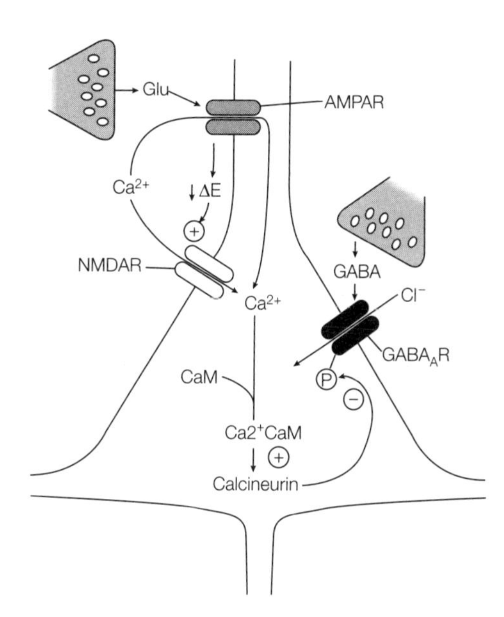

Figure 8.19. NMDA receptor activation causes decreased inhibition by favoring $GABA_A$ receptor dephosphorylation.

A further difference between the two processes is that for kindling to occur the stimulus must evoke afterdischarges. In LTP, the stimulus trains do not produce these spontaneous bursts. Afterdischarges result from the fact that hyperexcitability requires activation of many and widespread synapses whereas LTP is highly localized. It is surmised that afterdischarges account for the greater durability of kindling over LTP, though it is not known how.

8.11.6 *What produces epilepsy* in vivo?

Kindling is normally produced by tetanic stimulus trains or a number of drugs, none of which are endogenous, so the question arises about what it is that initiates any kindling-like process in epilepsy. One possibility is enhanced glutamate secretion. This may play a part in kindling. (1s,3s)-ACPD is a fairly selective agonist of presynaptic group II mGluRs. It inhibits veratridine-induced release of glutamate and aspartate from cortical synaptosomes and reversibly blocks kindling epileptogenesis.

There is currently considerable controversy as to whether glutamate release is increased in epilepsy. Microdialysis of human brains during complex partial seizures shows large increases in glutamate concentration in epileptic compared with nonepileptic hippocampus. Interestingly, seizures also caused increases in GABA concentration, but this was more marked in the nonepileptic hippocampus. This may relate to the reduced number of GABA transporters in epileptic tissue via which Ca^{2+}-independent GABA release may occur during seizures. The synaptic concentrations of glutamate inferred by these studies are above those shown to be excitotoxic to neurons in culture. This is supported by the common finding that the hippocampus in chronic epilepsy shows loss of pyramidal cells, accompanied by a reactive gliosis; a phenomenon first recognized some 150 years ago and originally described as Ammon's horn sclerosis.

However, many *in vivo* dialysis studies in animals in which seizures are induced with convulsants such as kainate, bicuculline or picrotoxin have failed to provide evidence for increased release of glutamate. One such, done at the Institute of Neurology in London, found no change in extracellular glutamate concentration from the hippocampus during picrotoxin-induced seizures. Moreover, whilst blockade of glutamate reuptake with L-*trans* pyrrolidine-2,4-dicarboxylate in normal rats produced massive increases in extracellular glutamate concentration for up to 10 min, it failed to produce any significant electrographic changes.

If epileptogenesis does not require elevated glutamate concentrations, what might be the necessary trigger? Possible candidates include an abnormal up-regulation of glutamate receptors, or an increase in their sensitivity. Direct hippocampal injection of herpes simplex virus vector containing the GluR6 subunit causes overexpression of kainate receptor and this results in limbic seizures in the rat. Focal epilepsy in encephalitis is associated with production of autoantibodies directed at GluR3 subunits and which appear to activate a population of glutamate receptors. Some 20% of all epilepsies are inherited, and linkage analysis has shown that one mutation, on chromosome 11, is localized to the same region as that thought to harbor the GluR4 gene.

Any successful theory of a human epilepsy must account for how commonly associated risk factors (brain trauma, tumors or infections) initiate the epileptogenesis, how the hyperexcitable state develops over time, what initiates, allows spread of, and terminates individual seizures, and will certainly be advanced by molecular biological approaches.

Box 8.1. Electrophoretic mobility shift assay (EMSA)

EMSA, also known as the gel retardation or gel shift assay, is a method for measuring the ability of a given protein to bind to a specific DNA fragment. The assay is based on the fact that when the DNA fragment and the protein are bound they form a complex which moves more slowly than the unbound DNA on an electrophoresis gel. The DNA fragments are usually manufactured on automated synthesizers using the known sequences of the putative protein-binding sites and then labeled, whilst the protein may be a mixture of proteins isolated from a cell nucleus or purified or recombinant pure protein. Even small amounts of protein binding will lead to the appearance of an extra shifted band on the gel, hence the name gel shift assay.

Box 8.2. Floxed receptor knockouts

Floxed NMDAR1 receptor knockouts are transgenic animals engineered so that the NMDAR1 subunit gene is excised from a single cell type (CA1 pyramidal cells) 3 weeks postnatally.

The technique uses a bacterial recombinase, Cre (causes recombination) an enzyme which recognizes a short sequence of DNA called the *loxP* (location of crossing over) site. Cre will excise DNA located between two *loxP* sites if they are in the same orientation. The key idea is to introduce loxP sites into a mouse so they flank the gene of interest. The gene is said to be floxed (flanked by lox*P*). The gene will be excised only in cells which express active Cre.

To produce floxed NMDAR1 knockouts two transgenic mouse lines were generated:

(i) fNR1 mice which are homozygous for the *lox P–NMDAR1–loxP* sequence;
(ii) Cre mice which are heterozygous for Cre expression only in hippocampal cells.

Production of homozygous fNR1 mice. For engineering of the fNR1 mice, a targeting vector was constructed in which two *loxP* sites were inserted into the NMDAR1 gene (*Figure 1a*). One was placed into intron 10 and the other, together with a *neo*^{r} gene, was inserted downstream of the last exon. In this way, the *loxP* sites did not interfere with coding regions. The vector DNA was introduced into mouse ES cells by electroporation. Neomycin-resistant ES cell colonies were picked, grown up and homologous recombinants selected. These were injected into mouse blastocysts. Mice homozygous for fNR1 were bred using standard selection methods (see ***Box 7.2***, p. 148).

Production of heterozygous Cre mice The plasmid vector used to make the Cre mice contained four elements (*Figure 1b*). The *Cre* gene is placed under the control of the αCamKII promoter which is known to be expressed in restricted regions of the forebrain but only 3 weeks postnatally. It was hoped this would permit the expression of Cre to be confined to the hippocampus and neocortex and to occur after brain development was complete. The nuclear localization signal transports the Cre recombinase to the nucleus, and the inclusion of the splicing

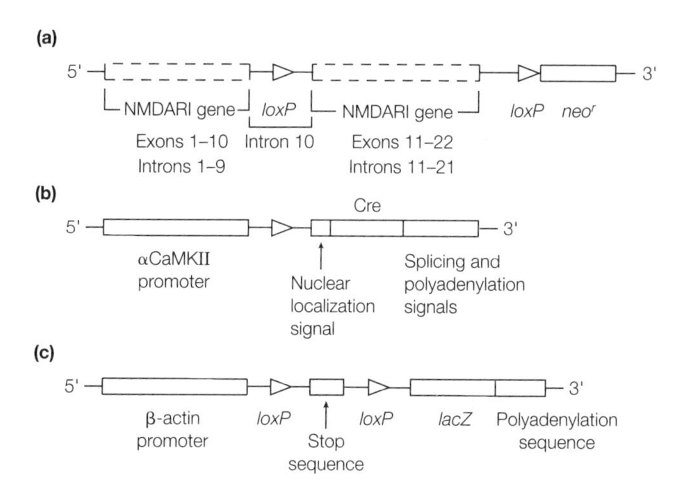

Figure 1. Targeting vectors used in the development of NMDAR1 knockout mice. (a) fNR1 targeting vector; (b) the Cre plasmid vector; (c) the β-actin/lacz targeting vector.

Continued

and polyadenylation signals at the 3′ end of the vector ensures that the resulting transcript is recognized by the protein synthetic machinery of the host cell. The vector was injected into the pronucleus of fertilized oocytes which were implanted into foster mice. The resulting founder mice were backcrossed with normal mice, and those founder mice with germline transmission were identified by Southern blot analysis of the DNA.

These Cre transgenic animals are heterologous recombinants so the position of the transgene varies with each animal. Since the pattern of expression of the αCamKII promoter is affected by where the transgene is inserted, the subsequent distribution of Cre expression was examined for each Cre transgenic mouse line. This was done by crossing the Cre transgenic animals with lacZ transgenic animals who carried the *lacZ* gene preceded by the β-actin promoter, which is widely activated in post-mitotic neurons, and a stop signal flanked by two *loxP* sites (*Figure 1c*). In Cre/lacZ double transgenic mice, in cells expressing the Cre recombinase, the stop sequence will be deleted and so β-galactosidase is expressed. These cells show blue by X-gal staining of brain sections. Cells not expressing Cre will appear white. In this way, a Cre transgenic mouse line was identified which expressed Cre only in post-natal CA1 pyramidal cells.

The CA1-NMDAR1 knockout mice were then generated by mating those mice carrying the *Cre* transgene (restricted to CA1) and mice homozygous for fNR1. Although this method was used to produce a specific knockout, it is virtually certain that this method could be applied to other genes. By placing Cre expression under the control of different promoters, it should be possible to produce animals in which floxed genes are deleted from different cell types and at various times.

Chapter

Molecular mechanisms in neurodegenerative disease

9.1 Introduction

The discovery, using molecular biology, of an ever increasing number of genes implicated in a variety of neuropathologies has stimulated the development of models of disease which may eventually form the basis for possible prevention or treatment.

For those diseases which have a clear genetic basis, an important step is to locate and sequence the gene responsible. In those cases where there is a known defective protein, then a synthetic oligonucleotide probe could be used to locate the gene. However, when the biochemistry of the underlying defect is not known, then the techniques of **linkage analysis** using **restriction fragment length polymorphisms** (RFLPs) allow the gene to be identified.

9.2 Huntington's disease causes a specific pattern of neuronal loss

Huntington's disease (HD) is an autosomal dominant disease characterized by involuntary movements, which may start as facial ticks which gradually spread and become more pronounced. These dyskinesias are responsible for the other names for the disease, Huntington's chorea (chorea is derived from the Latin choreus or dance) and St. Vitus' dance. There are also psychiatric symptoms, including hallucinations, delusions and dementia. The onset of the disease is usually between 40 and 50 years of age, and results in death in 15–20 years. The similarity of some of the symptoms with those of schizophrenic patients and the observation that L-3,4-dihydroxyphenylalanine (L-DOPA) aggravates the symptoms, and drugs which block dopamine activity reduce the symptoms, has led to the idea that HD is due to an excess of dopamine. The pathology of HD includes a severe loss of neurons in the caudate nucleus and putamen, which may be reduced to 10% of their original size, leading to an enlargement of the lateral ventricle. Later, cells in the cortex may be affected, and in advanced cases brain weight may be reduced by 30%. The cell loss in the basal ganglia leads to a loss of the projections to the substantia nigra and the globus pallidus. These pathways are GABAergic with co-localized neuropeptides, enkephalin and substance P. As dopaminergic projections to the striatum remain intact, with a loss of GABAergic inhibition there is an increase in excitatory activity to the cortex.

9.2.1 The gene for HD has been located and sequenced

HD occurs in 5–10 per 100 000 of the population but, because it is almost always familial, in some parts of the world it can be relatively common. In the Lake Maracaibo region of Venezuela, there exists the largest family known with HD; almost 12 000 individuals of whom over 250 have HD, with a further 4000 at some risk of being affected later in life. It was using blood samples and family trees from many of these individuals that the HD gene was first mapped to a specific locus on chromosome 4 and, finally, in 1993, the Huntington's Disease Collaborative Research Group identified and sequenced the gene itself, interesting transcript 15 (*IT-15*).

The discovery of the gene depended on the fact that the human genome contains a large number of variants or **polymorphisms** which are passed from generation to generation. Many of these polymorphisms may be responsible for the differences between individuals and/or may be phenotypically silent, if they produce no change in amino acid sequence or fall outside the coding regions of genes. However, these polymorphisms can be recognized by restriction enzymes, and this polymorphism forms the basis of linkage analysis using RFLPs (***Box 9.1***, p. 212)

9.2.2 HD is caused by an excess number of trinucleotide repeats

The *IT-15* gene codes for a protein of approximately 348 kDa called huntingtin. This protein, whose normal function is not known, is expressed in all tissues and is obviously essential for life, as homozygous knockout mice which produce no huntingtin protein do not survive beyond 9 days of gestation. The gene is characterized by a pattern of repeated trinucleotides CAG, coding for polyglutamine, in the first exon, followed by a smaller stretch of repeated CGG (polyproline). When the gene sequences of normal and HD individuals are compared, the length of the polyglutamine tract varies widely, but the pattern is clear. In normal individuals, there are between nine and 39 (mean ~18) repeats, whilst in HD the number of repeats varies from 36 to 121 (mean ~44). These studies showed that those individuals with more than 40 repeats will almost certainly develop HD, whilst anyone with less than 30 repeats is usually normal, allowing the number of repeats to have a predictive value for individuals who may suspect that they have inherited the gene but are as yet symptomless. For those people falling in the range 31–39, this is more problematic.

Interestingly, the number of repeats has an inverse correlation with the age of onset of the disease. In those individuals with more than 50 repeats, the age of onset may be before 30 years, although there is a wide variation possibly due to difficulties in deciding the exact onset of clinical symptoms. The number of repeats may change between generations. When the HD gene is inherited from the mother, the number of repeats tends to vary by about four, with a tendency towards an increase. However, when the gene is inherited from the father, the number of repeats varies much more and tends to increase. When inherited in this way, the number of repeats may double. This suggests that events during the formation of gametes may allow a change in the number of repeats, with spermatogenesis being particularly affected. This is substantiated by the large variation in repeat numbers seen in the sperm of patients with HD.

In sporadic HD, where there is no family history of HD, the typical expanded CAG repeats have also been observed. On examination of the unaffected father of one of these cases, his repeat number was in the intermediate 30–40 range, suggesting that an increase in repeat number during spermatogenesis was sufficient to increase the number of repeats into the affected range.

9.2.3 *The role of huntingtin is unknown*

Huntingtin lacks homology with any previously known protein, but it contains some recognizable motifs. Part of the gene codes for a leucine zipper motif (see Chapter 8), which would allow binding of other proteins with the same motif such as the transcription factors, Fos and Jun. It is also of note that the CAG trinucleotide is a sequence found in the TF-binding sites of both AP-1 and CRE. As huntingtin has no sequences which would suggest a membrane-spanning or membrane-associated protein and has been located both in the cytoplasm, particularly around synaptic vesicles, and in the nucleus, it is possible that the normal role of huntingtin is in the regulation of transcription of other genes. A brain-specific protein, huntingtin-associated protein (HAP-1), which binds to huntingtin with an affinity which varies with the number of CAG repeats, recently has been cloned, but whether this protein is only involved in the normal gene function or has a role in HD is not known.

The role of the expanded huntingtin in the HD-affected individual is not known, but there is evidence to suggest that the mutation is of the type termed 'gain of function'. This is where the affected locus gains a new function and the effects are not due simply to the loss of one allele. This is strongly suggested by the fact that individuals who are homozygous for the extended HD gene are no more affected and the age of onset is not greater than in heterozygotes. Also, loss of a normal HD allele does not cause HD, as patients suffering from Wolf–Hirschhorn syndrome do not suffer HD-like pathology yet their disease is caused by the loss of a region of chromosome 4 which includes the *IT-15* gene.

Various hypotheses have been proposed to explain how the extended CAG repeats could cause the pathology of HD. In light of the possible binding of TFs to the CAG sites, it has been suggested that HD is caused by a disruption of normal gene regulation. Simpler explanations involve the long polyglutamine sequences acting as substrates for transglutaminase enzymes which could gradually produce abnormal cross-linking of huntingtin. A similar explanation involves the aggregation of polyglutamine, producing insoluble, toxic precipitates. However, these explanations do not explain why homozygotes are no more affected than heterozygotes.

9.2.4 *HD can be mimicked experimentally*

The pattern of cell death seen in HD can be mimicked by the injection into the striatum of excitatory amino acid analogs, in a process which is known as **excitotoxicity**. Apart from its involvement in any other pathologies, the importance of excitotoxicity is that it is a mechanism responsible for killing neurons during the brain ischemia that accompanies a stroke. Before examining the possible role of excitotoxicity in HD, it is necessary to examine how excitotoxicity occurs and how it may kill cells.

9.3 Excitotoxicity involves inappropriate activation of glutamate receptors

The term excitotoxicity was first coined by John Olney who observed that a number of excitatory amino acids (EAAs), including kainate, killed neurons in the hypothalamus, with a potency that correlated with their efficacy in depolarizing cells. The lesion is selective in that it kills neurons with their cell bodies exposed to the EAAs but spares those whose axons only have been dosed. Glial cells seem unaffected.

Two distinct phases of glutamate receptor-mediated neuronal degeneration have been described, acute and delayed. The acute phase seems to be an osmotically driven water influx causing cell swelling and lysis following Na^+ entry into the neuron via iGluRs. The delayed phase has increases in intracellular free Ca^{2+} concentration as the trigger for excitotoxic cell death, which may arise in several ways. Calcium influx can occur through activated AMPA/kainate receptors (recall that some GluR isoforms have a significant Ca^{2+} conductance), or via NMDARs and VDCCs provided that the neuronal membrane is sufficiently depolarized. Even Na^+ influx via AMPA/kainate will result in increased Ca^{2+} entry by activating Na^+/Ca^{2+} exchange transporters. Alternatively, Ca^{2+} can be released from internal stores. Group I mGluRs are G protein-linked receptors (see *Table 6.2*). When stimulated by glutamate, they activate a pathway which results in the release of Ca^{2+} from the endoplasmic reticulum into the cytoplasm (Section 6.7.3).

9.3.1 *Cell death may be due to an increase in intracellular Ca^{2+} concentration*

The link between increased intracellular Ca^{2+} and cell death is not established unambiguously. Three hypotheses are in vogue, but they should not be regarded as mutually exclusive.

(i) Firstly, Ca^{2+} activates degradative enzymes, endonucleases and proteases (e.g. calpain I) the activity of which terminally disrupts cell function.

(ii) Secondly, Ca^{2+} activates mechanisms which generate highly reactive **free radical species** which oxidize numerous macromolecules, but polyunsaturated fatty acids are particularly vulnerable to peroxidation. The peroxy radicals generated propagate the lipid peroxidation in a positive feedback cascade, producing irreversible changes in physical and chemical properties of cell membranes. Several processes are candidates for free radical generation in excitotoxicity.
NMDAR activation has been shown to cause stimulation of Ca^{2+}-activated PLA_2 which catalyzes the hydrolysis of membrane phospholipids to release AA. The action of cyclooxygenase on AA produces a peroxylipid and subsequent generation of free radicals including superoxide anions (O_2^-) (*Figure 9.1*).

Under normal circumstances, mechanisms to inactivate these free radicals operate. These are particularly important in the brain, which relies exclusively on aerobic metabolism, and involve α-tocopherol (vitamin E), superoxide dismutase (SOD)

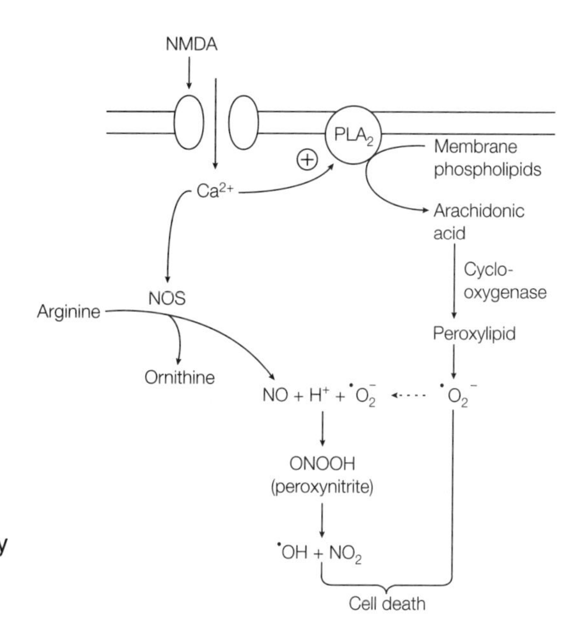

Figure 9.1. NMDA excitotoxicity is probably mediated by mechanisms involving PLA_2 and NOS.

and reduced glutathione. Since cyclooxygenase is itself stimulated by free radicals, the action of free radical inactivation mechanisms limits cyclooxygenase-mediated free radical generation. Presumably excessive NMDAR activation results in swamping of these inactivation mechanisms and so an uncontrolled climb in concentrations of free radicals. NMDAR-mediated Ca^{2+} influx also activates Ca^{2+}-inducible NOS. NO rapidly diffuses to neighboring neurons and partakes in free radical generation with resulting cell death.

Kainate-mediated excitotoxicity seems not to involve NOS activation, and the role of AA metabolism is ambiguous. It may be due to the activation of xanthine oxidase, as cerebellar granule cells are protected from kainate excitotoxicity by allopurinol, a xanthine oxidase inhibitor. Xanthine oxidase reacts xanthine with oxygen to produce uric acid, superoxide anions and peroxide. Interestingly, calpain I, a Ca^{2+}-activated peptidase (see above), converts xanthine dehydrogenase to xanthine oxidase, thus possibly linking kainate-induced Ca^{2+} influx and increased xanthine oxidase activity.

(iii) A third hypothesis which attempts to link the rise in intracellular calcium to cell death has it that Ca^{2+} causes persistent activation of immediate-early response genes (IEGs) which results in delayed neuronal death (DND). IEGs are genes transcribed, normally briefly, within minutes by the arrival of a variety of extracellular signals. IEGs encode proteins, such as the TFs, Fos and Jun, which regulate the expression of late response genes by binding to specific regulatory DNA sequences, such as AP-1 sites.

IEGs are expressed at low levels in nonstimulated neurons, and physiological stimuli result in expression of AP-1-containing genes over the following 8–12 h. By contrast, transient global ischemia or kainate in adult rats, which result in DND, is accompanied by increased levels of TFs over a much longer period (16–72 h). The morphology of cells stricken by excitotoxins suggests that DND is due to **apoptosis** (genetically programed cell death) rather than **necrosis**.

9.3.2 *Excitotoxic cell death may be the cause of the damage in HD*

The time course of all these excitotoxicity experiments is such that they can inform us about acute events such as strokes, but their relevance to HD is still under investigation. The normal striatum receives a large number of glutamatergic inputs from the cortex, so there is the potential for a large amount of local glutamate release. Although both kainate and ibotenate can cause the loss of striatal neurons, it is the injection of quisqualate which most closely mimics the pattern of changes seen in HD. Particularly, there is a selective loss of the medium-sized spiny GABAergic projection neurons, which themselves receive glutamatergic input from the cortex. There is a relative sparing of the large neurons which contain ACh, somatostatin and neuropeptide Y. Now, quisqualate is an agonist at both AMPA and group I mGlu receptors, and activation of both these types of glutamate receptor can cause excitotoxicity. However, in the presence of AMPA antagonists, it has been shown that quisqualate enhances NMDA-mediated toxicity. Group I mGluRs (mGluR1 and mGluR5) stimulate hydrolysis of PIP_2 and hence the IP_3-mediated release of Ca^{2+}. Activation of PKC, by Ca^{2+} and DAG, and the subsequent phosphorylation of NMDA receptors (Section 5.10.3) decreases the voltage-dependent Mg^{2+} block and allows activation of NMDA receptors at less depolarizing potentials. The distribution of GluRs in the striatum also suggests that group I mGluRs may be important; whilst there is no definite correlation between NMDA receptor distribution and the cells which die in HD, the mGluR5 is only found on the medium-sized striatal neurons and is not found on the neurons that are spared. The appearance of cells killed in HD shows evidence of apoptosis, as detected by the characteristic pattern of internucleosomal DNA fragmentation. This same pattern is also observed in early excitotoxic cell death, although later cell death is by necrosis. Huntingtin has been linked to apoptosis, by studies which showed that two enzymes which are involved in the cascade of reactions which produces apoptosis can interact with huntingtin.

Brain-scanning techniques such as PET and nuclear magnetic resonance (NMR) imaging have shown reduced glucose metabolism and elevated levels of lactate in HD patients. There is also evidence of a disruption in mitochondrial electron transport, all of which suggests that there may be an impairment of neuronal energy metabolism which plays a role in HD. The lack of energy could produce a build up of free radicals which could compound the effect of radicals produced by excitotoxic mechanisms.

9.3.3 *Other neurological diseases may involve excitotoxicity*

The evidence that excitotoxicity is an important component of strokes comes from animal studies. That brain ischemia causes the extracellular concentration of glutamate to rise is well documented by *in vivo* microdialysis studies. NMDAR antagonists, including 2-amino-7-phosphonoheptanoate (D-AP7), dizocilpine and dextrorphan, have been shown to reduce loss of neurons in hypoxic or ischemic models of stroke. Glutamate-induced death of neurons in culture is curtailed by removing Ca^{2+} from the bathing medium.

The link between ischemia and increased extracellular glutamate is easy to appreciate. During hypoxia, ATP concentrations are depleted so that energy-requiring processes are compromised. This will include the Na^+/K^+-ATPase. Decreased activity of the cation pump, by increasing intracellular Na^+ concentration, will promote osmotic cell swelling and Ca^{2+} influx via the Na^+–Ca^{2+} antiport. Loss of sodium pump activity will result in an increased extracellular K^+ concentration, causing membrane depolarization. Depolarization, by activating VDCCs, will promote the increased glutamate release that causes excitotoxicity. It is worth noting that the conditions needed for activation of NMDARs have been met; both synaptic glutamate and membrane depolarization to lift the Mg^{2+} blockade. Termination of the action of glutamate at synapses involves reuptake by a Na^+/K^+-dependent transporter (Section 7.16). Since this depends indirectly on metabolic energy, it too will be compromised in ischemia, and glutamate concentrations will remain high. In fact, it has been suggested that the intracellular Na^+ and extracellular K^+ concentrations can become high enough so that the Na^+/K^+-dependent glutamate transporters work in reverse and glutamate is actually transported out of the neuron, making a bad situation worse.

Excitotoxic cell death has also been implicated in amyotrophic lateral sclerosis (ALS), also known as motor neuron disease, although whether this is the primary or secondary event is not known. ALS involves the selective loss of spinal cord and cortical motor neurons, which leads to gradual paralysis with eventual respiratory depression and death. The involvement of glutamate-mediated excitotoxicity in ALS was suggested by the observation that two diseases in which there is degeneration of upper and lower motor neurons are both possibly caused by excitatory amino acid analogs. These are lathyrism, caused by ingestion of the toxin, β-*N*-oxalylamino-L-alanine, from the chickling pea, *Lathyrus sativus*, and Guamanian ALS–PD–dementia complex, where β-*N*-methylamino-L-alanine (L-BMMA) has been proposed as the responsible agent. L-BMMA is found in the flour produced from cycad nuts which are a traditional food source in the islands of Guam and Rota. In macaque monkeys, ingestion of L-BMMA can cause many effects which are similar to those seen in Guam disease, although the pattern of neuronal loss is not identical. However, more recently, the involvement of L-BMMA has been questioned. Firstly, it has been noted that during preparation of the cycad flour it is washed extensively, which removes most of the toxin and, secondly, the disease can occur up to 20–30 years after exposure to the toxin and there is no evidence that L-BMMA can act latently. However, circumstantial evidence points to the fact that the decline in incidence of Guam disease since the 1940s has coincided with a decrease in the use of the nut flour.

There have been reports of varied changes in glutamate and aspartate concentrations in the plasma, brain and spinal cord of ALS patients, but none of these are conclusive. There have also been reports of reduced glutamate transport in some but not all brain regions in ALS. A failure in the removal of released glutamate from the synaptic cleft could lead to inappropriate activation of AMPA/kainate receptors and, subsequently NMDARs and mGluRs, all of which could lead to excitotoxicity.

Most cases of ALS are sporadic, but there are a small number of familial forms where the disease has been linked to mutations of a gene on chromosome 21 which encodes the Cu/Zn-containing form of SOD (*SOD1*). Initially it was thought that the cell death in these forms of ALS was therefore due to a loss of SOD activity resulting in enhanced damage by free radicals. This is supported by the fact that inhibition of *SOD1* expression in transgenic mice leads to apoptosis in spinal motor neurons, indicating that SOD activity is required for the normal protection of motor neurons from damage. However, it has been suggested that there are additional gain-of-function effects which are not related to loss of normal SOD activity. There are patients with normal SOD activity who nonetheless have ALS, and this has also been seen in transgenic mice, generated by heterologous recombination (see ***Box 7.2***, p. 148), who possess both the original *SOD1* genes and the human mutant *SOD*, and have normal SOD activity, but develop ALS which closely resembles human ALS. The severity of the mouse ALS is correlated with the levels of expression of the mutant protein, strongly suggesting that the mutated SOD protein has a novel and deleterious activity which is unrelated to normal enzyme function. Interestingly, mutations in the *SOD* gene also occur in some cases of sporadic ALS.

Disruption of axonal transport, by the abnormal accumulation of neurofilaments, has also been suggested as a possible factor in the degeneration seen in ALS, although it is thought that this may be a secondary event. Accumulation of neurofilaments has been observed in familial ALS with SOD mutations and in the transgenic mice expressing mutant SOD, but more interestingly in the majority of sporadic cases. Transgenic mice expressing a mutated form of the human low molecular weight neurofilament protein (NF-L) display a pathology which closely resembles human ALS and, in three cases of ALS, mutations in the high molecular weight neurofilament (NF-H) have been observed.

So, in ALS, although there is the possible involvement of excitotoxic mechanisms and free radical damage, the involvement of other, as yet unknown mechanisms cannot be ruled out.

9.4 Parkinson's disease also causes degeneration of specific neurons

Parkinson's disease (PD) is a well-defined movement disorder, which involves the loss of a specific population of cells, the dopaminergic neurons in the pars compacta of the substantia nigra, which project from the substantia nigra to the neostriatum (caudate and putamen) and globus pallidus. Although there are losses of other cells, for example the noradrenergic neurons of the locus coeruleus and cholinergic neurons of the nucleus basalis of Meynert, the major lesion is in the substantia nigra. In the substantia nigra and the locus coeruleus, and to a lesser extent in other brain areas, there are Lewy bodies, these are intraneuronal proteinaceous structures with radiating filaments, which are thought to be diagnostic for PD (*Figure 9.2*).

The major symptoms of PD are tremor and rigidity. There is an inability to initiate voluntary movement (akinesia) and a slowness of movement (bradykinesia). There are some early subtle cognitive changes, which may be related to disruption in the pathways linking the caudate nucleus and the cortex, and there is an increased risk of dementia. PD is generally a disease of the elderly, with a mean age of onset of about 60, although in some individuals it can occur much earlier. L-DOPA is a dopamine precursor, is very successful in reducing many of the symptoms of PD, although it is not without side effects and becomes less efficacious over time. L-DOPA, unlike

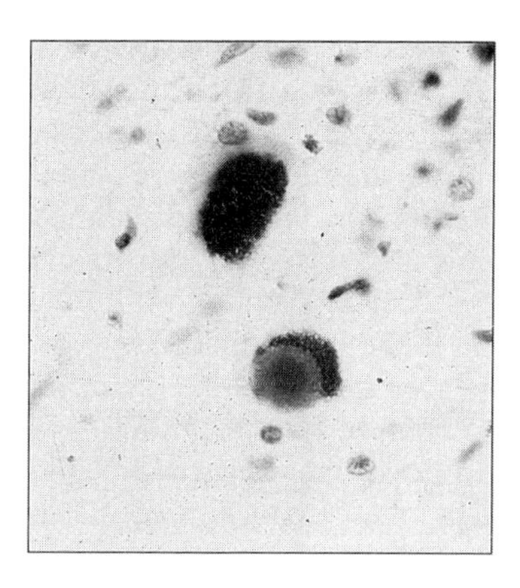

Figure 9.2. Lewy bodies. Reprinted from Allen, S.J. and Dawbarn, D. (1995) Alzheimer's disease: current controversies – an overview. In: *Neurobiology of Alzheimer's Disease* (eds D. Dawbarn and S.J. Allen). Reprinted by permission of Oxford University Press.

dopamine, is transported across the blood–brain barrier and converted to dopamine by decarboxylation. The exact mode of action of L-DOPA is not certain, but some evidence suggests that it may be taken into the remaining cells of the substantia nigra and released specifically onto target cells. Other cells may also convert L-DOPA and release dopamine, which bathes the striatum. The efficacy of this dopamine initially may be increased by the up-regulation of the dopamine receptors but, after L-DOPA treatment, receptor numbers are reduced to normal. After chronic treatment with L-DOPA there are often a number of side effects, especially abnormal movements or dyskinesias, such as those seen in HD, which may be due to excessive stimulation of dopamine receptors. There are also some psychiatric effects, which have some resemblance to those seen in schizophrenia (Section 6.13). These side effects may be due to the progressive down-regulation of dopamine receptors in the presence of continual stimulation by agonist (Section 6.4.4). As well as these side effects, after chronic L-DOPA treatment there is a marked reduction in clinical effectiveness, which is thought to be related to a continuing loss of dopaminergic neurons. This reduces both the storage of dopamine for release by dopaminergic neurons and the ability to decarboxylate L-DOPA.

9.4.1 *PD may be caused by an environmental factor*

Although there is a large amount of discussion on the issue of the heritability of PD, it seems that there is a clear genetic link in a particular form of early-onset PD, although the gene responsible has not been identified.

A number of factors have suggested as environmental cause for PD.

(i) Although a disease similar to PD had been described in ancient Indian medical texts, and despite the very visible and well-defined nature of the disease, the first full account of the disease was in 1817 by James Parkinson. Although there are no accurate figures for the prevalence of the disease in those times, it has been proposed that the disease is increasing in frequency and may thus be a disease caused by industrialization.

(ii) The viral disease, encephalitis lethargica, affected many people between 1915 and 1926. At least half of those who survived the disease, which had a 40% mortality, developed PD-like symptoms and were shown to have reduced dopamine levels in the substantia nigra and striatum. Although not identical to PD, some of these patients responded well to L-DOPA treatment.

(iii) In Guam disease (see above) there are PD-like symptoms.

(iv) In California in 1982, a number of heroin addicts who used a synthetic heroin substitute which had been illicitly manufactured, rapidly developed a syndrome which very closely resembles PD. Chemical analysis of the material showed that it contained a contaminant, **MPTP**, which was responsible for the PD-like symptoms. When injected into elderly primates, it causes a neuropathology closely resembling that of PD, including the appearance of Lewy bodies. This chance discovery has given a great impetus to the study of PD in two ways, firstly, the action of MPTP may give clues as to the underlying cause of the disease and, secondly, the use of MPTP in producing an animal model of the disease will allow better research into possible treatments.

9.4.2 PD may be due to free radical damage

The toxic effects of MPTP injection are not due to MPTP itself, but to a metabolite, MPP$^+$ (*Figure 9.3*). As mentioned in Section 7.16, MPTP can cross the blood–brain barrier, as it is lipophilic and uncharged, and is taken up by astrocytes. MPTP is then oxidized to MPDP$^+$ which rearranges to MPP$^+$. The oxidation reaction is carried out by MAO-B (monoamine oxidase-B). This enzyme, which is found in glial cells, such as astrocytes, has a higher affinity for MPTP than the neuronal MAO-A. Specific inhibitors of MAO-B can block the toxic effects of MPTP, whilst MAO-A inhibitors cannot. MPP$^+$ is then taken up by neurons via the high-affinity dopamine transporter. This accounts for the specific damage to dopaminergic neurons as only dopaminergic neurons will take up the toxin. MPP$^+$ is also bound in cells of the substantia nigra by neuromelanin. This dark pigment is present in relatively high concentrations in the substantia nigra and is what gives this brain region both its dark appearance and its name. In the nigral neurons, MPP$^+$ is concentrated in mitochondria where it inhibits complex I of the electron transport chain and hence oxidative phosphorylation. This results in a depletion in ATP which could cause cell death. Evidence from transgenic mice suggests that the proximate cause of cell death is due to **oxidative stress**, from excess free radical production. Transgenic mice which overexpress Cu/Zn-SOD are resistant to the effects of MPTP, suggesting that if there are sufficient levels of free radical inactivation then cells are not damaged by MPP$^+$. A link can also be established with excitotoxic cell death, as NMDA antagonists can also protect against the effects of MPP$^+$. A possible source

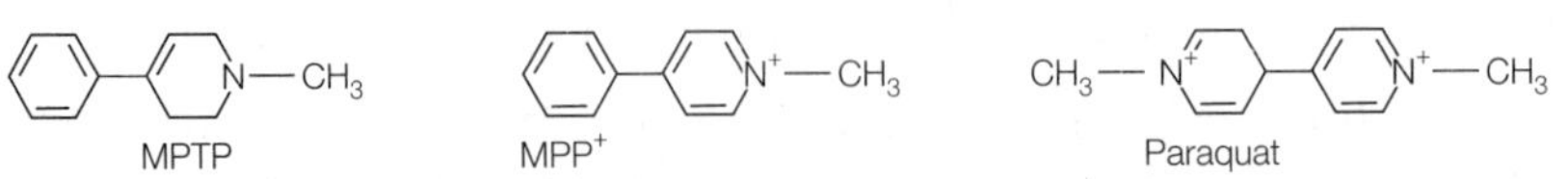

MPTP MPP$^+$ Paraquat

Figure 9.3. Structures of MPTP, MPP$^+$ and paraquat.

could be the glutamatergic neurons which project from the subthalamic nucleus to the pars compacta of the substantia nigra.

Speculation that there is an environmental equivalent to MPTP which can cause PD has focused attention on a number of chemicals, especially paraquat (*Figure 9.3*). A survey in the Canadian province of Quebec showed a significant correlation between usage of herbicides and the incidence of PD, although it was not possible to determine whether this included paraquat or not. Other possible candidates include pyridines, isoquinones and β-carbolines. Some of these compounds can become involved in the generation of free radicals by futile cycling involving single electron reduction by enzymes such as NADPH-cytochrome P_{450} (c) reductase, which may increase the oxidative stress.

In normal aging, there is a normal loss of nigral cells which may be related to the specific biochemistry of the nigral cells. Dopaminergic neurons have several features which make them susceptible to oxidative damage. MAO-A can catabolize intracellular dopamine in the presence of oxygen to produce hydrogen peroxide (H_2O_2). Then, in the Fenton reaction, H_2O_2 reacts with ferrous ions (Fe^{2+}) to produce ferric ions (Fe^{3+}), hydroxyl ions (OH^-) and a hydroxy radical (·OH). In the substantia nigra, there are high concentrations of iron which binds to neuromelanin, and there is some evidence that iron metabolism is disturbed in PD. Dopamine can also react nonenzymatically with oxygen, in reactions which also produce substrates for neuromelanin synthesis. Neuromelanin can be seen as an end-product of catecholamine metabolism which accumulates over time. This is shown by the fact that the substantia nigra only becomes visible after the age of six.

This all adds up to the situation in which nigral cells may be at greater risk of oxidative damage than many other types of neuron because of the presence of both dopamine and neuromelanin. Neurons normally do not divide, and so any loss of cells will be permanent. The normal loss of nigral cells appears to be relatively slow until about 60 years of age when there is a marked acceleration in the rate of cell loss (*Figure 9.4a*). It is only when the levels of dopamine in the striatum have been reduced by about 80% that there are clinical symptoms of PD, so any situation which reduces the number of remaining neurons to below the critical threshold could lead to PD.

It is possible to hypothesize how a number of different situations could lead to PD.

(i) There could be a gradual cell loss at a greater rate than normal, hence the threshold is reached earlier, resulting in PD. This could be due to a number of causes, including chronic toxic insult, infection, less effective protective enzymes, etc. (*Figure 9.4b*).
(ii) There could be an event, such as an acute toxic insult, which reduces the cell number to a lower level. Continued cell loss could then result in PD (*Figure 9.4c*). Note that in the original MPTP poisoning the subjects lost a very large number of nigral cells and were thus immediately below the threshold and developed PD-like symptoms.
(iii) Individual variations in the initial number of nigral cells. During the development of the nervous system, large numbers of cells die by apoptosis. If there were a larger die back at this time, those individuals would be at greater risk of PD (*Figure 9.4d*).

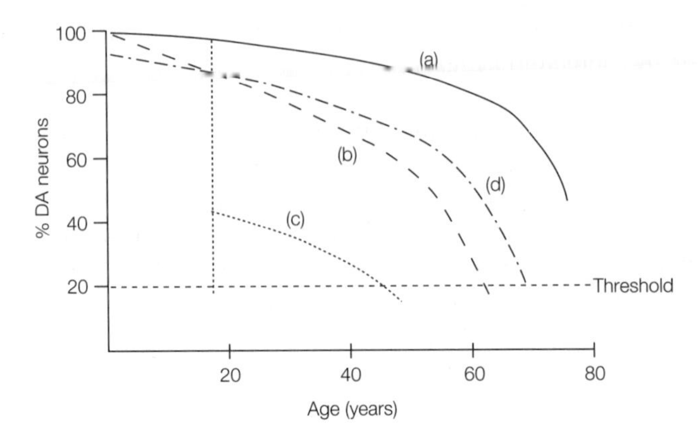

Figure 9.4. Different patterns of cell loss could lead to PD.

Whilst option (ii) could only be caused by an environmental agent, the other situations could occur due to either an environmental or a genetic cause. The genetic cause need not be a defective mutation but simply a natural variation which could confer a tendency towards PD.

9.5 Alzheimer's disease causes severe dementia

Alzheimer's disease (AD) is the commonest form of dementia in the elderly. It manifests as a progressive loss of memory, particularly recent memories, and other cognitive functions, until patients eventually become completely dependent on carers. Hallucinations and confusion are common; aggression, depression and Parkinsonism also occur in some but not all patients. Death occurs inevitably after 3–15 years. AD is predominantly a disease of the elderly; it occurs in about 5% of 70 year olds and in over 20% of people over 80. However, it can be found in younger people, in some particular cases from 30 years onwards (see below).

9.5.1 Alzheimer's disease has a wide pattern of neuronal degeneration

In AD there is gross atrophy of brain, with a large loss of neurons, in both cortical and subcortical regions. Brain weight is reduced by 30–40%, and this can be seen clearly in PET scans where there are enlarged ventricles and wider sulci. There is a wide individual variation in the exact cell losses, but in many patients there is a large reduction in cortical activity of the marker for cholinergic neurons, choline acetyltransferase (CAT), which is caused by the loss of basal forebrain cholinergic nuclei. This loss of cholinergic neurons is what underlies the rationale for the use of the only drug licensed for the treatment of AD, tacrine. This is an acetylcholinesterase inhibitor and the intention is to increase cholinergic transmission by preventing the breakdown of ACh liberated from the remaining cholinergic neurons. However, the drug can have severe side effects (liver toxicity) and many patients respond poorly, so in no way can it be considered a satisfactory treatment.

There are also significant losses in other neurotransmitter systems (NA, 5-HT, Glu). Particularly, there are major losses in the predominantly glutamatergic pathways leading to and from the hippocampus. This leads to a functional isolation of hippocampus and probably underlies the inability to form long-term memories which is so common in AD.

Although the exact pattern of neuronal loss may vary from patient to patient, the appearance of the brain at the cellular level shows two features, **neuritic plaques** and **neurofibrillary tangles**, which together are characteristic of most forms of AD (*Figure 9.5*). In fact, the only certain way of confirming a provisional diagnosis of AD, at present, is by the post-mortem examination of the brain for these features.

Neuritic plaques. Neuritic plaques are extracellular structures which consist predominantly of insoluble deposits of β-amyloid peptide (βA), hence the alternative name, amyloid plaques. However, there are also many other proteins associated with plaques including apolipoprotein E (apoE), components of the complement cascade and cytokines. The plaque core is surrounded by dystrophic neurites and reactive astroglia and microglia. Whilst neuritic plaques are particularly prevalent in areas of the brain showing substantial neuronal loss, neuritic plaques are also seen, albeit usually at a lower frequency, in normal elderly people without AD. However, the number of plaques does not seem to be invariably related to AD as there are some normal elderly people with large numbers of plaques and some AD patients who have few, if any, plaques.

Neurofibrillary tangles. Neurofibrillary tangles (NFTs) are dense aggregates of long unbranched filaments in the cytoplasm of cortical pyramidal cells. Under electron microscopy, they have been shown to consist of two 10-nm filaments which are twisted in a helix with a period of about 160 nm to form paired helical filaments (PHFs). These

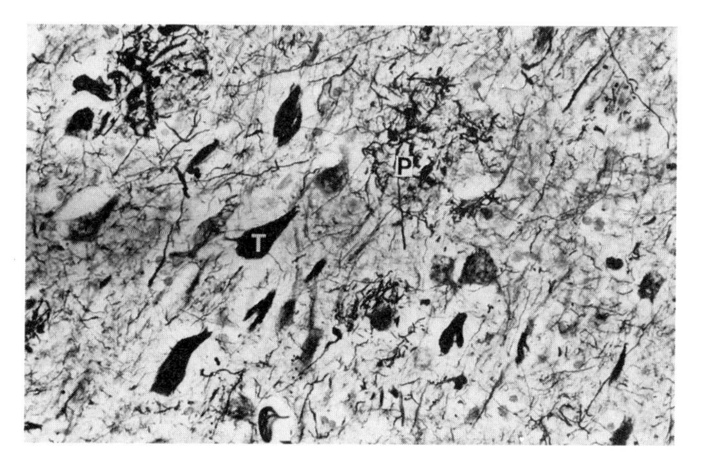

Figure 9.5. Neuritic (or senile) plaques (SP) and neurofibrillary tangles (T). Reprinted from Wischik, C.M., Edwards, P.C. and Harrington, C.R. (1995). The tau protein amyloidosis of Alzheimer's disease: it's mechanisms, potential trigger factors and consequences. In: *Neurobiology of Alzheimer's Disease*, pp. 89–148. Reprinted by permission of Oxford University Press.

filaments are composed of the microtubule-associated protein (MAP), tau, in an abnormally phosphorylated form which self-assembles to form the PHFs. Although NFTs are also seen in other neurodegenerative diseases, the number of NFTs is well correlated with the severity of the dementia. Remnants of dead cells which contained NFTs can also be seen as extracellular 'ghost' tangles.

Studies on the main components of plaques and tangles have given rise to two theories of the etiology of AD. These are:

(i) the β-amyloid hypothesis, which centers around the abnormal deposition of βA in plaques;
(ii) the tau hypothesis in which the key event is the formation of PHFs in tangles.

There are also a number of theories of AD based on linkages between specific genes and various forms of AD. Hypotheses on the cause of AD also have to account for the association between serious head injuries and later development of AD, and the subacute damage suffered by boxers and the later appearance of dementia pugilistica which appears to be a form of AD.

9.6 AD may be caused by a number of different genes

Although AD is very probably caused by a number of different mechanisms, these all seem to converge on a common neuropathology, as both the general symptoms and neuronal damage seem to be fairly consistent from case to case. Although, the majority of AD cases are classified as sporadic, late onset (>60 years), some sporadic cases are early onset (<60 years old). However, of great interest to researchers seeking clues as to the cause of AD are the cases which are clearly familial. Linkages have now been made between familial AD and genes on four different chromosomes. Before going on to look at the different models of AD, it is worth looking at the possibilities suggested by genetic linkages.

9.6.1 APP is associated with early-onset AD

AD can occur with a very early onset (30–40 years), particularly in people with Down's syndrome. Almost all individuals with Down's syndrome who live beyond 30 show some symptoms of AD, and neuritic plaques can be seen from much earlier. Down's syndrome is caused by the presence of an extra chromosome 21, hence the name trisomy 21, and this suggested that there may be a link between chromosome 21 and AD. It has been discovered subsequently that the gene for amyloid precursor protein (APP) is located on chromosome 21, so in Down's syndrome it is possible that an excess of APP can lead to AD. Mutations in the APP gene have been shown to be associated with a number of very rare, early-onset familial forms of AD which occur in less than 20 families.

9.6.2 Two homologous proteins can be linked to early-onset AD

Linkages have been found between chromosomes 14 and 1 and cases of early-onset familial AD. The two genes involved, named respectively presenilin-1 (PS-1) and presenilin-2 (PS-2), have now been located and sequenced, and have been found to code for homologous proteins. PS-1 mutations are associated with about 2% of all AD cases, whilst PS-2 mutations are linked to several families of Volga German descent. Both these proteins have a similar predicted structure, with 7–9 TM segments and two hydrophilic domains. It is not known what their function is and whether they are receptors or channels, but there is some similarity between these proteins and two proteins of *C. elegans* which may have a function in cell trafficking.

9.6.3 *Different alleles of the* APOE *gene are associated with susceptibility to late-onset AD*

Although there is no linkage between a single genetic mutation and late-onset AD, there is a clear association between susceptibility to AD and a particular allelic genotype at the *APOE* gene locus on chromosome 19. The *APOE* locus has three common alleles, ε2, ε3 and ε4. The susceptibility to AD depends on the combination of alleles inherited from both parents, not on any mutation in the genes themselves. The risk of developing AD increased and the mean age of onset decreased with each copy of the ε4 allele inherited. Inversely, the inheritance of the ε2 allele decreased the risk and increased the mean age of onset (*Table 9.1*).

As well as occurring in cases of familial AD, this correlation has been shown to occur in cases classified as sporadic AD. Estimates vary, but it has been suggested that between 60 and 90% of total AD cases can be accounted for by association with *APOE* genotype. Although exact allelic frequencies vary between racial and ethnic groups, the most common allele in the American population as a whole is ε3 (~74%) followed by ε4 (~16%) and ε2 (~10%). This means that the rarest combination is ε2/ε2, which is present in less than 1% of the population, and the most deleterious combination, ε4/ε4, is present in about 2.5%. When *APOE* genotypes were compared in two US groups of 176 AD patients and 91 nonAD controls, 17% of AD patients had the ε4/ε4 genotype compared with 2% of the controls. The ε3/ε4 genotype was also more prevalent in the AD patients, 43% compared with 21% of the controls. Similar results were shown in a study of Japanese AD patients, although the frequency of ε4 alleles is lower (~9%) in Japan. This leads to a predicted smaller number of ε4/ε4 and ε3/ε4 individuals in the Japanese population and correlates with a lower prevalence of AD in Japan with a higher mean age of onset.

Table 9.1. *APOE* genotypes alter the mean age of onset of AD

APOE genotype	Mean age of onset of AD
2/2	? (rare genotype)
2/3	>90
2/4	80–90
3/3	80–90
3/4	70–80
4/4	<70

Two points are worth making here. Firstly, if the mean age of onset of AD could be increased by somehow mimicking the effects of the ε2 allele or knocking out the effects of ε4, then AD would become a rare disease as the number of individuals over 90 is much lower than those in the younger age groups. Secondly, the possession of the ε2 allele confers no selective advantage as its effects are obviously not apparent until late in life. If it were advantageous before and during reproductive age then its frequency would be higher.

9.7 The amyloid hypothesis depends on the abnormal deposition of βA

The main component of neuritic plaques, βA, is protein containing between 39 and 42 amino acids which is cleaved from APP. APP is an integral membrane protein with a single TM segment, a C-terminal cytoplasmic domain and an N-terminal extracellular domain. The gene coding for APP on chromosome 21 has 18 exons, and alternative splicing can produce three variants of differing lengths, APP_{695}, APP_{751} and APP_{770}. The normal function of APP is not known, but all forms, especially the more abundant APP_{751}, can act as an adhesion molecule and can promote neurite outgrowth. It has also been reported that APP interacts with the G protein α_o subunit.

βA is formed from a part of APP which starts with 28 amino acids in the extracellular domain and ends with 11–14 amino acids embedded in the membrane (*Figure 9.6*). APP is processed by at least three, as yet unidentified, proteases called α, β and γ secretases.

α secretase cleaves membrane-associated APP at a site (Lys687) which is just outside the membrane and is in the middle of the βA domain. The N-terminal portion is secreted, sAPPα, leaving a 10-kDa C terminus portion (688–770) which may be further processed or degraded in lysosomes. In order to produce intact βA, the APP can also be cleaved by β secretase at Met671, producing the N terminus of the βA domain (Asp672). This results in a smaller sAPPβ. This processing may occur in Golgi or late endosomes. The remaining 12-kDa fragment (672–770) is then cleaved further by γ secretase in the region Val711 to Thr714 to produce the βA peptide (672–711 to 714). The 10-kDa fragment is also cleaved by γ secretase to produce a smaller 3-kDa peptide called p3 (*Figure 9.6*). Although βA occurs in a range of sizes, some forms, particularly the longer forms, are more able to form plaques. Cleavage of APP by the α pathway cannot produce βA, and the p3 peptide does not appear to form plaques. Thus, processing of APP down this α pathway does not appear to be involved in AD. However, the processing of APP by alternate pathways seems to be interrelated, and it has been suggested that increased processing via the β pathway could lead to excess production of βA and plaque formation. In cells transfected with one of the mutations of APP known to cause familial AD, known as the Swedish mutation, there was a 6- to 8-fold increase in the production of βA which was accompanied by a decrease in the amount of p3, suggesting that APP processing via the β pathway was increased, with corresponding reductions in processing via the α route.

Processing of APP can be regulated by receptor-mediated events, through ligands such as carbachol (mAChr agonist), glutamate, 5-HT and bradykinin. Stimulation of M1 and M3 mACh receptors, which are both coupled to PLC activation, increases the amount

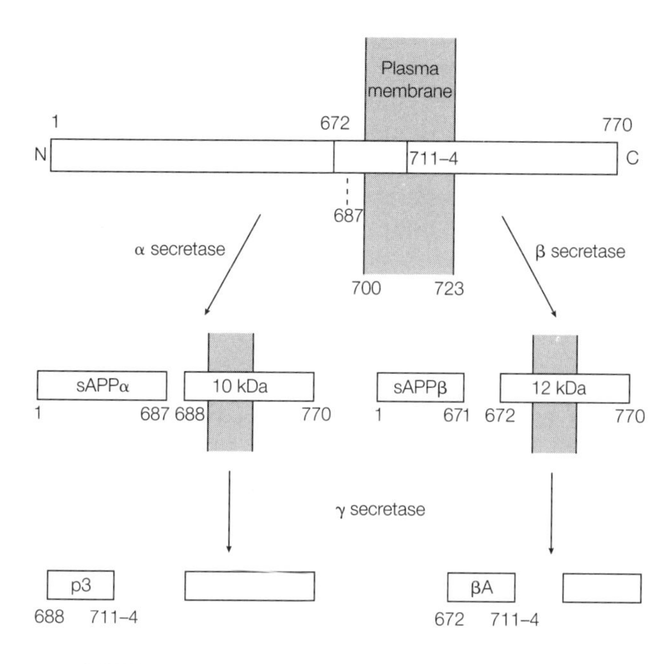

Figure 9.6. Structure of APP and its processing by α, β and γ secretases.

of both sAPPα and p3, whilst reducing the amounts of sAPPβ and βA. This seems to be controlled through the activation of PKC, although this does not appear to be the case with all ligands.

As the βA peptide contains part of the membrane-anchoring domain of APP, it was thought originally that βA was only produced in response to cell injury. However, it has since been shown that βA is secreted by normal cells and has been detected in normal human cerebrospinal fluid. No clear role for this secreted βA has been found, although a limited homology with the tachykinins has suggested that βA may act as a receptor ligand.

In normal conditions, the βA which is secreted is soluble and seems to have no adverse effects on cells. However, in AD, the protein forms aggregates of insoluble β-pleated sheets of fibrillar βA protein (hence the name β-amyloid). The conformational change from soluble to fibrillar forms seems to be a spontaneous event, and no specific trigger is known. The fibrillization is increased with higher concentrations of βA and with the longer forms of βA, so any production of larger amounts of βA than normal or production of the larger, less soluble forms of βA will tend to increase plaque formation. Once the βA plaque has started to form, other molecules can interact with the nascent plaque to produce eventually the mature plaque with its associated areas of cell death.

Although the manner in which the formation of plaques could cause cell death is unproven, it has been shown that aggregates of βA can disrupt Ca^{2+} homeostasis in neurons. In cultured human neurons, chronic βA exposure resulted in raised resting intracellular Ca^{2+} concentrations and increased intracellular Ca^{2+} in response to depolarization and EAAs. This action of βA was prevented by removal of extracellular Ca^{2+}. Immunostaining of these neurons showed aggregates of βA accumulating on the outside of the plasma membrane. Incorporation of βA in lipid bilayers showed that βA can form Ca^{2+}-permeable channels. Even a relatively small reduction in the ability of cells to regulate intracellular Ca^{2+} could leave cells vulnerable to damage by excitotoxicity, ischemia and free radical damage. In contrast, the secreted sAPPα has been shown to activate K^+ currents and reduce intracellular Ca^{2+} concentrations through cGMP production and protein dephosphorylation. In this way, processing of APP through the α pathway would be neuroprotective whilst production of βA by the β pathway is neurotoxic. If stimulation of metabotropic cholinergic and glutamatergic receptors linked to PKC is lost due to a degeneration of local synapses, this could cause an increase in the processing of APP via the amyloidogenic β pathway.

9.8 Aggregation of tau proteins underlies the formation of tangles

Whilst the amyloid hypothesis is seen by some to be central to the damage seen in AD, there is a competing hypothesis which says that it is the formation of fibrillary tangles and the subsequent disruption of axonal transport which is the reason for cell death. This is substantiated by the good correlation between the degree of dementia and the frequency of tangles.

Although tangles have been shown to contain many different proteins, the core of the tangle is made up very largely of tau protein. Physiologically, tau protein binds to microtubules via a C-terminal binding domain which contains three or four tandem repeats of a 31–32 amino acid sequence. There are six human isoforms (A–F), which are all generated by alternative splicing of a single gene. All the isoforms are expressed in adult brains, but only tau-A is expressed in fetal brain. Tau proteins may form cross-bridges between microtubules which stabilize the microtubules, promoting tubulin polymerization and microtubule bundling which are essential for axon elongation and maintenance. Tau is a phosphoprotein, and the ability of tau to bind to microtubules is reduced by phosphorylation. Tau can be phosphorylated by a number of different kinases including MAPK (see Section 6.10.2) and can be dephosphorylated by protein phosphatase 2A.

The tau protein isolated from PHFs is hyperphosphorylated (tau-HP) when compared with normal tau, and this greatly reduces its ability to bind to microtubules. Comparison of phosphorylation patterns between normal tau and tau-HP shows that there are 17 potential phosphorylation sites. Normally, only a few of these are phosphorylated, but in tau-HP a large number of these sites are phosphorylated.

It would be nice to find a relationship between the ability of tau to form PHFs and its phosphorylation state, but in fact phosphorylation seems to inhibit both tau–tau and tau–tubulin binding. Tau–tau (and tau–tubulin) binding occurs at a site in the tandem repeat region, and phosphorylation of the tau protein seems to mask the site, thus preventing binding. In fact, it has been suggested, from *in vitro* experiments, that it is only normal tau that can self-aggregate, by undergoing a critical conformational change which allows self-association, and that it is an autocatalytic event where the presence of altered tau can promote the alteration of other tau molecules. This has been compared with the action of **prion proteins** where the normal α-helical form is thought to be converted to the pathogenic β-sheet form by the presence of the β-form which induces the switch.

However, none of these hypotheses explain why PHFs do not accumulate in normal brain if there are no abnormal modifications necessary to produce tau aggregation. Also, they do not explain the presence of tau-HP in PHFs. Other factors may be needed to explain the paradox.

The level of soluble tau protein in the frontal cortex is highly predictive of the degree of cognitive impairment, and some studies have shown that levels of soluble tau in cerebrospinal fluid have been found to be raised in AD patients. This has raised the possibility of a predictive test for AD, but there is at least a 20% overlap between cerebrospinal fluid levels in normal and AD patients, so the test is not definitive.

9.9 ApoE interacts with both βA and tau in an isoform-specific manner

The close association between particular alleles of the *APOE* gene and susceptibility to AD has stimulated much research on the possible interactions between apoE and the components of plaques and tangles.

ApoE is involved in the recycling of cholesterol during membrane repair and remodeling. ApoE binds to both lipoproteins and to the low-density lipoprotein (LDL) receptor. The affinity of binding to the different types of lipoprotein and to the LDL receptor varies with the apoE isoform, but the significance of these different interactions is not obvious, although it has been suggested that the ε4 isoform is less effective than ε2 and ε3 at repairing membrane breakdown. Neurons may be particularly susceptible to reductions in the effectiveness of apoE, as the other apolipoproteins, apoB and apoA1, are not found in the brain.

ApoE has also been shown to bind to both βA and tau proteins both *in vivo* and *in vitro*. Incubation of apoE with βA *in vitro* has shown that ε4 binds more rapidly than ε3, although this reaction requires a large molar excess of βA. After a long period of incubation (days), apoE and βA form fibrils, but these are not the same as the types of fibrils seen with βA alone or with *in vivo* plaques. ApoE binding to plaques is extracellular, but apoE has also been shown to enter the neuronal cytoplasm, a necessary property if it is to interact with tau proteins *in vivo. In vitro* experiments have shown that the ε3, but not the ε4, isoform binds to tau in a manner that can only be reversed by boiling under reducing conditions. Neither ε3 nor ε4 can interact with tau-HP. The interaction between tau and apoE ε3 is via the LDL receptor-binding site on apoE and the tandem repeat sequences on tau. A similar interaction has also been shown between ε3 and MAP2c. This suggests a mode of action of apoE where ε3 can bind to the tandem repeat sequences of tau and MAP2c and prevent tau–tau interactions, whilst ε4 is unable to do this and therefore there is a tendency, in the presence of ε4 rather than ε3, for PHFs to form, with the subsequent disruption of the axonal cytoskeleton. However, in order for this model to be viable, it would be necessary for tubulin to be able to displace ε3 from tau in order to stabilize microtubules. Again, the question is left open as to the role of tau-HP. One possible hypothesis is that the accumulation of tau-HP is an event which occurs after the formation of PHFs and has no significance to the disease process. It has been shown recently that some of the phosphorylation sites on tau which have been considered to be abnormal can also be seen in normal tau. The reason for the previous failure to observe these was due to the fact that post-mortem there is a rapid dephosphorylation of these sites.

9.10 Other hypotheses have been suggested for the cause of AD

Although much research effort has been targeted at the role of βA, tau and apoE in the pathogenesis of AD, there are number of other hypotheses which have been proposed as a cause of AD, and some of these are described briefly below.

9.10.1 *AD may be an inflammatory disease*

Microglia are activated and proliferate locally at the sites of AD lesions. Microglia then produce cytokines and free radicals which could produce a chronic inflammatory response. In both head injury and ischemia, there has been shown to be activation of microglia and increases in APP expression which appear very rapidly. βA is known to activate microglia, although whether activated microglia secrete βA or not is a contentious issue. The complement receptor, C1q, binds to βA, provoking the complement cascade. In this way, the inflammatory process may be both initiated and potentiated by βA. The role of cytokines in the development of AD was examined in a retrospective study of the effects of drugs which inhibit cytokines. It showed a lower incidence of AD in people who had been taking anti-inflammatory drugs.

9.10.2 *Abnormal glycation and accumulation of AGEs could produce AD*

With increasing age, there is an accumulation of abnormally glycated proteins, called advanced glycosylation end-products (AGEs), and an increase in the production of free radicals, consistent with a permanent state of oxidative stress. Due to the reduced rate of protein turnover with age, this leads to an increased accumulation of damaged proteins and lipids. A number of features have suggested that these mechanisms are exacerbated in AD, leading to abnormal modifications of proteins including βA and tau which can stimulate their aggregation. It has been shown that AGE-modified βA can act as a template for further deposition of soluble βA, and tau from AD but not normal brains is also AGE modified and undergoes oxidation-induced cross-linking. However, why some aged individuals develop AD and not others is not known, although if apoE ε4 is significantly less able to repair damaged neurons this could lead to greater oxidative stress and the development of AD. In injury-related AD, it is possible that an increase in oxidative stress due to high levels of cell damage, both acute and chronic, could act as a triggering factor for the development of AD.

9.11 New therapies may be based on understanding the molecular basis of AD

As mentioned above, there is as yet no satisfactory test for the early presence of AD, and it has been estimated that by the time there are recognizable symptoms of AD, 70% of the neurons have already been lost. The presence of the apoE ε4 allele is not in itself a test but maybe if it can be associated with raised levels of tau this could reduce the number of false-positive diagnoses. Whether it is then feasible to treat all susceptible individuals will depend on both the benefits of treatment and the cost. As in many other brain diseases, there is always the problem of access to the brain tissue. Experiments in rats and primates on the effects of nerve growth factor (NGF) in promoting the survival of neurons in the aging brain have proved to be positive, but the difficulties in introducing NGF into the brain will probably preclude its use as a general therapy. It has been suggested that implantation, into the brain, of neurons which secrete NGF might be appropriate, but the diffuse nature of the lesion in AD makes this a more difficult task than in PD where the cell loss is both anatomically and biochemically specific.

Looking at the models of AD presented above there are some obvious targets for possible new therapies.

(i) If excess βA production causes AD, then a possible therapy could involve the inhibition of the β secretase which would result in the production of less βA. However, the role of the normally secreted βA would have to be examined.
(ii) If tau–apoE interactions are important in preventing the formation of tangles, then the introduction in some way of more ε3 or ε2 would prevent abnormal tau–tau interactions. However, how this would be done is not obvious. Knocking out the effects of ε4 would not, on the face of it, improve the situation.
(iii) If AD is an inflammatory process, then anti-inflammatory drugs might be useful in prevention. In fact, a recent double-blind trial of the effects of indomethacin showed clearly that cognitive decline in treated patients was halted. Whilst placebo patients declined by 8.4% over the 6 months of the trial, the indomethacin patients improved by 1.3%.
(iv) If oxidative stress is a component of AD, then antioxidants and free radical scavengers might have some therapeutic value.

Common paths to oxidative stress and free radical damage in neurodegenerative disease

In all the neurodegenerative diseases discussed in this chapter, a common theme emerges, in that many different mechanisms can lead to the activation of free radical generation and subsequent cell damage. The involvement of glutamate-mediated excitotoxicity in HD and stroke, the selective vulnerability of dopaminergic neurons in PD and the involvement of SOD in ALS all point to oxidative stress as an important component in the death of neurons. Even if oxidative stress is not the cause of AD, it is probable that it plays a role in the final degeneration associated with plaques and tangles.

Box 9.1. Linkage analysis of DNA using RFLPs

When human DNA is cut with restriction enzymes and the fragments separated by gel electrophoresis, the pattern of fragments will be different for each individual and for each chromosome pair of that individual. This is due to the presence of individual **polymorphisms** in the DNA which ensure that some of the restriction enzyme recognition sites will be present on some chromosomes and not on others. It has been estimated that the variation is one nucleotide in every 100–500 in noncoding DNA sequences. The parts of the DNA which encode transcribed regions are more highly conserved, presumably because if the function of the resulting protein was disturbed the individual would be less likely to survive.

In order to locate a gene of interest, it is necessary first to locate a marker gene which is close to the gene locus. If the marker gene is close to the disease gene then it is less likely that they will be separated during the recombination which occurs during meiosis. Samples of DNA from a number of individuals in the same family, both with and without the disease gene, are cut with a specific restriction enzyme. The DNA is separated on a gel and Southern blotted using a labeled probe which will label the marker gene. If the region of the marker and disease genes is polymorphic with respect to the restriction enzyme sites, then different patterns of fragments will be produced. The aim is to find a combination of restriction enzymes which produces a pattern of fragments which vary between chromosomes carrying the normal and disease gene alleles.

A simple example (*Figure 1*) shows how this type of analysis can be used to distinguish affected individuals and carriers of an autosomal recessive disease. Both parents are unaffected but have a daughter who has developed the disease and three other children (*Figure 1a*). The mother (I-1) is a carrier, and cutting her DNA produces two fragments (a and b), one from each chromosome. The father (I-2) is also a carrier, and cutting his DNA also produces two fragments (a and c) (*Figure 1b*). Now, either fragment a corresponds to the affected allele in both parents, or each parent has the affected allele on a different restriction fragment,

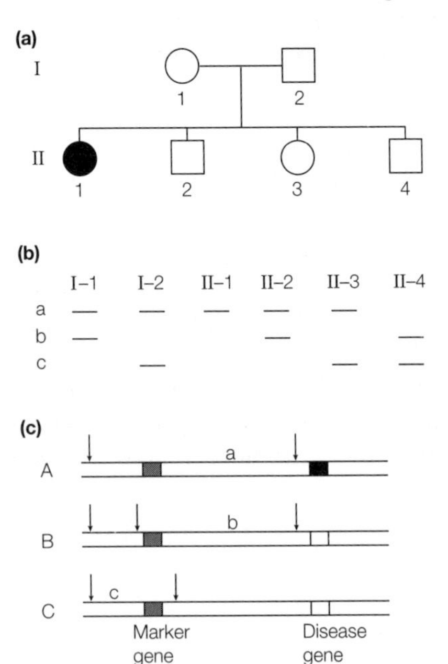

Figure 1. Example of linkage analysis using RFLP. (a) Family tree (females = circles; males = squares; affected individuals = filled). (b) Pattern of restriction fragments a, b and c produced by each individual. (c) Chromosomes (A, B and C) with restriction sites (shown by arrows) which would produce the marked fragments a, b and c.

Continued

either b or c. By examining the patterns produced by their children, it may be possible to identify which possibility is correct and therefore predict the status of the other children. Both II-2 and II-3 have patterns similar to their parents and are therefore unaffected, but are both carriers. The affected child (II-1) has inherited fragment a from both parents, therefore fragment a is associated with the disease gene and the youngest son (II-4) in an unaffected noncarrier. Thus, in this example, the restriction enzyme cuts the chromosomes (A–C) in three different ways (*Figure 1c*) to produce the three different sized marked fragments.

The following example of an RFLP linkage analysis for HD illustrates how the method works in practice. The DNA from five members of a family with a living HD-affected individual (II-2) and an affected deceased grandparent (I-1) was cut with two different restriction enzymes, *Hin*dIII and *Bgl*II. The six possible fragments produced can be used to assign a **haplotype**, corresponding to a particular set of alleles, to each chromosome and to determine whether a given individual is at risk of developing HD.

Figure 2a shows the family tree for these individuals and *Figure 2b* the pattern of fragments produced. The grandparent I-2 has the haplotype which we can call A1/A1 and the unaffected offspring (II-1) has A1 (from the A1/A1 parent) and B2, which is associated with a normal allele, from the affected grandparent (I-1) whose DNA sample was unavailable. The affected individual has the A1 haplotype plus an A2 haplotype which must be associated with the HD gene. Confusingly, the unrelated parent, II-3, also has an A2 haplotype (as well as B2), but one which must be normal as this individual is unaffected by HD. Examination of the haplotype of III-1 shows the pattern A1/A2. He must have inherited his A1 haplotype from his affected father since his mother's haplotype is A2/B2. Therefore, his A2 haplotype must be the normal A2 type from his mother and he is not at risk of developing HD. It is now possible to infer that the affected grandparent must have had the haplotype A2(affected)/B2.

This example shows how this type of analysis can help to predict whether an individual is at risk of HD and other genetic disorders. Whether this information is beneficial to the person concerned, if they are found to be affected when there is no available cure, is debatable.

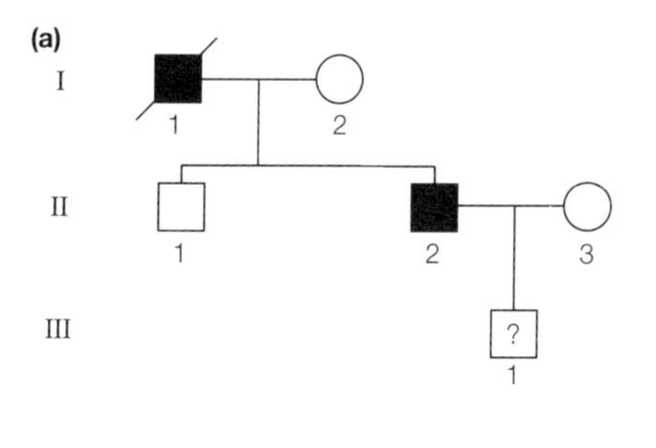

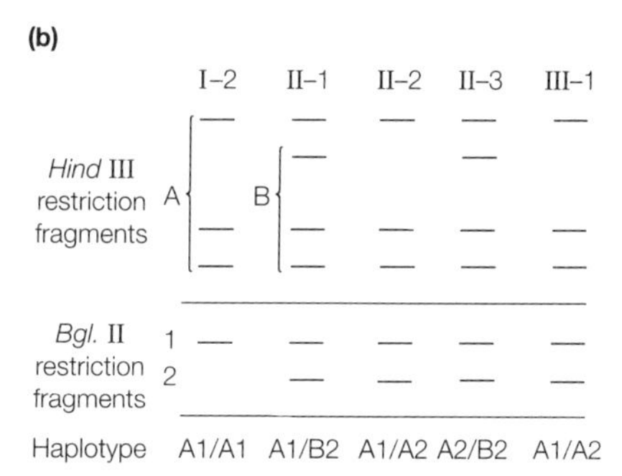

Figure 2. Example of linkage analysis in the identification of individuals at risk from HD. (a) Family tree; (b) patterns of restriction fragments for each individual

Appendices

Appendix 1. Genetic code

First position	Second position				Third position
	U	C	A	G	
U	Phe	Ser	Tyr	Cys	U
U	Phe	Ser	Tyr	Cys	C
U	Leu	Ser	*Stop*	*Stop*	A
U	Leu	Ser	*Stop*	Trp	G
C	Leu	Pro	His	Arg	U
C	Leu	Pro	His	Arg	C
C	Leu	Pro	Gln	Arg	A
C	Leu	Pro	Gln	Arg	G
A	Ile	Thr	Asn	Ser	U
A	Ile	Thr	Asn	Ser	C
A	Ile	Thr	Lys	Arg	A
A	Met/*Start*	Thr	Lys	Arg	G
G	Val	Ala	Asp	Gly	U
G	Val	Ala	Asp	Gly	C
G	Val	Ala	Glu	Gly	A
G	Val	Ala	Glu	Gly	G

Appendix 2. Single-letter code and three-letter abbreviations for amino acids

A	Ala	alanine
C	Cys	cysteine
D	Asp	aspartate
E	Glu	glutamate
F	Phe	phenylalanine
G	Gly	glycine
H	His	histidine
I	Ile	isoleucine
K	Lys	lysine
L	Leu	leucine
M	Met	methionine
N	Asn	asparagine
P	Pro	proline
Q	Gln	glutamine
R	Arg	arginine
S	Ser	serine
T	Thr	threonine
V	Val	valine
W	Trp	tryptophan
Y	Tyr	tyrosine

Appendix 3. Some second messenger-activated enzymes

Enzyme	Type	Target	Activator(s)
Kinases			
cAMP-dependent kinase (PKA)	I	Serine	cAMP
	II	Serine	cAMP
Protein kinase C	α,β,γ	Serine	DAG/Ca^{2+}
	δ,ϵ,ζ	Serine	DAG
Ca^{2+}/CaM kinase	I	Serine	Ca^{2+}
	II	Serine	Ca^{2+}
	III	Threonine	Ca^{2+}
cGMP-dependent kinase (PKG)		Serine	cGMP
Phospholipases			
Phospholipase A_1		sn-1 position on membrane lipid[a]	
Phospholipase A_2	I,II,III,IV	sn-2	Ca^{2+}
Phospholipase C	β	sn-3	G proteins
	γ		Tyrosine kinases
	δ		
Phospholipase D		Terminal phosphodiester bond	Small G proteins and PKC
Protease			
Calpain	I		Ca^{2+} (μM)
	II		Ca^{2+} (mM)
Phosphatase			
Phosphatase-2B (calcineurin)		Serine Threonine	Ca^{2+}/CaM

[a]sn-1 through 3 correspond to the three hydroxyl groups of glycerol.

Glossary

Adaptors are short synthetic double-stranded oligonucleotides which have one blunt end and one sticky end. They can be ligated to blunt-ended DNA in order to provide sticky ends appropriate for ligating with similar sticky-ended DNA.

Active zones are the regions at the synapses, particularly obvious at the neuromuscular junction, where arrays of calcium channels are situated next to vesicles docked in preparation for release.

Aequorin is a chemiluminescent protein from the jellyfish, *Aequora forkalea*, which emits light on binding calcium. It has been used to measure changes in calcium during events such as muscle contraction and neurotransmitter release.

Affinity chromatography is used to separate molecules such as receptors from a mixture based on their high-affinity binding (see *Figure 2.15*).

Afterdischarges are prolonged synchronized bursts of action potentials of hippocampal neurons which occur more or less spontaneously after the tetanic conditioning of kindling.

Alpha helix is a common type of regular secondary protein structure. It consists of a helical structure stabilized by hydrogen bonding between every fourth peptide bond. The α-helix has 3.6 amino acids per turn with a pitch of 0.54 nm.

Alternative splicing is the *in vivo* transcription of two or more different mRNAs, or splice variants, from the same gene. This can be achieved by various methods including the use of different promoters to initiate transcription or the inclusion of different exons in the final transcripts.

Annealing refers to the formation of double-stranded nucleic acids from complementary single strands by Watson–Crick base pairing. Because this occurs via hydrogen bonding, annealing will not occur at high temperatures.

Antigene suppression is a variant of the antisense knockout technique in which the cell is transfected with the target gene in the reverse orientation so that antisense mRNA will be produced in the cell (see ***Box 6.1***, p. 126).

Antiport is membrane transport in which two or more substrates are cotransported across the membrane in opposite directions. Usually the movement of one substrate down its concentration gradient is used to move a second substrate in the opposite direction.

Antisense knockouts are organisms in which the production of a particular protein has been blocked by introducing oligonucleotides, called antisense oligonucleotides, into the cell which are complementary to the mRNA coding for that protein (see ***Box 6.1***).

Antisense oligonucleotides are oligonucleotides of about 15–18 bp which are complementary to the mRNA of the protein of interest. They block production of the protein and are thought to work either by forming duplexes with the normal mRNA or by forming triplexes with the DNA (see ***Box 6.1***)

Apoptosis is genetically programed cell death which is characterized by fragmentation of the cellular DNA. It occurs naturally during development but can also occur in neurodegenerative disease.

Associative learning occurs by the association, or pairing, of two stimuli in time. Classical conditioning and operant conditioning are two associative learning paradigms.

Autonomous activation is the continued activation of an enzyme after the concentration of the activator has fallen below baseline levels.

Autoradiography is a method used to detect the position of radiolabeled molecules on a gel or filter by laying an X-ray film over the gel until the radioactivity causes bands or spots to be formed on the film, when developed, corresponding to the positions of the molecules.

Autoreceptors are presynaptic receptors which are activated by the neurotransmitter being released by the neuron. Usually they regulate neurotransmitter release.

Autosomal dominant forms of inheritance occur when the gene is situated on one of the non-sex chromosomes and has to be inherited from only one parent in order for the trait to occur.

Autosomal recessive forms of inheritance occur when the gene is situated on one of the non-sex chromosomes and has to be inherited from both parents in order for the trait to occur.

Bacteriophages are viruses which infect bacteria.

Biotin is a protein which is used to label antibodies. It can be detected using the binding of labeled streptavidin.

Blunt ends are ends of double-stranded DNA which are cut at the same position in each strand.

Botulinum toxins are a group of clostridial neurotoxins which can block neurotransmitter release by cleaving the SNAREs, synatobrevin, syntaxin and SNAP-25.

Caged molecules are modified chelators which on exposure to UV light change their conformation and release the chelated molecules. By introducing the caged molecules into an experimental setup and then supplying a flash of UV light using an arc lamp, a controllable amount of the required molecule can be delivered very rapidly to the site of action.

cAMP response element-binding protein (CREB) is a transcription factor which, when phosphorylated by cAMP-dependent protein kinase (PKA) binds to specific sequences of DNA, called CREs (cAMP response elements) which activate gene transcription.

cDNA libraries are collections of vectors containing cDNA derived from the mRNA expressed in a particular cell.

Chemiluminescence is the property of some chemical reactions to produce light. An example of this is the reaction between luminol, horseradish peroxidase and hydrogen peroxide which produces light in proportion to the concentration of hydrogen peroxide.

Chimeras refers to an entity which is composed of items from more than one source. A chimeric protein may be derived from a gene which has been spliced from different isoforms. A chimeric channel can be composed of subunits from different species or different subunit subtypes. A chimeric animal is one whose blastocyst cells are derived from more than one individual blastocyst.

Clones are individuals containing identical copies of DNA.

Cloning is the manufacture of multiple copies of a fragment of DNA, often by insertion of the selected DNA fragment into bacteria which then make multiple copies.

Co-agonist is a molecule which enhances the responses to an agonist.

Concatamers are long DNA molecules formed from multiple copies of DNA which can anneal through their cos sites.

Consensus sequences are parts of a protein which have an amino acid sequence which in other proteins in known to have a particular function. For example, certain enzymes will only phosphorylate proteins at specific residues with particular flanking sequences. The presence of similar regions on novel proteins suggests that these regions may be similarly phosphorylated.

Conservative substitutions are when differences in amino acid sequences of related peptides are due to replacing amino acids with others having similar side chains, for example the substitution of aspartate for glutamate. When this occurs, the overall hydropathy of the peptide chain is relatively unchanged.

Constitutive activation of receptors occurs when they are continually active even in the absence of agonist.

Copy number is the number of copies of a plasmid which are produced within a bacteria. This depends on the nature of both the host cell and the particular plasmid.

Cos sites on linear double-stranded DNA molecules are cohesive ends that can anneal with each other to form either a circular molecule or an extended concatemer.

Degenerate refers to a property of the genetic code whereby a given amino acid can be coded for by more than one triplet codon.

Dihydropyridines are a group of L-type calcium channel ligands, some of which, like nifedipine, are antagonists, whilst others, like BAY K 8644, act as agonists.

Docking in the context of neurotransmitter release refers to the step in which the vesicle becomes closely bound to the presynaptic membrane.

Edman degradation is a chemical method for sequencing short protein sequences. The N-terminal amino acid is labeled with phenylisothiocyanate which is then cleaved from the protein chain using a mild acid, giving a cyclic phenylthiohydantoin derivative of the amino acid which can be identified using HPLC. This process is then repeated to identify the new N-terminal amino acid until the entire peptide is sequenced.

Electromobility shift assay is a method for measuring the ability of a given protein to bind to a specific DNA fragment (see ***Box 8.1***, p. 189).

ELISA assays are a form of immunoassay in which either the antigen or antibody is adsorbed on to a rigid surface. The reaction between antibody and antigen is detected by an enzyme, hence the name enzyme-linked immunosorbent assay. This method is now widely used to detect a huge range of molecules.

Endocytosis is the means by which membrane inserted into the plasma membrane during exocytosis is retrieved. It is also the method by which some ligands are internalized by the cell, in a process called receptor-mediated endocytosis. An area of plasma membrane forms an invagination which is internalized to form an intracellular vesicle.

Epilepsy is a neurological disease which is characterized by periodic spontaneous seizures with associated loss of consciousness, and motor and sensory disturbances.

Epileptogenesis refers to the processes which underlie the development of the capacity to generate epileptic seizures.

Episodic memory is a form of declarative memory – memory which is associated with factual recall – which refers to the remembering of specific past events.

Epitope is the exact site on an antigen which is recognized by antibody. A molecule may have many independent epitopes which will be recognized by different antibodies.

Equilibrium potential also called the Nernst potential or reversal potential. This is the membrane potential at which there is no net current flow through an ion channel. It can be predicted by the Nernst equation which for a monovalent cation at 37°C is given by $E = -61 \log [C]_o/[C]_i$, where E is the membrane potential and $[C_i]$ and $[C_o]$ are the intracellular and extracellular concentrations of a single ion species. Because of this, it is also known as the Nernst potential. Changes in membrane potential from E produce either net inward or outward currents; thus, E is the potential at which the current reverses and hence the alternative name, reversal potential.

Excitotoxicity is the death of cells due to the presence of excess levels of excitatory amino acids and their analogs.

Exocytosis is the process by which the cell can secrete molecules across the cell membrane. It involves the fusion of an intracellular vesicle with the plasma membrane leading to the opening of the vesicle to the external medium. It is the method by which the regulated secretion of neurotransmitters occurs.

Fluorescence imaging is a technique which allows the visualization of a fluorescent signal in living tissue. It can be used to monitor such things as changes in intracellular calcium during stimulation (see ***Box 7.1***, p. 147).

Free radical species are molecules which contain an unpaired electron which makes them highly reactive. They can oxidize a wide range of biologically active molecules, damaging membranes, enzymes and DNA.

Fura-2 is a fluorescent analog of the calcium chelator EGTA which changes it absorption spectrum on binding calcium. It has been widely used to monitor intracellular calcium concentrations.

Fusion in the context of neurotransmitter release refers to the final step of exocytosis during which the vesicle becomes fully fused with the plasma membrane allowing the release of the contents into the synaptic cleft.

Fusion proteins are foreign proteins which are expressed as part of another protein in order to provide a known antigenic marker.

Genomic DNA is the total cell DNA.

Genomic libraries are collections of vector containing the entire genomic DNA of an organism.

Haplotype is the pattern of restriction fragments derived from a single chromosome which can be used to follow the inheritance of an individual chromosome. When a haplotype can be associated with a particular phenotype it can be used to determine the pattern of inheritance of genetic diseases.

Hebb's rule states that if two neurons are excited at the same time, then any active synapses between them will be strengthened. An often-used phrase to summarize the rule: 'What fires together, wires together'.

Hemifusion is a partial fusion of the lipid bilayer which is thought to occur during the priming step of neurotransmitter release.

Hetero-oligomers are multisubunit proteins which are made up of a number of different subunits.

Heterologous gene probe is used to identify genes which have similar sequences to a previously identified gene. Under conditions of low stringency, a probe may anneal with sequences that are less homologous. It can be used to identify other members of a gene family.

Heterologous recombination inserts a recombinant transgene into the genome at a random position. This form of recombination may lead to multiple copies of the transgene being inserted at many different positions (see ***Box 7.2***, p. 148).

Heterosynaptic LTD is where tetanic synaptic activation at one synapse induces long-term depression (LTD) in other synapses on the same neuron (see *Figure 8.15*).

Homo-oligomers are multisubunit proteins which are made up of a number of identical subunits.

Homologous recombination inserts a recombinant transgene at the position in the genome corresponding to the original gene. This occurs because of homology between the transgene and the target gene (***Box 7.2***).

Homosynaptic LTD is where appropriate synaptic activation induces long-term depression (LTD) in the same synapses (see *Figure 8.15*).

Hydropathy analysis is a method of determining the average hydrophobicity of a stretch of amino acids. It can be used to indicate the distribution of a protein in a membrane and is used to develop models of the secondary structure of transmembrane proteins.

Immunocytochemistry uses the binding of antibodies raised against molecules of interest such as enzymes and receptors to identify their location and number.

Inhibitor peptides are short fragments of the full-length protein of interest, which when microinjected into a cell act as pseudosubstrates, binding to the protein's normal intracellular target and preventing access by the native protein. Any process blocked by an inhibitor peptide is likely to be mediated by the full-length protein, just as a competitive antagonist will block the effects of a neurotransmitter.

Interictal spikes are fast spike-like potentials seen in the electroencephalogram (EEG) between seizures of humans with epilepsy. They correspond to the paroxysmal depolarizing shifts and burst firing of cortical neurons within the epileptic foci. An increase in the frequency of these interictal spikes may lead to a generalized seizure (or ictus).

Inward rectifier is a channel which allows current to flow into the cell but not out (see *Figure 4.10a*).

Ionotropic receptors are receptors which contain an intrinsic ion channel which is opened on binding of the receptor ligand.

Isoforms are different forms of the same gene or protein. For example, many receptor subunits are found to exist in slightly different versions and these are called isoforms.

Kindling is an animal model of certain forms of epilepsy. It can be induced either chemically, by the administration of low doses of either excitatory amino acid agonists or inhibitory amino acid antagonists, or electrically, via the tetanic stimulation of chronically implanted electrodes once or twice each day for about 2 weeks.

Knockouts are organisms in which the function of a particular protein has been removed by a variety of genetic engineering manipulations including the use of antisense oligonucleotides (see ***Box 6.1***, p. 126) and by the production of transgenic animals (see ***Box 7.2***).

Ligation is the method by which DNA fragments are joined into vectors. Ligation, carried out by DNA ligases, may join either complementary sticky ends or blunt ends. DNA ligases can also seal nicks in a single strand of double-stranded DNA.

Linkage analysis is a method for estimating the distance between two genes on the same chromosome based on the frequency of recombination events which occur during meiosis. If two genes or markers are close together then the probability of their coinheritance will be high.

Linkers are synthetic double-stranded oligonucleotides containing one or more restriction enzyme recognition sites which can be ligated to blunt ends in order to produce sticky ends.

Metabotropic receptors are receptors which are linked to changes in intracellular metabolism, for example usually via the activation of a G protein and the generation of second messengers.

Miniature end-plate potential (mepp) is the small postsynaptic depolarization which occurs due to the arrival of a single quantum of neurotransmitter at the neuromuscular junction. These may occur spontaneously at the rate of about 1 per sec and have an amplitude of about 0.5 mV.

MPTP is a neurotoxin which produces a syndrome resembling Parkinson's disease. *In vivo* it is converted to its active form MPP^+ which can damage mitochondria, particularly in dopamine-containing cells.

mRNA differential display is a method which uses a number of arbitrary primers to amplify by PCR the total mRNA in a given cell. It can be used to compare changes in gene expression under different conditions.

Necrosis is cell death which occurs due to adverse conditions. It can be distinguished from apoptotic cell death by differences in the morphology of the dying cell.

Neuritic plaques are extracellular structures which consist predominantly of insoluble deposits of β-amyloid; hence the alternative name, amyloid plaques. They contain other proteins and are surrounded by dystrophic neurites and reactive glial cells. They are thought to be characteristic of Alzheimer's disease although they do not occur in all cases.

Neurofibrillary tangles are dense aggregates of paired helical filaments, composed of a microtubule-associated protein, tau. They occur in a number of neurodegenerative diseases but are particularly prevalent in Alzheimer's disease.

Neuromuscular junction is the synapse which occurs between the motor neuron and the skeletal muscle. In vertebrates, the neurotransmitter is acetylcholine, acting on nicotinic acetylcholine receptors.

Nonassociative learning occurs when an animal is exposed once or repeatedly to a single type of stimulus, which allows an animal to learn about the properties of that stimulus. Habituation and sensitization are two forms of nonassociative learning.

Nonrectifying refers to a current where the membrane resistance is constant with voltage. This leads to a linear current/voltage relationship and the current is referred to as Ohmic.

Northern blotting is a technique which detects RNA molecules on an agarose gel by the hybridization of a labeled oligonucleotide probe which is complementary to the sequence(s) of interest. It is named for its similarity to Southern blotting (which detects DNA).

Open channel blockers are ion channel blocking agents which only bind to the channel when it is in the open conformation.

Origin of replication (ori) is a short sequence of DNA (50–100 bp) which binds enzymes from the host cell in order to initiate replication.

Orphan receptors are gene sequences which apparently code for receptors which have no known endogenous ligand.

Outward rectifier is a channel which allows current to flow out of but not into the cell (see *Figure 4.10b*).

Oxidative stress occurs when the production of free radicals is greater than the cell's mechanisms for their inactivation. It may occur due to either excess production of free radicals or the lack of enzymes such as superoxide dismutase.

Paroxysmal depolarizing shifts are large depolarizations of the membrane potential which underlie the burst firing of cortical neurons during interictal spikes.

Patch clamping encompasses a number of electrophysiological techniques. In the single-channel mode it enables the current flow through single ion channels to be measured. In whole-cell mode, it enables voltage clamping to be applied to cells which are usually too small to be examined using conventional voltage clamping (see ***Box 4.2***, p. 72)

Patch sniffing is a method derived from patch clamping in which an outside-out patch containing a receptor for an endogenous agonist is used to detect that molecule in the extracellular fluid.

Perforated synapses seen on mushroom-shaped dendritic spines have discontinuities in both presynaptic grids and postsynaptic densities produced by local narrowing of the cleft or the presence of a spinule, a finger-like projection of postsynaptic elements into the presynaptic bouton. Such synapses have two or more functionally independent active zones. They might have greater efficiency by virtue of the fact that a single action potential has the chance to trigger release at more than one active zone simultaneously and independently; failures will be fewer.

Periplasmic space is the space between the cell wall and the outer membrane in Gram-negative bacteria.

Phages *see* Bacteriophages.

Phenylalkylamines are a group of calcium channel antagonists which act predominantly on L-type calcium channels. The most well known is verapamil.

Photoaffinity labeling is a technique for identifying the position of a particular binding site or amino acid residue in a protein. It involves the incubation of protein with a photoreactive compound which will either bind to the site of interest or will react with the required amino acid. On exposure to UV light, the compound will react covalently with the site. The protein can then be isolated, digested and fractionated, and the position of the label identified.

Place cells are cells in the hippocampus which fire at higher frequencies when the animal is in a particular location.

Place fields are the regions in an animal's environment which elicit a response in the appropriate place cell.

Plasmids are double-stranded circular DNA molecules which occur naturally in bacteria. Genetically engineered plasmids are used as vectors in recombinant DNA experiments.

Polylinker is a short stretch of DNA containing the recogntion sequences for several different restriction enzymes.

Polymerase chain reaction (PCR), invented by Kerry Mullis, is a method for synthesizing large amounts of a specific DNA starting from very small amounts of starting material; theoretically, a single copy of the required DNA is sufficient.

Polymorphisms are the individual variations in the genome which are passed from generation to generation. Many of these do not occur in the part of the genome coding for expressed genes but will produce a variation in the restriction fragments produced.
Positron emission tomography is a technique which allows imaging of tissues, particularly of the brain, by measuring the emission of positrons derived from radioactive precursors, with a spatial resolution of about 5–10 mm. Using arrays of detectors and computer algorithms, it is possible to reconstruct a three-dimensional image of the pattern of emission.
Post-translational modification of newly translated proteins can consist of cleavage of peptides in order to produce the correct folding pattern or in order to ensure the correct cellular processing. It can also consist of the chemical modification of the termini or the side chains. These modifications can cause the addition of a variety of chemical groups including acetyl, hydroxyl and methyl groups, as well as the addition of more complex fatty acids and sugars.
Postsynaptic density consists of electron-dense material which is found associated with the postsynaptic membrane at the regions under the active zones. It consists of many proteins including receptor proteins, ion channels, enzymes and cytoskeletal elements.
Primers are short strands of single-stranded DNA, used in PCR, which are complementary to the 3′ and 5′ ends of the DNA strand to be amplified.
Priming in the context of neurotransmitter release refers to the step during which the vesicle becomes partly fused with the plasma membrane prior to release, an event called hemifusion.
Prion proteins are naturally occurring proteins which in abnormal forms are responsible for diseases such as sheep scrapie, bovine spongiform encephalopathy (BSE) and Creutzfeldt–Jakob disease (CJD).
Quanta is the name given to the discrete packets of neurotransmitter which are contained within a single vesicle.
Quantal analysis is a method which measures the average amplitude of responses generated by the release of a single quantum of neurotransmitter and the frequency with which different numbers of quanta are released simultaneously. It has been used to determine whether changes in neurotransmitter release are due to changes in pre- or postsynaptic events.
Quantal content ($m = np$) is taken as a measure of the number of quanta released by a single action potential.
Quantal size (q) is the amplitude of the miniature excitatory postsynaptic potential (mepsp) which is classically thought to be determined by the number of postsynaptic receptors.
Receptive fields are the regions on a sensory surface which, if stimulated, will elicit a change in the activity of the receptive neuron.
Recombinant DNA (rDNA) is a combination of the vector, such as a plasmid, with the foreign DNA to be cloned.
Restriction fragments are pieces of DNA which have been cut by restriction enzymes.
Recruitment in the context of neurotransmitter release refers to the initial calcium-dependent process which allows synaptic vesicles to move from regions of the synapse, where they are tethered to the cytoskeleton, to the active zones.
Rectification is the property by which membrane resistance varies with voltage. This means that the voltage/current relationship (I–V plot) for that current is not linear. Rectifying channels can show inward or outward rectification.

Relaxed control refers to plasmids which replicate to produce high numbers (10–200) of copies per cell.

Restriction enzymes are enzymes which recognize specific base sequences in DNA and cut the DNA at defined sites within the recognition sequence.

Restriction fragment length polymorphisms are fragments of DNA produced when genomic DNA is digested with restriction enzymes. Because of the presence of polymorphisms in individual DNA, the pattern of fragments will vary between individuals. However, similar patterns of fragments will occur in closely related individuals, allowing the use of this technique in DNA fingerprinting and in the identification of disease-causing genes (see *Box 9.1*, p. 212).

Retrograde amnesia is the failure to recall events which occurred prior to brain trauma or lesion.

Retrograde messenger is a diffusible second messenger which is generated postsynaptically and crosses the synapse where it alters presynaptic neurotransmitter release.

Reverse transcriptase is an enzyme which catalyzes the synthesis of DNA from an RNA template and is found in retroviruses. It is used to make cDNA from mRNA.

RNA editing is the post-transcriptional modification of mRNA prior to translation leading to a change in the nucleotide sequence.

RNA fingerprinting is based on PCR, using mRNA as a template and a variety of random primers to amplify expressed sequences.

Sanger sequencing is a method for sequencing DNA which uses labeled dideoxynucleotides to act as chain terminators in the enzymatic synthesis of DNA. In four different reactions, each containing a different ddNTP, DNA fragments are produced of all different chain lengths corresponding to the position of the ddNTP. The fragments produced in the four different reactions can then be separated on gels and the sequence deduced from the position of the fragments.

Second messengers are intracellular molecules which are produced via the activation of a receptor and act to change the activity of either other receptors, channels or intracellular enzymes.

Seizures are the abnormal, synchronized rhythmic firing of large populations of neurons.

SH2 (Src homology 2) domains are conserved regions in several molecules involved in intracellular signal transduction which bind short amino acid sequences containing phosphotyrosine. These include the nonreceptor tyrosine kinase, Src and the β-adrenergic receptor kinase.

Silent synapses are synapses which when stimulated produce no postsynaptic response; this may be due to the absence of functional postsynaptic receptors.

Single-channel recording is a type of patch clamping which allows the current flowing through single ion channels to be measured.

Site-directed mutagenesis is a technique for mutating DNA in a precise manner (see *Box 4.3*, p. 74).

SNAPs are soluble *N*-ethylmaleimide-sensitive attachment proteins which are involved with SNAREs in the docking and priming steps of neurotransmitter release.

SNAREs are a group of proteins associated with neurotransmitter vesicles and the plasma membrane which are involved with SNAPs in the docking and priming steps of neurotransmitter release. They include the proteins, synaptobrevin, syntaxin and SNAP-25.

Southern blotting is a technique which detects DNA molecules on nitrocellulose filters by the hybridization of a labeled oligonucleotide probe which is complementary to the sequence(s) of interest. The DNA is separated on agarose gels and transferred from the gel on to the filter by capillary action. It is named after its inventor Edwin Southern.

Spatial navigation learning involves animals learning the features of their immediate environment by using positional cues.

Splice cassettes are sections of mRNA which can be included or excluded during alternative splicing (see *Figure 5.12*).

Status epilepticus is continuous maintained epileptic seizure.

Sticky ends are ends of double-stranded DNA which are asymmetric, leaving protruding 5′ or 3′ ends.

Stringent control refers to plasmids which only replicate a few (< 20) times per cell.

Synaptic tagging is the process by which a recently tetanized synapse acquires a marker which allows it to be targeted by proteins synthesized in the cell body. It is a mechanism for ensuring specificity of LTP.

Tetanus toxin is a clostridial neurotoxin which can block neurotransmitter release by cleaving the SNARE, synatobrevin.

Transfection is the incorporation of DNA into eukaryotic cells (*see* Transformation).

Transformation is the introduction of recombinant DNA into bacteria. It can be carried out by incubation of cells with a hypotonic calcium solution followed by a heat shock. This increases the efficiency with which the rDNA is taken up.

Transgenic animals are those which have had a synthetic gene, called a transgene, inserted into the genome, allowing the function of that gene to be examined in the context of the whole animal (***Box 7.2***).

Vector is a nucleic acid molecule which has the ability to replicate in the host. Vectors are usually derived from bacterial plasmids or bacteriophage viruses.

Voltage clamping is an electrophysiological technique which injects the current required to maintain a constant membrane potential during the movement of ions across the membrane. The injected current thus gives a measure of the current flowing across the membrane at different potentials. It enable fast events such as the action potential to be studied (see ***Box 4.1***, p. 71)

Voltage-gated ion channels are ion channels which open in response to a change in the local membrane potential. Their opening may also be affected by phosphorylation or dephosphorylation by intracellular kinases and phosphatases.

Weighting refers to the strength of a particular synapse. It is the efficacy with which presynaptic activation can affect the postsynaptic cell.

Western blotting is a technique which detects proteins on an SDS–polyacrylamide gel by the binding of specific antibodies. It is named for its similarity to Southern blotting which identifies DNA.

X-ray crystallography is a technique in which beams of X-rays are passed through a crystal of protein. The diffraction pattern produced by scattering of the X-rays allows deductions to be made as to the three-dimensional structure of the protein.

Further reading

We have chosen a number of key articles for each chapter as well as reviews of each subject.

General books

Davies, R.W. and Morris, B.J. (eds) (1997) *Molecular Biology of the Neuron.* Oxford University Press, Oxford.

Hall, Z.W. (ed.) (1992) *An Introduction to Molecular Neurobiology.* Sinauer Associates Ltd, Sunderland, MA, USA.

Nicholls, D.G. (1994) *Proteins, Transmitters and Synapses.* Blackwell Scientific Publications, Oxford.

Chapter 2

Noda, M., Takahashi, H., Tanabe T., Toyosato, M., Furutani, Y., Hirose, T., Asai, M., Inayama, S., Miyata, T. and Numa, S. (1982) Primary structure of alpha-subunit precursor of *Torpedo californica* acetylcholine receptor deduced from cDNA sequence. *Nature* **299**: 793–797.

Watson, J.D. Gilman, M. Witkowski, J. and Zoller, M. (1992) *Recombinant DNA.* Scientific American Books, W.H. Freeman and Co., New York.

Chapter 3

Kerr, M.A. and Thorpe, R. (eds) (1994) *Immunochemistry Labfax.* BIOS Scientific Publishers, Oxford.

Livesey, F.J. and Hunt, S.P. (1996) Identifying changes in gene expression in the nervous system; mRNA differential display. *Trends Neurosci.* **19:** 84–88.

Mullis, K.B. (1990) The unusual origin of the polymerase chain reaction. *Sci. Am.* **April**: 36–43.

Chapter 4

Aidley, D.J. and Stanfield, P.R. (1996) *Ion Channels: Molecules in Action.* Cambridge University Press, Cambridge.

Catterall, W.A. (1995) Structure and function of voltage-gated ion channels. *Annu. Rev. Biochem.* **64:** 493–531.

Heinemann, S.H., Terlau, H., Stühmer, W., Imoto, K. and Numa, S. (1992) Calcium channel characteristics conferred on the sodium channel by single mutations. *Nature* **356:** 441–443.

Jan, L.Y. and Jan, Y.N. (1997) Cloned potassium channels from eukaryotes and prokaryotes. *Annu. Rev. Neurosci.* **20:** 91–123.

Noda, M., Shimizu, S., Tanabe, T. *et al.* (1984) Primary structure of *Electrophorus electricus* sodium channel deduced from cDNA sequence. *Nature* **312**: 121–127.

Receptor and Ion Channel Nomenclature Supplement (annual). *Trends Pharmacol. Sci.* Elsevier Science, Cambridge.

Stühmer, W., Conti, F., Suzuki, H., Wang, X., Noda, M., Yahagi, N., Kubu, H. and Numa, S. (1989) Structural parts involved in activation and inactivation of the sodium channel. *Nature* **339**: 597–603.

Chapter 5

Karlin, A. (1993) Structure of nicotinic acetylcholine receptors. *Curr. Opin. Neurobiol.* **3**: 299–309.

McKernan, R.M. and Whiting, P.J. (1996) Which $GABA_A$ receptor subtypes really occur in the brain? *Trends Neurosci.* **19**: 139–143.

Scoepfer, R., Monyer, H., Sommer, B. *et al.* (1994) Molecular biology of glutamate receptors. *Prog. Neurobiol.* **42**: 353–357.

Smith, G.B. and Olsen, R.W. (1996) Functional domains of $GABA_A$ receptors. *Trends Pharmacol. Sci.* **16**: 162–168.

Unwin, N. (1993) Neurotransmitter action: opening of ligand-gated ion channels. *Cell* **72**: 31–41.

Zukin, R.S. and Bennett, M.V.L. (1995) Alternatively spliced forms of the NMDAR1 receptor subunit. *Trends Neurosci.* **18**: 306–313.

Chapter 6

Albert, P.R. and Morris, S.J. (1994) Antisense knockouts: molecular scalpels for the dissection of signal transduction. *Trends Pharmacol. Sci.* **15**: 250–254.

Hepler, J.R. and Gilman, A.G. (1992) G proteins. *Trends Biochem. Sci.* **17**: 383–387.

Masu, M., Tanabe, Y., Tsuchida, K., Shigemoto, R. and Nakanishi, S. (1991) Sequence and expression of a metabotropic glutamate receptor. *Nature* **349**: 760–765.

Sokoloff, P. and Schwartz, J.-C. (1995) Novel dopamine receptors half a decade later. *Trends Pharmacol. Sci.* **16**: 270–275.

Sternweis, P.C. (1994) The active role of $\beta\gamma$ in signal transduction. *Curr. Opin. Cell Biol.* **6**: 198–203.

Chapter 7

Amara, S.G. and Arriza, J.L. (1993) Neurotransmitter transporters: three distinct gene families. *Curr. Opin. Neurobiol.* **3**: 337–344.

Capecchi, M.R. (1994) Targeted gene replacement. *Sci. Am.* **March:** 34–41.

Dunlap, K., Leubke, J.I. and Turner, T.J. (1995) Exocytotic Ca^{2+} channels in mammalian central neurons. *Trends Neurosci.* **18**: 89–98.

Söllner, T., Whiteheart, S.W., Brunner, M., Erdjument-Bromage, H., Geromanos, S., Tempst, P. and Rothman, J.E. (1993) SNAP receptors implicated in vesicle targeting and fusion. *Nature* **362**: 318–324.

Sudhof, T.C. (1995) The synaptic vesicle cycle: a cascade of protein–protein interactions. *Nature* **375**: 645–653.

Walch-Solimena, C., Jahn, R. and Sudhof, T.C. (1993) Synaptic vesicle proteins in exocytosis: what do we know? *Curr. Opin. Neurobiol.* **3**: 329–336.

Chapter 8

Bartsch, D., Ghirardi, M., Skehel, P.A., Karl, K.A., Herder, S.P., Chen, M., Bailey, C.H. and Kandel, E.R. (1995) *Aplysia* CREB2 represses long-term facilitation: relief of repression converts transient facilitation into long-term functional and structural change. *Cell.* **83:** 979–992.

Bliss, T.V.P. and Collingridge, G.L. (1993) A synaptic model of memory: long-term potentiation in the hippocampus. *Nature* **361**: 31–39.

Fazeli, S. and Collingridge, G.L. (eds) (1996) *Cortical Plasticity: LTP and LTD.* Oxford University Press, Oxford.

Linden, D.J. (1994) Long-term synaptic depression in the mammalian brain. *Neuron* **12**: 457–472.

Wilson, M.A. and Tonegawa, S. (1997) Synaptic plasticity, place cells and spatial memory: study with second generation knockouts. *Trends Neurosci.* **20:** 102–106.

Chapter 9

Coyle, J.T. and Puttfarcken, P. (1993) Oxidative stress, glutamate and neurodegenerative disorders. *Science* **262**: 689–695.

Dawbarn, D. and Allen, S.J. (eds) (1995) *Neurobiology of Alzheimer's Disease.* Oxford University Press, Oxford.

Haque, N.D.S., Borghesani, P. and Isacson, O. (1997) Therapeutic strategies for Huntington's disease based on a molecular understanding of the disorder. *Mol. Med. Today* **April**: 175–183.

The Huntington's Disease Collaborative Research Group (1993). A novel gene containing a trinucleotide repeat that is expanded and unstable on Huntington's disease chromosomes. *Cell* **72**: 971–983.

Strittmatter, W.J. and Roses, A.D. (1996) Apolipoprotein E and Alzheimer's disease. *Annu. Rev. Neurosci.* **19:** 53–77.

Index

LIVERPOOL
JOHN MOORES UNIVERSITY
AVRIL ROBARTS LRC
TITHEBARN STREET
LIVERPOOL L2 2ER
TEL. 0151 231 4022